Ultra-High Performance Liquid Chromatography and Its Applications

Ultra-High Performance Liquid Chromatography and Its Applications

Edited by

Quanyun Alan Xu

Library of Congress Cataloging-in-Publication Data:

Ultra-high performance liquid chromatography and its applications / edited by Quanyun Alan Xu
   pages cm
   Includes bibliographical references and index.
   ISBN 978-0-470-93842-3 (hardback)
   1. Liquid chromatography.   I. Xu, Quanyun A., 1961–
   QD79.C454U48 2013
   543′.84–dc23   2012035740

Printed in the United States of America

10 9 8 7 6 5 4 3 2 1

Contents

Preface

High performance liquid chromatography (HPLC) is widely used in analytical chemistry and biochemistry to separate, identify, and quantify compounds. For decades, it has been the method of choice in drug discovery and development both in the pharmaceutical and biopharmaceutical industries and also in academia. With the advent of ultra-high performance liquid chromatograph (UHPLC) a few years ago, the science of liquid chromatography (LC) has come to a new era. UHPLC provides the separation science with high speed, high resolution, high sensitivity, and short runtime. The ultra-high performance of liquid chromatography is achieved either by ultra-high pressure liquid chromatographic instrumentation which couples with sub-2 μm particle column packing or by conventional high performance liquid chromatography which capitalizes on core-shell (or fused-core) particle column packing. Ultra-high pressure liquid chromatographic system can withstand extremely high pressures, up to 1,000 bars, which a sub-2 μm particle column generates. On the other hand, a core-shell particle-packing column usually produces column back pressure up to 400 bars. Two different packing materials, one goal.

Ultra-High Performance Liquid Chromatography and Its Applications will help readers develop skills in UHPLC method development; transfer methods from HPLC to UHPLC and from UHPLC to HPLC; maximize the performance of their UHPLC; understand benefits and limitations of UHPLC; and apply UHPLC in pharmaceutical research, clinical research, food safety, and environmental services. The first half of this book describes method development procedure that includes defining method goals, scouting columns and mobile phases, selecting separation conditions, optimizing and validating the method, presents the approaches of method transfer between HPLC and UHPLC platforms that include selecting the right column and properly scaling flow rate, injection volume, and gradient conditions (if applicable), and discusses practical aspects of UHPLC, including its benefits and limitations. It reviews the recent development of the coupling of UHPLC with mass spectrometry (UHPLC-MS) in terms of its advantages and challenges.

As an alternative to sub-2 μm particles in ultra-high pressure liquid chromatography, core-shell particle column packing balances column efficiency and back pressure and is used on conventional high performance liquid chromatography. Chapter 5 presents an insight into the theory (van Deemter equation) behind the success of core-shell particle packing and reviews advantages and applications in the fields of bioanalysis and environmental analysis.

The second half of this book provides examples of applications of UHPLC in bioanalysis, including analysis of drugs of abuse in human biological matrices, analysis of isoflavones and flavonoids in natural products and biological samples, characterization and analysis of therapeutic proteins, and studies of pharmacokinetics, drug metabolism, and metabonomics of Tranditional Chinese Medicine.

I would like to acknowledge all of the authors who found time in their busy schedules to write these exceptional chapters and who made this book possible. My thanks also go to Mr. Bob Esposito and Michael Leventhal at John Wiley & Sons for their much valued assistance throughout the preparation of this book.

QUANYUN ALAN XU

Contributors

Irena Barnowska, Professor, Department of Analytical Chemistry, Silesian University of Technology, Gliwice, Poland

Tilak Chandrasekaran, MS, Merck & Co., Inc., Rahway, NJ

Ray Chen, PhD, Thermo Fisher Scientific, San Jose, CA

Shujun Chen, PhD, API Chemistry and Analysis, Product Development, Platform Technology and Science, GlaxoSmithKline, King of Prussia, PA

Thomas F. Cullen, PhD, Abbvie, North Chicago, Illinois

Ying Deng, MS, Department of Analytical Chemistry, School of Pharmacy, Shenyang Pharmaceutical University, Shenyang, Liaoning, China

Diab Elmashni, Marketing Communications, Thermo Fisher Scientific, Sunnyvale, CA

Szabolcs Fekete, PhD, School of Pharmaceutical Sciences, University of Geneva, Geneva, Switzerland

Jeno Fekete, Professor, Department of Inorganic and Analytical Chemistry, Budapest University of Technology and Economics, Budapest, Hungary

Hong Gao, PhD, Merck & Co., Inc., Rahway, NJ

Maria Carla Gennaro, Department of Science and Technological Innovation, University of Piemonte Orientale, Alessandria, Italy

Fabio Gosetti, PhD, Department of Science and Technological Innovation, University of Piemonte Orientale, Alessandria, Italy

Davy Guillarme, PhD, School of Pharmaceutical Sciences, University of Geneva, Geneva, Switzerland

Guifeng Jiang, PhD, Applied Market, Thermo Fisher Scientific, San Jose, CA

Alireza Kord, PhD, API Chemistry and Analysis, Product Development, Platform Technology and Science, GlaxoSmithKline, King of Prussia, PA

Laila Kott, PhD, Analytical Development Small Molecules, Millennium (The Takeda Oncology Company), Cambridge, MA

Famei Li, PhD, Department of Analytical Chemistry, School of Pharmacy, Shenyang Pharmaceutical University, Shenyang, Liaoning, China

Sylwia Magiera, PhD, Department of Analytical Chemistry, Silesian University of Technology, Gliwice, Poland

Eleonora Mazzucco, PhD, Department of Science and Technological Innovation, University of Piemonte Orientale, Alessandria, Italy

Theresa K. Natishan, MS, Merck & Co., Inc., Rahway, NJ

Jennifer C. Rea, PhD, Protein Analytical Chemistry, Genentech, Inc., South San Francisco, CA

Serge Rudaz, PhD, School of Pharmaceutical Sciences, University of Geneva, Geneva, Switzerland

Julie Schappler, PhD, School of Pharmaceutical Sciences, University of Geneva, Geneva, Switzerland

Jason R. Stenzel, PhD, Crime Laboratory, Washington State Patrol, Cheney, WA

Jean-Luc Veuthey, PhD, Professor, School of Pharmaceutical Sciences, University of Geneva, Geneva, Switzerland

Yajun Jennifer Wang, PhD, Protein Analytical Chemistry, Genentech, Inc., South San Francisco, CA

Gregory K. Webster, PhD., MBA, Abbvie, North Chicago, IL

Christopher J. Welch, PhD, Merck & Co., Inc., Rahway, NJ

Naijun Wu, PhD, Analytical R&D, Celgene Corporation, Summit, NJ

Zhili Xiong, PhD, Department of Analytical Chemistry, School of Pharmacy, Shenyang Pharmaceutical University, Shenyang, Liaoning, China

Quanyun Alan Xu, PhD, Pharmaceutical Development Center, Department of Experimental Therapeutics, Division of Cancer Medicine, UT MD Anderson Cancer Center, Houston, TX

Li Zhang, MS, Merck & Co., Inc., Rahway, NJ

Taylor Zhang, PhD, Protein Analytical Chemistry, Genentech, Inc., South San Francisco, CA

Chapter 1

UHPLC Method Development

Shujun Chen and Alireza Kord

1.1 INTRODUCTION

UHPLC has been gradually adopted in industrial labs, especially the pharmaceutical industry due to its high resolution, high speed, and solvent saving, since its introduction in early 2004 (1). A UHPLC method using a sub-2 μm column could reduce the analysis time by up to 80% and save the mobile phase consumption by at least 80% compared with an HPLC method using a conventional 3.5 μm column without sacrificing separation performance (2). In addition, the much shorter run time significantly reduces UHPLC method development scouting time (1).

HPLC method development principles can be applied to UHPLC method development, although detailed procedures may differ. In addition, many existing HPLC methods used in the pharmaceutical industry can be converted to UHPLC methods. In practice, a UHPLC method may need to be converted to HPLC when a UHPLC system is not available.

This chapter provides an overview of the UHPLC method development process and the conversion process of an HPLC method to UHPLC or vice versa. It mainly focuses on analytical reversed phase UHPLC method development of small molecules. A general process and detailed steps are discussed as well as practical examples given.

1.2 METHOD DEVELOPMENT

There are many publications on HPLC method development strategies (3–10). These strategies can be applied to UHPLC method development, although instrumentation and columns are different. The UHPLC method development process includes gathering sample information and defining method goals, scouting columns and mobile phases, analyzing scouting results and selecting separation conditions, optimizing the method, and validating the method. Steps in common UHPLC method development

Ultra-High Performance Liquid Chromatography and Its Applications, First Edition. Edited by Quanyun Alan Xu.
© 2013 John Wiley & Sons, Inc. Published 2013 by John Wiley & Sons, Inc.

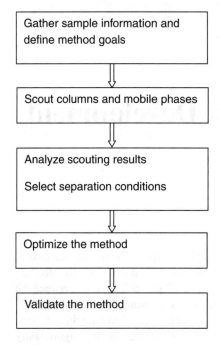

Figure 1.1 General process of UHPLC method development.

processes are summarized in Figure 1.1. Each step is described and discussed in the following sections.

1.2.1 Gather Sample Information and Define Method Goals

1.2.1.1 Gather Sample Information

Sample information is very useful for method development. The sample information includes the process used to generate the sample, chemical structures and physical and chemical properties of sample components, and their toxicity.

Understanding the process is helpful to method development and is achieved by talking to a chemist or formulator who provides the sample. In the pharmaceutical industry, there is rarely a totally unknown sample. In process chemistry, active pharmaceutical ingredient (API), intermediate (IM), and starting material (SM) are made from synthetic schemes. Most impurities in the sample are known, although some impurities may be unpredictable. In the pharmaceutical process, drug products are manufactured by mixing API with excipients. The major components of the sample are known, although some degradation products may not be predictable.

In addition, knowledge of chemical structures and physicochemical properties of sample components is useful for method development. The chemical structures of possible sample components provide data on molecular weights and functional groups. Special attention should be paid to acidic, basic, aromatic, and other functional groups from which pK_a, solubility, chromophore, and stability can be inferred if data of physical and chemical prosperities is not available. Based on pK_a, a suitable column and mobile phase buffer pH can be selected for a robust method. The solubility and stability of major components can be used to select a suitable diluent and mobile phase. The UV chromophore can be used to select a suitable detection wavelength.

It is very important to read material safety data sheet (MSDS) before working on samples. Some samples may be toxic and should be handled in a safe manner recommended by MSDS.

1.2.1.2 Define Method Goals

There are four major types of tests in the pharmaceutical industry. These are identification tests, quantitative tests for impurities' content, limit for control of impurities, and quantitative tests of active moiety (11). A quantitative method, for example, an assay method, can be used as an identification method. In addition, many methods can be used for quantitative tests for both assay and impurities' contents.

Different types of methods have different goals. The method goals are usually defined as specificity, accuracy, precision, sensitivity, and robustness, or more specifically as resolution, linearity, recovery, repeatability, and quantitation limit for an assay and impurity method. The method goal for an identification test is often defined as "specificity." This chapter mainly focuses on the assay and impurity method.

The same type of methods may have different goals for a different sample. For an API sample, its impurities must be separated from API and each other, and the limit of quantification (LOQ) for the impurities must be lower than the reporting threshold recommended in ICH Q3A(R2) guideline (12). For SM or IM, its impurities must be separated from SM or IM and from each other, and LOQs are determined by the criticality of the impurities. For a drug product, its impurities must be separated from its API, excipients, and each other, and LOQ must be lower than the reporting threshold recommended in ICH Q3B(R2) guideline (13).

In addition, method goals also change with the phase of each project. For an early-phase project, robustness and ruggedness are not required for an API method. However, for a late-phase project, robustness and ruggedness are required for an API method. The comparison of method goals between early- and late-phase projects is made in Table 1.1.

1.2.2 Scout Columns and Mobile Phases

1.2.2.1 Select a UHPLC Mode

After gathering sample information and defining goals, a suitable UHPLC mode can be selected. In HPLC, there are four major separation modes: reversed phase, normal phase, ion exchange, and size exclusion (14). However, there are only two UHPLC modes because of UHPLC column current availability. These are reversed-phase and normal-phase UHPLC (15, 16). Like HPLC, most pharmaceutical compounds can be separated with reversed-phase UHPLC, which is the main focus of this chapter.

Table 1.1   Comparison of API Method Goals Between Early- and Late-Phase Projects

	Early Phase	Late Phase
Specificity/resolution	Resolution ≥ 1.2	Resolution ≥ 1.5
	Discrimination factor ≥ 0.5	Discrimination factor ≥ 0.8
Accuracy/linearity	Quantitation limit (QL) to at least 120% of the analyte (main and specified impurities)	Quantitation limit (QL) to at least 120% of the analyte (main and specified impurities)
Accuracy/recovery	No	Yes
Repeatability	$\leq 1.0\%$	$\leq 1.0\%$
Sensitivity	Depends on maximum daily dose	Depends on maximum daily dose
Solution stability	Yes	Yes
Robustness and ruggedness	No	Yes

1.2.2.2   Select Columns, Mobile Phases, and Detection and Setting

1.2.2.2.1   Columns   There are many commercial UHPLC columns, most of which are based on silica gel packing material. The major limitation of the silica gel packing is its instability at basic pH. Some UHPLC columns are packed with hybrid particles such as bridged ethylene hybrid (BEH) technology. The advantage of the hybrid packing material is its stability at alkaline pH and less peak tailing for basic compounds compared with the silica packing materials (17, 18). In addition, some UHPLC columns are packed with core-shell particles (19–21).

There are different bonded phases for each packing material. For example, there are C18, C8, C4, phenyl, and polar-embedded C18 bonded phases for BEH packing material. The selected UHPLC columns are listed in Table 1.2

Table 1.2   Selected UHPLC Columns

Manufacturer	Type	Column
Waters Corp.	C18	BEH C18 1.7 μm
	Hydrophilic carbamate with C18	BEH Shield RP18 1.7 μm
	C8	BEH C8 1.7 μm
	Phenyl	BEH phenyl 1.7 μm
	Amide	BEH amide 1.7 μm
	HILIC	BEH HILIC 1.7 μm
	C18	HSS C18 1.7 μm
	C18	HSS C18 SB 1.7 μm
	T3	HSS T3 1.7 μm
Agilent technologies	C18	Zorbax Eclipse Plus C18 RRHD 1.8 μm
	C18	Zorbax SB-C18 RRHD 1.8 μm
	C8	Zorbax SB-C8 RRHD 1.8 μm

Different bonded phases are used during method development because they provide different selectivity. Their different selectivity gives confidence in method specificity. Typically C18, C8, and phenyl are selected. In addition, mobile phase pH use range should be considered when selecting UHPLC columns. As mentioned, the silica-based packing columns are stable under acidic conditions but not at basic conditions. Therefore, the hybrid-based packing columns are selected under basic conditions.

Column stability needs to be checked before column selection or during method validation. A column must be stable for at least 200 injections for routine use (22). Column manufacturers may provide this information. In addition, studies on some column stability were published (23, 24). However, the information is limited at specific conditions and needs to be confirmed experimentally.

1.2.2.2.2 Mobile Phases Mobile phases mainly consist of organic solvent and buffer. Acetonitrile and methanol are commonly used organic solvents. However, the disadvantages of using methanol are its low sensitivity at low UV wavelength and high back pressure. Tetrahydrofuran (THF) may also be used as a UHPLC organic solvent. However, it may cause a detection problem and damage to a UHPLC instrument. Consult with the UHPLC instrument manufacturer beforehand.

The other component of mobile phase is aqueous buffer. The buffer pH may be critical to some separation. pKa values of compounds of interest should be considered when selecting mobile phase pH. The buffer pH of a mobile phase should be outside pKa ± 2 to ensure method robustness. Table 1.3 lists the commonly used mobile phase buffers. The most commonly used mobile phase buffer is a trifluoroacetic acid (TFA) aqueous solution. One issue with TFA is its low sensitivity at wavelength 210 nm.

Different mobile phase organic solvents have different selectivity for neutral, acidic, and basic compounds. Therefore, different mobile phase organic solvents are tested as modifiers during method development. In addition, mobile phase buffer pH affects selectivity of acidic and basic compounds. Therefore, different buffer pHs are investigated during method development for acidic and basic compounds.

Ideal mobile phase provides excellent resolution as well as sensitivity (25–29). It needs to have a low UV cutoff to give sufficient sensitivity for UV detection. It is also preferable that it be compatible with mass spectrometry (MS).

Table 1.3   Commonly Used Mobile Phase Buffers

Buffer	pKa	pH Range
Trifluoroacetic acid (TFA)	0.2	1.5–2.5
Formic acid	3.8	2.8–4.8
Acetic acid	4.8	3.8–5.8
Ammonium acetate	4.8 and 9.2	3.8–5.8 and 8.2–10.2
Ammonium hydroxide	9.2	8.2–10.2

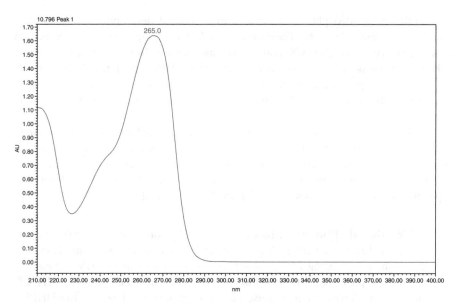

Figure 1.2   Example of a UV profile for wavelength selection.

1.2.2.2.3   Detection   Most pharmaceutical compounds have UV chromophores. UV detector is the most commonly used detector. Wavelengths at λ_{max} or λ_{val} are usually selected for method robustness. For example, three wavelengths at 265 (λ_{max}), 226 (λ_{val}), and 210 nm (λ_{max}) can be potentially used for a compound with the UV profile shown in Figure 1.2. However, sensitivity is the highest at 265 nm, which is why 265 nm should be selected for best detection. In addition, UV profiles of impurities in the sample should be considered. A detection wavelength should be selected to ensure that all components have acceptable sensitivity, not just the major component.

Beside wavelength selection, UV detection setting is also critical (30). An HPLC peak is usually broad, and its peak width is typically on order of several seconds. However, a UHPLC peak is very narrow, and its peak width is usually on order of one second. For UHPLC, at least 20 Hz data acquisition rate should be used. Low data acquisition rates cause peak distortion and broadening as shown in Figure 1.3.

For analytes with no UV chromophore or weak UV chromophore, detection methods such as Corona charged aerosol detection (CAD), evaporative light scattering detection (ELSD), or MS can be used. These detectors require that analytes be nonvolatile and mobile phase be compatible with detectors.

1.2.2.3   Prepare Sample Solution

Sample diluent must be appropriately selected. It must be able to dissolve all components in a sample. In addition, all sample components must be stable in the selected diluent. The selected diluent also does not interfere with separation. Typically, aqueous mobile phase or a mixture of aqueous mobile phase and organic mobile phases is

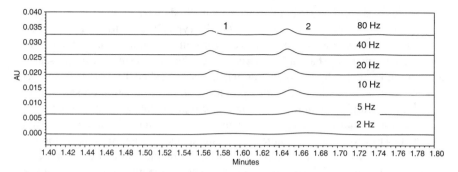

Figure 1.3   Example of the influence of the data acquisition rate on peak distortion. The filter time is 0.1 s. Conditions: Zorbax Bonus RP 100 × 2.1 mm, 1.8 μm; mobile phase A 0.1% TFA water; mobile phase B acetonitrile; gradient 20%–56% B for 2.06 min; flow rate 0.8 mL/min; wavelength 268 nm. Peaks 1 and 2 are impurities.

used as the sample diluent. A high percentage of organic solvent in the diluent may reduce resolution and sensitivity.

For drug products, sample preparation may involve grinding, extraction, and filtration (31–35). For samples with a more complicated matrix, solid phase extraction (SPE) and other sample preparation techniques can be used to remove interfering species and to increase sensitivity (36–41). The procedure needs to ensure sufficient recovery of target analytes for accuracy and sensitivity.

1.2.2.4   Run a Scouting Experiment

Once columns, mobile phase organic solvents and buffers, and method of detection are selected and the sample is prepared, a scouting experiment can be performed. There are many HPLC method strategies published (5–10, 42). However, these can be classified into two main approaches. One, called a trial-and-error approach, involves trying one mobile phase and column at a time. The other, called an automated screening approach, involves screening multiple columns and mobile phases using an automated system. These HPLC strategies can be applied to UHPLC.

1.2.2.4.1   Trial-and-Error Approach   This strategy has often been used in the past and is still used widely today. A method can be developed very quickly by carefully selecting a column and mobile phase for a sample with simple composition. However, for a very complicated sample, an analyst needs to analyze the scouting results and decide what to do next, which usually is most effective when the analyst has extensive experience. In addition, the approach lacks efficiency due to manual column changing as well as good understanding due to limited information.

1.2.2.4.2   Automated Screening Approach   The prerequisite for this strategy is to have an automated system. After selecting columns, mobile phases, and detection and settings, it is necessary to build instrument methods and a sample set. All columns and mobile phases can be screened automatically. Typically a photo diode array (PDA) detector is used, and the automated system runs all conditions. This approach does not require an analyst to have a lot of experience. The analyst simply runs preselected columns and mobile phases on an automated system, called Toolkit. This approach is highly efficient. In addition, it may provide a lot of information on selectivity of different bonded phase columns, mobile phase organic solvents, and buffer pHs. It facilitates better understanding of the method and gives more insurance on method specificity. However, when a sample is not complicated, this approach takes longer to screen all columns and mobile phases than when using the trial-and-error approach.

In practice, both trial-and-error and automated screening approaches are used. Both have pros and cons. Analysts can choose an appropriate approach based on instrument availability and complexity of a sample.

1.2.3 Analyze Scouting Results and Select Separation Conditions

1.2.3.1 Analyze Scouting Results

After a sample set run is complete, an analyst needs to analyze the scouting results. Depending on the extent of sample complexity, it may have several conditions that provide desirable separations. However, it may turn out that no tried conditions can give desirable separation. The analyst needs to understand the results and find a clue as to what conditions, columns, or mobile phases may work for the separation. Then new conditions, columns, or mobile phases must be tried, until a desirable separation is achieved.

1.2.3.2 Select Separation Conditions

Once separation is achieved for all peaks of interest, a condition needs to be selected from several potential conditions. The goals of a method must be considered against candidate conditions. At this point, resolution and tailing factor must be evaluated for candidate conditions. For a small impurity peak on shoulder of a major peak shown in Figure 1.4, US pharmacopeia (USP) resolution at equal to or greater than 1.5 is not adequate to ensure accurate integration. The integration result of the small peak varies with a way of integration, which introduces uncertainty to the analytical results. In this case, a discrimination factor needs to be used (14). The discrimination factor is defined in the following:

$$d_o = \frac{h_p - h_v}{h_p} \tag{1.1}$$

where d_0 = discrimination factor, h_p = impurity peak height, and h_v = height of the valley by drop integration.

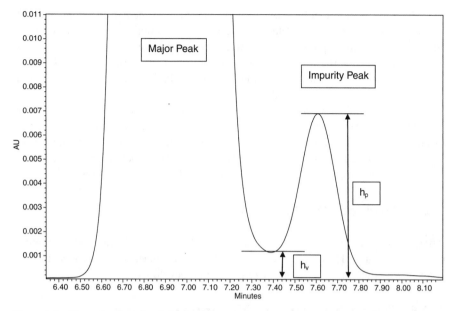

Figure 1.4   Illustration for a discrimination factor.

For an early-phase project, the discrimination factor is usually equal to or greater than 0.5. However, for a late-phase project, it needs to be equal to or greater than 0.8 for high accuracy.

1.2.4   Optimize the Method

Once separation conditions are selected, column, mobile phase organic solvent, and buffer are known. However, column dimension, flow rate, column temperature, mobile phase organic solvent gradient range and gradient time, and buffer pH need to be optimized for better resolution, robustness, and shorter run time. If necessary, buffer concentration needs to be optimized or triethylamine (TEA) needs to be added for better resolution or tailing factor.

There are several software and automated systems for HPLC method development and optimization, such as Drylab®, Chromsword®, and ACD/AutoChrom MDS, and others (43–47). Their principles can be applied to UHPLC. In addition, Waters Corp. (Milford, MA) has recently promoted Fusion Method Development™ software. Fusion Method Development™ software from S-Matrix integrates seamlessly with Water's ACQUITY UPLC and Empower 2 Chromatography software to automate method development. The software automatically generates instrument methods and sample sets. Another feature of this software is to visualize data by statistically fitting the results. However, it cannot generate simulated chromatograms at predicted conditions, like Drylab® can.

The optimization can be efficiently performed with method development software such as Drylab® using data generated from the scouting experiment or a few additional runs. This approach can effectively use the data generated during the scouting and necessitates the limited number of additional runs. Furthermore, Drylab® provides useful information on robustness and ruggedness based on the limited number of runs (48). This is discussed in Examples 1.1 and 1.2.

In addition, injection amount determined by sample concentration and injection volume may need to be optimized. Lower injection amount may result in less sensitivity and higher efficiency and resolution. However, higher injection amount may cause higher sensitivity and lower efficiency and resolution. Therefore, the injection amount needs to be optimized to meet requirements for both sensitivity and resolution.

1.2.5   Validate the Method

After carefully selecting separation conditions and optimizing the method, the method must be validated. However, if any parameter cannot meet the predefined method goals, the method must be re-optimized or even redeveloped.

1.2.6   Phase-Appropriate Method Development

Appropriate analytical methods are required for drug development at different phases, which span from drug discovery to new drug application (NDA) or marketing authorization application (MAA) filing to launch and manufacture, as shown in Figure 1.5.

API method development effort during drug discovery is limited. In phase I, all impurities must be separated from a major peak to ensure peak purity. In phase II, a synthetic route needs to be selected and optimized. The method needs to be challenged when the route or process is changed. In phase III, the route is selected and the process is finalized. Critical Quality Attribute (CQA) impurities are defined and their references are typically prepared for method development. Method robustness and ruggedness must be demonstrated, and relative response factors of all impurities CQAs are determined before regulatory filing such as NDA and MAA. The validated definitive methods are transferred to manufacturing sites.

The API method can be modified for pharmaceutical dosage forms that consist of API and excipients throughout the drug development. The modified method should be able to separate API, degradants, impurities, and excipients from each other to accurately determine assay and impurity in the pharmaceutical dosage forms. An appropriate sample preparation is needed to ensure method accuracy and sensitivity for the pharmaceutical dosage forms (31–35).

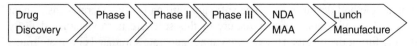

Figure 1.5   Drug development process.

1.2.7   Very High Pressure and Frictional Heating in UHPLC

Very high pressure in UHPLC alone can increase the retention factor compared with HPLC (49–53). For neutral compounds, a relatively small increase in the retention factor was observed over a pressure increase of 500 bar. However, for polar solute, or solutes with large molecular weight, acids, or bases, a much larger increase in the retention factor was observed. The differential increase in retention factors can provide significant selectivity effect for a mixture of different types of analytes in UHPLC.

Frictional heating becomes more pronounced in UHPLC, although smaller column diameters are used compared with HPLC. Both theoretical and experimental results indicate that significant radial and longitudinal temperature gradients are formed along the column (54). The frictional heating manifests itself in the retention factor reduction with pressure increase (2, 55). The change in the retention factor with pressure may provide an additional tool to manipulate selectivity in UHPLC.

Very high pressure and frictional heating in UHPLC may provide different selectivity than HPLC. Therefore, they may make HPLC method conversion to UHPLC or vice versa more complicated. Their effects on separation need to be considered when an HPLC or UHPLC method is converted.

1.2.8   Relevance of Various Instrumentation to Method Development

Currently, there are several UHPLC instrument vendors; Waters Corp and Agilent Technologies are two major suppliers. Ideally, a method developed in one instrument works in another instrument. However, these instruments may perform differently because their specifications or even their designs may be different. For example, Agilent 1290 uses a Jet weaver to mix mobile phases, whereas Waters Acquity uses a regular volume mixer to mix mobile phases. The dwell volume of the former may be smaller than that of the latter. In addition, the Jet Weaver has much better mixing efficiency than the regular volume mixer. The effect is pronounced when mobile phase contains high concentration TFA, e.g., 0.1%. Therefore, the difference in instrumentation needs to be considered when a method is developed on one instrument but validated or used on another instrument.

1.2.9   Method Resolution and Speed Requirements

For most methods, resolution is required for method specificity, and high speed is desirable for short turnaround time and high throughput. In addition, high speed also contributes to great solvent saving. However, for some methods, only a major component is required to be determined. For example, dissolution testing only determines API concentration in dissolution media at different time points. It is unnecessary to determine impurities. In this case, high speed is required, especially for real-time monitoring.

There are different approaches to increasing UHPLC speed. These include using a short column of small particle sizes, high flow rate, and high temperature. The small particle sizes make it possible to use a short column and high flow rate without sacrificing efficiency or resolution. In addition, high temperature reduces mobile phase viscosity, resulting in high diffusion coefficient and flow rate without significant loss of efficiency and increase in column back pressure (23).

1.2.10 Example 1.1

Method development and simultaneous optimization of gradient program and column temperature for UHPLC separation

This example describes method development for an intermediate. Four columns, two organic solvents, and six buffers were scouted at UV wavelength 265 nm. Four columns were BEH C18 50 × 2.1 mm, 1.7 μm, BEH phenyl 50 × 2.1 mm, 1.7 μm, Zorbax SB-C8 50 × 2.1 mm, 1.8 μm, and Bonus RP 50 × 2.1 mm, 1.8 μm. Two organic solvents were acetonitrile and methanol. Six buffers were 0.1% TFA in water (pH ∼2.0), 0.1% acetic acid (pH 3.3), and ammonium acetate (pH 4, 4.8, 5.8, and 6.8).

The scouting results indicate that the compound was unstable at pH close to 7, and it was difficult to elute the intermediate and one of its impurities out of the columns with methanol. Therefore, 0.1% TFA buffer, acetonitrile, and BEH phenyl column were selected. However, separation of its impurities needs to be further optimized.

Two basic gradients with different gradient times (2.5 and 7.5 min) were run at two different column temperatures (20 and 50°C) on BEH phenyl 100 × 2.1 mm, 1.7 μm to optimize the separation of an API intermediate and its impurities. The data was input to Drylab®, and a resolution map was obtained and shown in Figure 1.6, where the smallest value of resolution (R_s) of any two critical peaks in

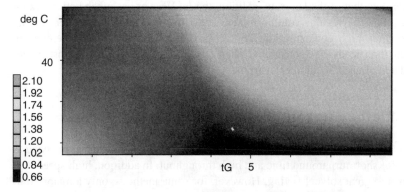

Figure 1.6   Two-dimensional resolution map of the column temperature (°C) against gradient time (t_G, min) for the separation of an intermediate and its impurities 1–4.

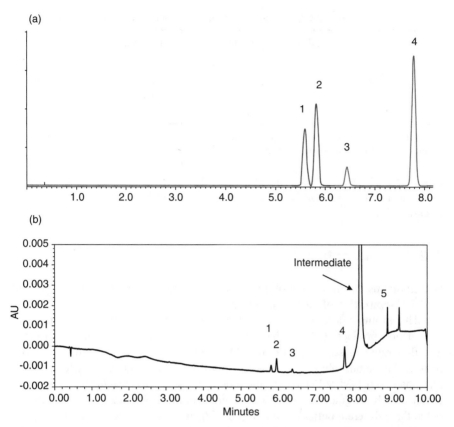

Figure 1.7   Predicted (a) and experimental (b) chromatograms of an intermediate were optimized by 2.5 and 7.5 min gradient basic runs at two different column temperatures (20 and 50°C). Conditions: BEH Phenyl 100 × 21 mm, 1.7 μm; mobile phase A 0.1% TFA in water; mobile phase B acetonitrile; gradient program 32%–60% B for 6.9 min, 60%–100% B for 1.1 min and hold at 100% B for 1 min; flow rate 0.6 mL/min; column temperature 50°C; injection volume 0.7 μL; detection wavelength 265 nm. Peaks 1–5 are impurities.

the chromatogram is plotted as a function of two simultaneously varied experimental parameters. In this case the parameters are gradient time and column temperature. Figure 1.6 indicates that higher temperature and longer gradient time can achieve better resolution. Gradient time (tG) of 6.9 min and column temperature of 50°C were selected with the consideration of resolution criterion, run time, and column life at high temperature.

The predicted optimum conditions were run, and the predicted and experimental chromatograms are shown in Figure 1.7. Only data of impurities 1–4 was input to Drylab® because the intermediate and impurity 5 eluted after the gradient. The experimental chromatogram is similar to the predicted chromatogram for impurities 1–4. In addition, retention times of the predicted and experimental chromatograms were listed and compared in Table 1.4. The results indicate that the predicted retention

Table 1.4   Comparison of Experimental Retention Times with the Predicted by Drylab®

Peak	Experimental (min)	Predicted (min)	Difference (min)[a]	% Error[b]
1	5.803	5.58	0.223	4.0
2	5.951	5.82	0.131	2.3
3	6.37	6.43	−0.06	−0.9
4	7.778	7.78	−0.002	0.0

[a]Difference = experimental − predicted by Drylab.
[b]% error = [(experimental − predicted)/predicted] × 100.

times were close to the experimental ones. The relative retention time error was not greater than 4.0%. DryLab® demonstrated reasonable prediction accuracy for retention time.

1.2.11   Example 1.2

Simultaneous optimization of gradient program and mobile phase pH for UHPLC separation of basic compounds

This example describes method development for basic API, its impurities, and degradation products (48). Zorbax SB C18 50 × 2.1 mm, 1.8 μm was selected. The flow rate was set at 0.5 mL/min. Methanol and buffer were used as a mobile phase. The mobile phase A consisted of 5% methanol and 95% buffer (10 mM phosphate + 0.1% triethylamine); the mobile phase B was 80% methanol and 20% buffer. Two basic gradients with different gradient times (7 and 21 min) were carried out at three different buffer pH values (6.2, 6.6, and 7.0).

The data was input to Drylab®, and the resolution map has been obtained and is shown in Figure 1.8. In this case the critical resolution in the chromatogram is plotted as a function of gradient time (min) and mobile phase buffer pH. The resolution map

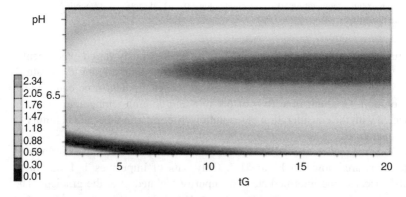

Figure 1.8   Two-dimensional resolution map of the gradient time (min) against mobile phase pH for the separation of basic API and its related impurities and degradation products. Reprinted from Fekete, S.; Fekete, J.; Molnar, I.; Ganzler, K.; *J. Chromatogr. A.* 2009, 1216: 7816–7823. Copyright (2009), with permission from Elsevier.

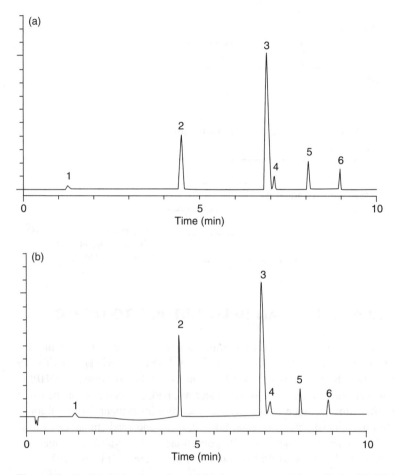

Figure 1.9   Predicted (a) and experimental (b) chromatograms. Column: Zorbax SB C18 50 × 2.1 mm, 1.8 μm, mobile phase A: methanol–buffer 5–95 V/V% (buffer: 10 mM phosphate + 0.1% triethylamine, pH 6.7), mobile phase B: methanol–buffer 80–20 V/V% (buffer: 10 mM phosphate + 0.1% triethylamine, pH 6.7), gradient elution (initial 0% B, at 0.7 min 0% B, at 3.1 min 65% B and 100% B at 10 min), flow: 0.5 mL/min ($p = 531$ bar), column temperature: 30°C, injection volume: 3 μL, detection: 230 nm, analytes: basic drug API and its related impurities and degradation products: (1) peak of light stress origin (unknown) (2) 1-naphtol (3) duloxetine (4) duloxetine-3-isomer impurity (5) dimethyl-duloxetine impurity, and (6) duloxetine impurity A. Reprinted from Fekete, S.; Fekete, J.; Molnar, I.; Ganzler, K.; *J. Chromatogr. A.* 2009, 1216: 7816–7823. Copyright (2009), with permission from Elsevier.

indicates buffer pH 6.7 would achieve enough resolution within 10 min. The predicted optimal condition was tested. The predicted and experimental chromatograms are shown in Figure 1.9. The experimental chromatogram was similar to the predicted chromatogram. The predicted retention times and resolution were in good agreement with the experimental ones. The average relative predicted retention time error was less than 2%, and the average relative predicted resolution (R_s) error was 6.5%.

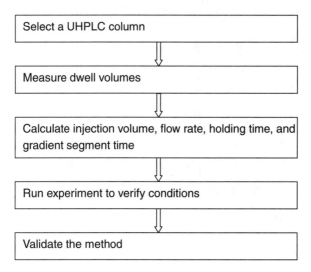

Figure 1.10   General process for conversion of an HPLC method to UHPLC.

1.3   CONVERSION OF AN HPLC METHOD TO UHPLC

There are many HPLC methods used for commercial drug products and new investigational drug products in the pharmaceutical industry. Some converted methods were reported in the literature (2, 56–62). HPLC methods can be converted to UHPLC systematically via the following steps: (i) select a UHPLC column with the same chemistry to maintain the same selectivity; (ii) measure instrument dwell volumes; (iii) calculate injection volume, flow rate, holding time, and gradient segment time for UHPLC; (iv) run an experiment to verify conditions; and (v) validate the method. Each step is described and discussed in detail and summarized in Figure 1.10.

1.3.1   Select a UHPLC Column with the Same Chemistry

There are a variety of HPLC columns used in the pharmaceutical industry. To convert HPLC methods to UHPLC, one should always contact the HPLC column vendors to see if they provide UHPLC columns with the same chemistry. The major UHPLC instrument manufacturers such as Waters Cor. (Milford, MA) and Agilent Technologies, Inc. (Santa Clara, CA) also provide some UHPLC columns with the same chemistry of their HPLC columns. Table 1.5 lists some examples.

If UHPLC columns with the same chemistry as HPLC columns are not available, UHPLC columns with a similar chemistry from other column vendors can be evaluated.

Once the column with the same or similar chemistry is selected, column dimensions such as length and diameter are chosen. The most common diameter of UHPLC

Table 1.5   Examples of HPLC and UHPLC Columns with the Same Chemistry

	HPLC	UHPLC
Waters	XBridge C18 3.5 or 5 μm	BEH C18 1.7 μm
	XBridge Phenyl 3.5 or 5 μm	BEH Phenyl 1.7 μm
	XBridge CN 3.5 or 5 μm	BEH CN 1.7 μm
Agilent	Zorbax Eclipse Plus C18 3.5 or 5 μm	Zorbax Eclipse Plus C18 RRHD 1.8 μm
	Zorbax SB-C18 3.5 or 5 μm	Zorbax SB-C18 RRHD 1.8 μm

columns is 2.1 mm, and the column lengths are 10 and 5 cm. Different column lengths can be selected, depending on the complexity of separation.

1.3.2 Measure Dwell Volumes of HPLC and UHPLC Systems

To calculate holding time, the dwell volumes of both the HPLC and UHPLC systems need to be determined. There are several ways to measure the dwell volume. The following is an example for an HPLC system.

Prepare 1% acetone in water as mobile phase B, and water alone as mobile phase A. A column under measurement is removed from the line, and the tubing is connected with a zero dead volume union. A gradient of 10 min is run from 100% A to 100% B at the flow rate of 2 mL/min and held for 10 min at 100% B with detection wavelength at 260 nm. The difference between 5 min and the time at half height between the initial baseline and the plateau times the flow rate is the dwell volume for the HPLC system. The typical dwell volume for most HPLC systems is approximately 1 mL.

For a UHPLC system, a similar procedure can be used. However, the flow rate needs to be adjusted according to maximum pressure of the UHPLC system. In addition, a restrictor may be needed if the back pressure is too low. The typical dwell volume for most UHPLC systems is 0.1 mL as shown in Figure 1.11. The approximate dwell volumes for Waters Acquity and Agilent 1290 are listed in Table 1.6.

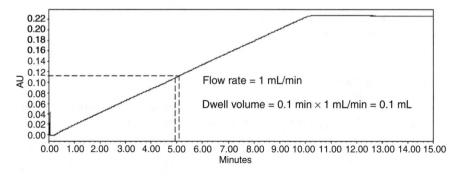

Flow rate = 1 mL/min

Dwell volume = 0.1 min × 1 mL/min = 0.1 mL

Figure 1.11   Measurement of UHPLC dwell volume.

Table 1.6   Approximate Dwell Volumes of Waters Acquity and Agilent 1290

System Name	Mixer (μL)	Approximate Dwell Volume (μL)
Waters Acquity	50	100
	250	300
Agilent 1290	35	100
	100	165

1.3.3 Calculate Injection Volume, Flow Rate, Holding Time, and Gradient Segment Time

A few calculation tools are available from instrument vendors and UHPLC practitioners. The basic principle is to keep the number of the elution column volume the same for HPLC as for UHPLC. In addition, appropriate flow rate and injection volume should be used (63, 64).

1.3.3.1   Injection Volume

Injection volume is adjusted to be proportional to UHPLC column dimensions to maintain its resolution and sensitivity. Sample diluent and concentration are kept the same. The following is used to calculate the UHPLC injection volume:

$$UHPLC\ injection\ volume = HPLC\ injection\ volume$$
$$\times \frac{UHPLC\ column\ volume}{HPLC\ colume\ volume}$$

$$Column\ volume = \pi \times \left(\frac{d}{2}\right)^2 \times l \times \varepsilon \qquad (1.2)$$

where d is the column inner diameter, l is the column length, and ε is the packing material porosity (generally 0.4)(2).

1.3.3.2   Flow Rate

The UHPLC flow rate needs to be adjusted according to column diameter (d) and optimal linear velocity (μ_{opt}) as expressed in the following equation:

$$Flow\ rate(mL/min) = 60 \times \pi \left(\frac{d}{2}\right)^2 \times \mu_{opt} \qquad (1.3)$$

The optimal linear velocity (μ_{opt}) depends on particle size (d_p) and the diffusion coefficient in the mobile phase of compounds to be separated (D_m). The relationship is shown in the following equation (52):

$$\mu_{opt} = 3\, D_m/d_p \qquad (1.4)$$

For small molecules with a diffusion coefficient of 6×10^{-6} cm²/s, the optimal flow rate on a 2.1 mm i. d. 1.7 μm UHPLC column is approximately 0.2 mL/min. However, higher flow rates can be used because the efficiency decreases little at higher flow rates (1). In practice, the flow rates of 0.4 mL/min or 0.6 mL/min or even higher are used.

1.3.3.3   Holding Time and Gradient Segment Time

When converting an HPLC method to UHPLC, one should express system dwell volumes and gradient segments in terms of column volume and maintain the same numbers of column volume for UHPLC and HPLC.

$$Number\ of\ column\ volume\ for\ system = \frac{System\ dwell\ volume}{Column\ volume} \qquad (1.5)$$

$$Number\ of\ column\ volume\ for\ gradient\ segment = \frac{Gradient\ volume}{Column\ volume} \qquad (1.6)$$

1.3.3.4   UHPLC Columns Calculator

Besides manual calculation, UHPLC conditions can be obtained with a UHPLC calculator, for example, the ACQUITY UPLC Columns Calculator, with the following procedure: (i) enter parameters and conditions such as column length, diameter, particle size, molecule weight of samples, flow rate, column temperature, injection volume, instrument dwell volume, and gradient as shown in Figure 1.12; (ii) click the Calculate button to open the Gradient Results window then check the Show Additional Options box as shown in the right of Figure 1.13; and (iii) click a gradient result, and the corresponding gradient will appear. Different gradient results are shown in Figure 1.13, Figure 1.14, Figure 1.15, and Figure 1.16. In addition, the ACQUITY UPLC Columns Calculator provides gradient results for different column dimensions as shown in Figure 1.17.

Usually, the UPLC conditions with scaled gradient (accounting for particle size) are selected for method verification; however, the UPLC conditions with scaled gradient (disregarding particle size) are not correct. The UPLC conditions for shortest analysis time at original peak capacity usually have column back pressure close to or over the instrument limit. The conditions with maximum peak capacity at original

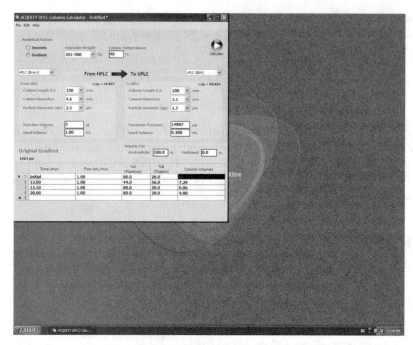

Figure 1.12   Enter parameters and conditions.

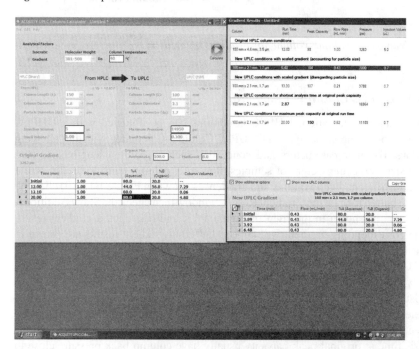

Figure 1.13   New UPLC conditions with scaled gradient (accounting for particle size).

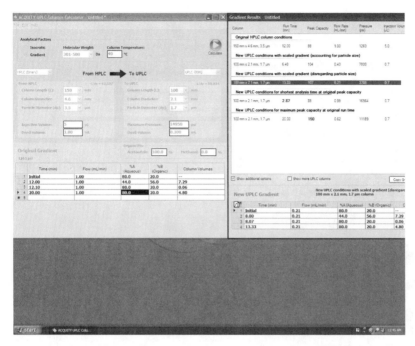

Figure 1.14   New UPLC conditions with scaled gradient (disregarding particle size).

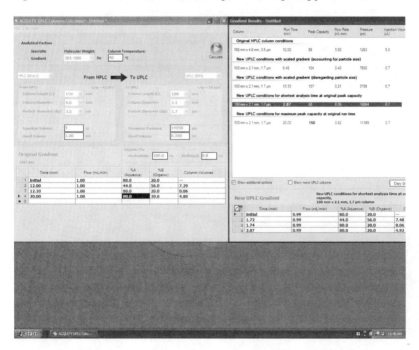

Figure 1.15   New UPLC conditions for shortest analysis time at original peak capacity.

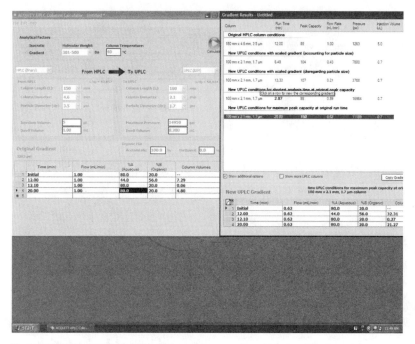

Figure 1.16   New UPLC conditions for maximum peak capacity at original run time.

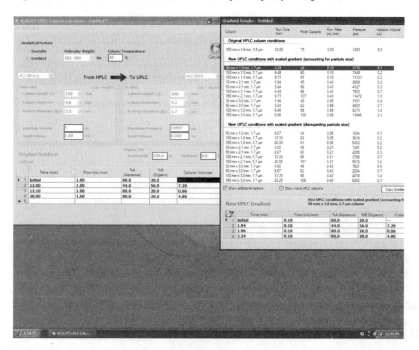

Figure 1.17   Gradient results for different column dimensions.

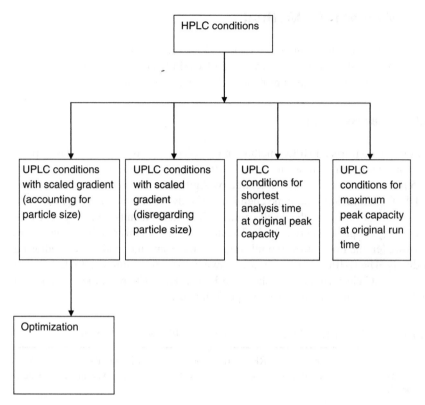

Figure 1.18   Flowchart for the use of the ACQUITY UPLC Columns Calculator.

run time have too long a run time. The flowchart for the use of the ACQUITY UPLC Columns Calculator is shown in Figure 1.18.

1.3.4 Perform Verification Experiment

Once a column is selected and the injection volume, flow rate, holding time, and gradient time are calculated, the UHPLC method conditions are known. The mobile phase, sample diluent and concentration, column temperature, and UV detection wavelength are the same as the HPLC method. However, the UV detection setting needs to be adjusted. The data acquisition rate must be at least 20 Hz to avoid peak distortion and broadening. Perform the verification experiment and compare the results with HPLC in resolution, tailing factor, sensitivity, and repeatability. Different results from HPLC may be obtained because of the effect of very high pressure and frictional heating in UHPLC as discussed in Section 1.2.7. If worse results are obtained, analyze the conditions and adjust accordingly. The conditions such as flow rate and gradient time can be optimized if necessary.

1.3.5 Validate the Method

Method validation is performed like HPLC. However, column lifetime needs to be carefully examined because the knowledge of UHPLC columns is very limited. In addition, an abridged validation protocol is possible if justified.

1.3.6 Example 1.3

This example converts an HPLC method in the left column of Table 1.7. The typical HPLC chromatogram is shown in Figure 1.19. The HPLC column is Zorbax Bonus-RP, 3.5 μm, 150 × 4.6 mm, manufactured by Agilent Technologies, Inc. It provides the UHPLC column with the same chemistry. The dwell volume of an Agilent 1100 HPLC system is 1 mL, and the dwell volume of a Waters Acquity UPLC system with a TFA mixer is approximately 0.3 mL. The flow rate is 0.43 mL/min based on Figure 1.13. However, the flow rate of 0.6 mL/min was used, and the gradient was adjusted accordingly. The UHPLC conditions are shown in the right column of Table 1.7. The typical UHPLC chromatogram is shown in Figure 1.20. The run time was reduced from 20 to 4.65 min without sacrificing performance.

Table 1.7 HPLC and UHPLC Method Conditions for API Assay and Impurities

Column details (column type, particle size and column dimensions)	Zorbax Bonus-RP, 3.5 μm, 150 × 4.6 mm, or validated equivalent			Zorbax Bonus-RP, 1.8 μm, 100 × 2.1 mm, or validated equivalent		
Column temperature	40°C			40°C		
Mobile phase A	0.1% (v/v) TFA in water			0.1% (v/v) TFA in water		
Mobile phase B	Acetonitrile			Acetonitrile		
Flow rate	1.0 mL/min			0.6 mL/min		
Gradient profile	Time (min)	%A	%B	Time (min)	%A	%B
	0	80	20	0	80	20
	12	44	56	2.78	44	56
	12.1	80	20	2.80	80	20
	20	80	20	4.65	80	20
Detector wavelength	268 nm			268 nm		
Injection volume	5 μL			0.7 μL		
Data collection time/reporting time	12 min			2.78 min		
Run time	20 min			4.65 min		
UV detection data acquisition rate	5 Hz			20 Hz		
Autosampler wash solvent	Diluent if needed			Weak washing solvent: mobile phase A Strong washing solvent: mobile phase B		

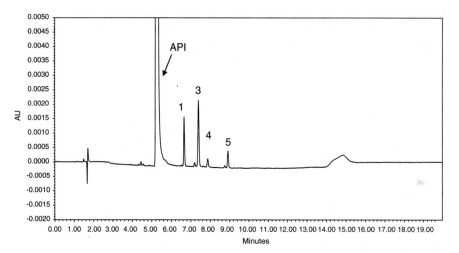

Figure 1.19   Typical HPLC chromatogram. Peaks 1–5 are impurities.

1.3.7   Example 1.4

This example converts an HPLC method for an intermediate to UHPLC. The HPLC conditions are listed in the left column of Table 1.8. The typical HPLC chromatogram is shown in Figure 1.21. The HPLC column is XBridge C18, 150 × 4.6 mm, 3.5 μm, manufactured by Waters Corp. It provided the UHPLC column with the same chemistry. The dwell volume of an Agilent 1100 system is 1 mL, and the dwell volume of a Waters Acquity UPLC system with a TFA mixer is 0.3 mL. The flow rate is 0.43 mL/min based on Figure 1.22. However, the flow rate of 0.6 mL/min was used, and the gradient was adjusted accordingly. The UHPLC conditions are shown in the right column of Table 1.8. The typical UHPLC chromatogram is shown in Figure 1.23. The run time was reduced from 25 to 5.80 min without sacrificing performance.

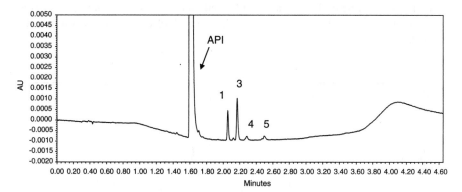

Figure 1.20   Typical UHPLC chromatogram. Peaks 1–5 are impurities.

Table 1.8 HPLC and UHPLC Method Conditions for an Intermediate

Column details(column type, particle size and column dimensions)	XBridge C18, 150 × 4.6 mm, 3.5 μm, or validated equivalent			BEH C18, 100 × 2.1 mm, 1.7 μm, or validated equivalent		
Column temperature	40°C			40°C		
Mobile phase A	10 mM NH_4OAc in water			10 mM NH_4OAc in water		
Mobile phase B	Acetonitrile			Acetonitrile		
Flow rate	1.0 mL/min			0.6 mL/min		
Gradient profile	Time (min)	%A	%B	Time (min)	%A	%B
	0	84	16	0	84	16
	20	53	47	4.63	53	47
	20.1	84	16	4.65	84	16
	25	84	16	5.80	84	16
Detector wavelength	242 nm			242 nm		
UV detection data acquisition rate	5 Hz			20 Hz		
Injection volume	5 μL			0.7 μL		
Data collection time/reporting time	20 min			4.63 min		
Run time	25 min			5.80 min		
Autosampler wash solvent	Not applicable			Weak washing solvent: mobile phase A Strong washing solvent: mobile phase B		

1.4 CONVERSION OF A UHPLC METHOD TO HPLC

UHPLC has many advantages over HPLC. It is faster and has large solvent saving. However, it is much more expensive than HPLC. The high cost of UHPLC is one of major reasons that prohibits wide adoption of UHPLC. Because it takes much less solvent and time to develop a UHPLC method than an HPLC, one strategy is to develop a UHPLC method and then convert it to an HPLC method and validate the method.

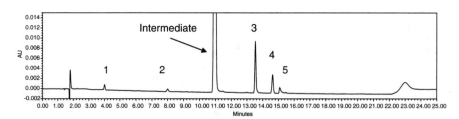

Figure 1.21 Typical HPLC chromatogram. Peaks 1–5 are impurities.

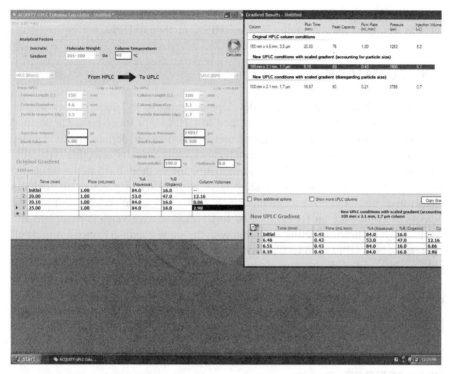

Figure 1.22   New UPLC conditions with scaled gradient (accounting for particle size).

In addition, outsourcing becomes more and more common to reduce the cost of drug development in the pharmaceutical industry. UHPLC methods developed in a company with UHPLC capability need to be transferred to contract research organizations (CROs), which may not have UHPLC capability.

Either case described earlier needs to convert a UHPLC method to HPLC. The principle of the conversion from a UHPLC method to HPLC is the same as the

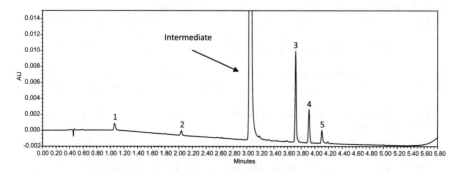

Figure 1.23   Typical UHPLC chromatogram. Peaks 1–5 are impurities.

conversion from an HPLC method to UHPLC. The examples in Section 1.3 can also serve as examples for conversion from a UHPLC method to HPLC.

1.5 SUMMARY

This chapter describes the method development process and strategies as well as the process of converting an HPLC method to UHPLC, or vice versa. The method development process includes gathering sample information, defining method goals, scouting columns and mobile phases, analyzing scouting results, selecting separation conditions, optimizing the method, and validating the method. Each step is discussed in detail, and the examples of simultaneous optimization of gradient and column temperature and gradient and mobile phase buffer pH are given. In addition, the process of converting an HPLC method to UHPLC or vice versa is described and discussed. It includes selecting a column with the same chemistry; measuring dwell volumes of both HPLC and UHPLC systems; calculating injection volume, flow rate, holding time, and gradient segment time; performing experiments for verification; and validating methods. API and intermediate HPLC methods are converted to UHPLC methods as examples using the ACQUITY UPLC Columns Calculator. The same principle, process, and examples can be used for conversion from a UHPLC method to HPLC.

REFERENCES

1. SWARTZ, M. E.; *Liq. Chromatogr. & Rel. Technol.* 2005, 28: 1253–1263.
2. CHEN, S.; KORD, A.; *J. Chromatogr. A.* 2009, 1216: 6204–6209.
3. SNYDER, L.R.; KIRKLAND, J.J.; GLAJCH, J.L.; *Practical HPLC Method Development, 2nd Edition.* New York: Wiley-Interscience, 1997.
4. DONG, M.W.; *Modern HPLC for Practicing Scientists.* [Online] New York: John Wiley & Sons, 2006.
5. WILSON, N.S.; GILROY, J.; DOLAN, J.W.; SNYDER, L.R.; *J. Chromatogr. A.* 2004, 1026: 91–100.
6. WYNDHAM, K.D., et al.; *Anal. Chem.* 2003, 75: 6781–6788.
7. LI, Y.; TERFLOTH, G.; KORD, A.; *American Pharm. Review.* May/June 2009.
8. RAO, R.N.; NAGARAJU, V.; *J. Pharm. Biomed. Anal.* 2003, 33: 335–377.
9. HEWITT, E.F.; LUKULAY, P.; GALUSHKO, S.; *J. Chromatogr. A.* 2006, 1107: 79–87.
10. LI, W.; RASMUSSEN, H.T.; *J. Chromatogr. A.* 2003, 1016: 165–180.
11. ICH Q2(R1); Validation of Analytical Procedures: Test and Methodology, Current Step 4 version. [Online] 2005.
12. ICH Q3A(R2); Impurities in New Drug Substances, Current Step 4 version. [Online] October 25, 2006.
13. ICH Q3B(R2); Impurities in New Drug Products, Current Step 4 version. [Online] June 2, 2006.
14. NEUE, U.D.; *HPLC Columns: Theory, Technology, and Practice.* New York: John Wiley & Sons, 1997.
15. RUTA, J.; RUDAZ, S.; MCCALLEY, D.V.; VEUTHEY, J.-L.; GUILLARME, D.; *J. Chromatogr. A.* 2010, 1217: 8230–8240.
16. CANCELLIERE, G.; CIOGLI, A.; D'ACQUARICA, I.; GASPARRINI, F.; KOCERGIN, J.; MISITI, D.; PIERINI, M.; RITCHIE, H.; SIMONE, P.; VILLANI, C.; *J. Chromatogr. A.* 2010, 1217: 990–999.
17. O'GARA, J.E.; WYNDHAM, K.D.; *J. Liq. Chromatogr. & Rel. Technol.* 2006, 29: 1025–1045.

18. WYNDHAM, K.D.; O'GARA, J.E.; WALTER, T.H.; GLOSE, K.H.; LAWRENCE, N.L.; ALDEN, B.A.; IZZO, G.S.; HUDALLA, C.J.; IRANETA, P.C.; *Anal. Chem.* 2003, 75: 6781–6788.
19. GRITTI, F.; LEONARDIS, I.; SHOCK, D.; STEVENSON, P.; SHALLIKER, A.; GUIOCHON, G.; *J. Chromatogr. A.* 2010, 1217: 1589–1603.
20. GRITTI, F.; GUIOCHON, G.; *J. Chromatogr. A.* 2010, 1217: 1604–1615.
21. GRITTI, F.; LEONARDIS, I., ABIA, J.; GUIOCHON, G.; *J. Chromatogr. A.* 2010, 1217: 3819–3843.
22. YE, C.; TERFLOTH, G.; LI, Y.; KORD, A.; *J. Pharm. Biomed. Anal.* 2009, 50: 426–431.
23. NGUYEN, D. T.-T.; GUILLARME, D.; HEINISCH, S.; BARRIOULET, M.-P.; ROCCA, J.-L.; RUDAZ, S.; VEUTHEY, J.-L.; *J. Chromatogr. A.* 2007, 1167: 76–84.
24. KAMENIK, Z.; HADACEK, F.; MARECKOVA, M.; ULANOVA, D.; KOPECKY, J.; CHOBOT, V.; PLHACKOVA, K.; OLSOVSKA, J.; *J. Chromatogr. A.* 2010, 1217: 8016–8025.
25. BIJLSMA, L.; SANCHO, J.V.; PITARCH, E.; IBANEZ, M.; HERNANDEZ, F.; *J. Chromatogr. A.* 2009, 1216: 3078–3089.
26. MARIN, J.M.: GRACIA-LOR, E.; SANCHO, J.V.; LOPEZ, F.J.; HERNANDEZ, F.; *J. Chromatogr. A.* 2009, 1216: 1410–1420.
27. WHELAN, M.; KINSELLA, B.; FUREY, A.; MOLONEY, M.; CANTWELL, H.; LEHOTAY, S.J.; DANAHER, M.; *J. Chromatogr. A.* 2010, 1217: 4612–4622.
28. AZNAR, M.; CANELLAS, E.; NERIN, C.; *J. Chromatogr. A.* 2009, 1216: 5176–5181.
29. VILLIERS, A.D.; CABOOTER, D.; LYNEN, F.; DESMET, G.; SANDRA, P.; *J. Chromatogr. A.* 2009, 1216 3270–3279.
30. FOUNTAIN, K.J.; NEUE, U.D.; GRUMBACH, E.S.; DIEHL, D.M.; *J. Chromatogr. A.* 2009, 1216: 5979–5988.
31. KRISHNAIAH, C.; REDDY, A.R.; KUMAR, R.; MUKKANTI, K.; *J. Pharm. Biomed. Anal.* 2010, 53: 483–489.
32. RAO, D.D.; SATYANARAYANA, N.V.; REDDY, A.M.; SAIT, S.S.; CHAKOLE, D.; MUKKANTI, K.; *J. Pharm. Biomed. Anal.* 2010, 51: 736–742.
33. DECONINCK, E.; SACRE, P.Y.; BAUDEWYNS, S.; COURSELLE, P.; BEER, J.D.; *J. Pharm. Biomed. Anal.* 2011, 56: 200–209.
34. DECONINCK, E.; CREVITS, S.; BATEN, P.; COURSELLE, P.; BEER, J.D.; *J. Pharm. Biomed. Anal.* 2011, 54: 995–1000.
35. KRISHNAIAH, C.; MURTHY, M.V.; KUMAR, R.; MUKKANTI, K.; *J. Pharm. Biomed. Anal.* 2011, 54: 667–673.
36. CHICO, J.; RUBIES, A.; CENTRICH, F.; COMPANYO, R.; PRAT, M.D.; GRANADOS, M.; *J. Chromatogr. A.* 2008, 1213: 189–199.
37. HAN, Z.; ZHENG, Y.; CHEN, N.; LUAN, L.; ZHOU, C.; GAN, L.; WU, Y.; *J. Chromatogr. A.* 2008, 1212: 76–81.
38. NURMI, J.; PELLINEN, J.; *J. Chromatogr. A.* 2011, 1218: 6712–6719.
39. BAKER, D.R.; KASPRZYK-HORDERN, B.; *J. Chromatogr. A.* 2011, 1218; 1620–1631.
40. KAMENIK, Z.; HADACEK, F.; MARECKOVA, M.; ULANOVA, D.; KOPECKY, J.; CHOBOT, V.; PLHACKOVA, K.; OLSOVSKA, J.; *J. Chromatogr. A.* 2010, 1217: 8016–8025.
41. HAN, Z.; ZHENG, Y.; LUAN, L.; REN, Y.; WU, Y.; *J. Chromatogr. A.* 2010, 1217: 4365–4374.
42. I, T.-P.; SMITH, R.; GUHAN, S.; TAKSEN, K.; VAVRA, M.; MYERS, D.; HEARN, M.T.W.; *J. Chromatogr. A.* 2002, 972: 27–43.
43. HOANG, T.H.; CUERRIER, D.; MCCLINTOCK, S.; MASO, M.D.; *J. Chromatogr. A.* 2003, 991: 281–287.
44. MOLNAR, I.; *J. Chromatogr. A.* 2002, 965: 175–194.
45. KRISKO, R.M.; MCLAUGHLIN, K.; KOENIGBAUER, M.J.; LUNTE, C.E.; *J. Chromatogr. A.* 2006, 1122: 186–193.
46. XIAO, K.P.; XIONG, Y.; LIU, F.Z.; RUSTUM, A.M.; *J. Chromatogr. A.* 2007, 1163: 145–156.
47. GARCIA-LAVANDEIRA, J.; LOSADA, B.; MARTINEZ-PONTEVEDRA, J.A.; LORES, M.; CELA, R.; *J. Chromatogr. A.* 2008, 1208: 116–125.
48. FEKETEA, S.; FEKETE, J.; MOLNÁR, I.; GANZLER, K.; *J. Chromatogr. A.* 2009, 1216: 7816–7823.
49. GUIOCHON, G.; SEPANIAK, M.J.; *J. Chromatogr.* 1992, 606: 248–250.

50. FALLAS, M.M.; NEUE, U.D.; HADLEY, M.R.; MCCALLEY, D.V.; *J. Chromatogr. A.* 2008, 1209: 195–205.
51. FALLAS, M.M.; NEUE, U.D.; HADLEY, M.R.; MCCALLEY, D.V.; *J. Chromatogr. A.* 2010, 1217: 276–284.
52. MACNAIR, J.E.; LEWIS, K.C.; JORGENSON, J.W.; *Anal. Chem.* 1997, 69: 983–989.
53. MACNAIR, J.E.; PATEL, K.D.; JORGENSON, J.W.; *Anal. Chem.* 1999, 71: 700–708.
54. GRITTI, F.; GUIOCHON, G.; *Anal. Chem.* 2008, 80: 5009–5020.
55. VILLIERS, A.D.; LAUER, H.; SZUCS, R.; GOODALL, S.; SANDRA, P.; *J. Chromatogr. A.* 2006, 1113: 84–91.
56. NOVAKOVA, L.; MATYSOVA, L.; SOLICH, P.; *Talanta.* 2006, 68: 908.
57. ALVI, K.A.; WANG, S.; TOUS, G.; *J. Liq. Chromatogr. & Rel. Technol.* 2008, 31: 941–949.
58. WU, T.; WANG, C.; WANG, X.; XIAO, H.; MA, Q.; ZHANG, Q.; *Chromatographia.* 2008, 68: 803.
59. FEKETE, S.; FEKETE, J.; GANZLER, K.; *J. Pharm. Biomed. Anal.* 2009, 49: 833–838.
60. PEDRAGLIO, S.; ROZIO, M.G.; MISIANO, P.; REALI, V.; DONDIO, G.; BIGOGNO, C.; *J. Pharm. Biomed. Anal.* 2007, 44: 665–673.
61. JERKOVICH, A.; MAKAROV, A.; LOBRUTTO, R.; VIVILECCHIA, R.; *American Pharm. Review.* 2007, 10: 32–37.
62. WREN, S.A.C.; TCHELITCHEFF, P.; *J. Chromatogr. A.* 2006, 1119: 140–146.
63. GUILLARME, D.; NGUYEN, D.T.-T.; RUDAZ, S.; VEUTHEY, J.-L.; *Eu. J. Pharm. Biopharm.* 2007, 66; 475–482.
64. GUILLARME, D.; NGUYEN, D.T.-T.; RUDAZ, S.; VEUTHEY, J.-L.; *Eu. J. Pharm. Biopharm.* 2008, 68: 430–440.

Chapter 2

Method Transfer Between HPLC and UHPLC Platforms

Gregory K. Webster, Thomas F. Cullen, and Laila Kott

2.1 INTRODUCTION

In its quest for faster, more efficient chromatographic separations, the pharmaceutical and environmental industries have welcomed the innovation of UHPLC (ultrahigh pressure liquid chromatography) technology into their analytical laboratories. The advent of UHPLC has challenged these industries to move toward improved capabilities in run time, selectivity, and detection. Interestingly, this movement has not been limited to the use of UHPLC but has extended its reach into challenging the realm of traditional liquid chromatography as well. In other words, the coming of age of UHPLC has also challenged those still using traditional HPLC (high performance liquid chromatography) platforms to optimize their methods to deliver near UHPLC efficiencies when possible, most notably accomplished through the use of superficially porous particle stationary phases (1).

With the introduction of the first commercial UHPLC system in 2005, pharmaceutical laboratories rushed to investigate whether these high-pressure systems would truly offer efficiencies that would result in a greater ability to resolve difficult-to-separate matrices. The systems were introduced because of their enhanced separation efficiencies, not solely for a reduction in mobile phase usage or analysis times. These benefits were welcome bonuses. Scientists looked to develop straightforward ways to convert existing methods validated on 3 μm, 4.6 mm i.d. columns to the new UHPLC standard. The use of UHPLC enabled improved efficiencies in pharma (2–13) as well as in environmental (14–16) applications.

In 2007, an unexpected turn came for the industry with the introduction of superficially porous particles (SPP), typically in the range of 2.7 μm. With these column packings, laboratories could achieve near-UHPLC separations without needing the capital investment of new instrumentation (17–21). The attractiveness of these

Ultra-High Performance Liquid Chromatography and Its Applications, First Edition. Edited by Quanyun Alan Xu.
© 2013 John Wiley & Sons, Inc. Published 2013 by John Wiley & Sons, Inc.

columns was that they eliminated the need for capital investments in equipment to transfer methods globally. In this regard, does transferring methods to SPP-based stationary phases make more sense than migrating to UHPLC platforms? Current trends in the near term market appear to indicate that UHPLC may have its niche in high throughput determinations and SPP-based stationary phases operating from 400–600 bar may be the optimum for routine testing if the separation can handle a 10%–20% efficiency loss from UHPLC (22, 23). This pressure range is easily achieved on most commercial LCs these days but will not likely be predominant until the Waters 2695, Agilent 1100/1200, and equivalent systems end their useful life in most laboratories. Today's analytical chemist needs to know when to apply each of these column formats to best suit their analytical and business needs.

The primary hindrance to the universal implementation of UHPLC is, for most laboratories, that the platform often requires new capital outlays to buy instrumentation capable of running at higher system pressures (typically above 400–1,500 bar). Access to properly qualified systems can create a roadblock to routine regulated testing. This limited availability of UHPLC becomes less of an issue each year as instrumentation comes to an end of its depreciation lifespan and gets replaced with newer LC systems. Yet, in the current state, pharmaceutical companies across the industry are less inclined to develop and transfer UHPLC analytical methods until they are convinced receiving laboratories have qualified UHPLC systems available on demand. In addition, not all UHPLC instruments perform to the same specifications (14), making transfer even more difficult as vendor-specific limitations can potentially cripple an analytical transfer.

Adapting methods from traditional HPLC to UHPLC has been the focus of many laboratories in recent years. In addition, early adoption of UHPLC and the subsequent inability to transfer to receiving laboratories led to investigations examining how to revert from UHPLC back to traditional HPLC conditions (24–31). Transferring methods from UHPLC to traditional LC platforms potentially addresses the analytical method transfer limitations of UHPLC within the industry and allows labs to execute Quality by Design robustness studies using UHPLC methods and then to scale the conditions back to more commonly available HPLC platforms. Because of this diversity of liquid chromatographic platforms available across the industry, there has been much interest in how to transfer HPLC methods to the UHPLC platform, as well as in returning to HPLC from a UHPLC platform. The adaptation of chromatographic method conditions from one platform to the other is the focus of this chapter.

2.2 TRANSFERRING HPLC METHODS TO UHPLC

As UHPLC instrumentation has become more readily available in routine quality control (QC) laboratories, the incentive to convert existing HPLC methods to utilize the enhanced capabilities of these instruments has increased. Accordingly, there has been an increased emphasis on the development of strategies and approaches for accomplishing this.

UHPLC is characterized by the use of chromatographic conditions that employ smaller particle size packing materials and instrumentation capable of operating at higher pressures and higher linear velocities than those typically encountered in HPLC. Most UHPLC columns also have smaller column diameters than those typical of HPLC methods.

2.2.1 Fundamental Physical Relationships

The fundamental relationships between the physical dimensions of a chromatographic column and the performance of separations performed using it are well known and have been extensively described in the literature. Of these, perhaps the most critical to understanding the implications of transferring HPLC methods to UHPLC are as follows:

1. The relationship between column length (L), the particle size (d_p) of the stationary phase packing material and the height equivalent of a theoretical plate (H) (and thus with the number of theoretical plates, N) (32):

$$H \propto \frac{d_p}{L}, \ H = \frac{N}{L}, \ N \propto d_p \qquad (2.1)$$

This relationship illustrates the fundamental advantage of smaller diameter packing materials, namely, higher chromatographic efficiency if column length is held constant or the ability to use shorter columns (and correspondingly shorter methods) if constant efficiency is maintained. Another benefit can be discerned by examining the van Deemter equation in more detail, particularly as it relates to linear velocity, u.

$$H = A + \frac{B}{u} + Cu \qquad (2.2)$$

Where A, B, and C are constants and $C \propto d_p^2$.

As the C term is the one that punishes high linear velocities, its proportionality to the square of particle size enables the use of higher linear velocities with a lower accompanying loss of plates. The combination of these effects (the ability to use shorter columns at higher flow rates) results in the shorter run times associated with UHPLC methods.

2. The relationship between particle size and the pressure drop (ΔP) across the column (33, 34):

$$\Delta P \propto \frac{1}{d_p^2} \qquad (2.3)$$

And so the fundamental limitation associated with smaller diameter packing material is clear: the lower practical limit for particle size is driven by the extremely high pressures associated with flow through them.

3. The relationship between resolution and plates (and therefore, between resolution and particle size) (35):

$$R \propto \sqrt{N} \propto \sqrt{d_p} \tag{2.4}$$

This relationship, when taken together with the practical pressure limitations described, illustrates the diminishing returns associated with ever diminishing stationary phase particles. As a practical matter, particles with diameters of less than 1.5 μm are uncommon.

2.2.2   Types of UHPLC Applications

Manipulation of the preceding relationships enables the practitioners of UHPLC to maximize resolution and/or reduce chromatographic run time. Chestnut and Salisbury (36) describe three distinct types of applications that can benefit from the application of UHPLC technology.

- High throughput methods are described as those that require few theoretical plates (typically <5,000), but where short run times are essential. Examples include dissolution assays and in-process controls. For these, there may not be a driving force to convert from UHPLC to HPLC platforms unless a critical separation forces it.

- High productivity separations are described as comprising the bulk of the work in a QC lab. These require moderate efficiencies (typically 10,000–15,000 plates) and can benefit from decreased run times. In these cases, the goal is to leverage the smaller values of H to reduce column lengths and thus shorten analyses.

- Finally, high-resolution methods are defined as requiring the maximum possible chromatographic efficiency, where run time is a secondary consideration.

Chestnut and Salisbury provide a case study where a conventional HPLC method was converted to UHPLC in both high-productivity and high-resolution modes. The resulting data for the number of theoretical plates observed and the analysis time achieved illustrate the benefits of UHPLC and the trade-offs that must be considered when employing it. Users of UHPLC have the choice of reducing run time by a significant factor while losing little efficiency, or of gaining significant efficiency without increasing run time.

Clearly, then, when discussing the transfer of HPLC methods to UHPLC, it is critical to define the purpose for the platform change. For the remainder of this section we will confine ourselves to the most common scenario: the development of high productivity methods to achieve separations equivalent to those achieved by HPLC, but with reduced analysis times.

2.2.3 Additional Fundamental Considerations

While the theoretical basis for and advantages of UHPLC can be inferred as described earlier, the practical implementation of these theories reveals additional complexities. Some of these involve relationships and phenomena that were previously of little importance to and thus largely ignored by chromatographers.

In early studies of the use of UHPLC for protein and peptide samples, Jorgenson and McNair (2, 37) found that the effects of frictional heating were, perhaps unsurprisingly, more pronounced in smaller particle columns. The power of heat generation (E) is strongly related to the particle diameter (34):

$$E \propto \frac{1}{d_p^3} \tag{2.5}$$

The higher resistance to mass transport inherent in more tightly packed beds of smaller particles leads to higher rates of frictional heating, which can lead to radial temperature gradients. These gradients result in nonuniform flow and increased dispersion with an accompanying loss of efficiency. Because the problem of radial temperature gradients had previously been solved for the case of capillary electrophoresis using smaller diameter columns by Jorgenson (38), the same approach was attempted with good success for UHPLC. The magnitude of the temperature gradient (ΔT) is related to the column radius (r) (34).

$$\Delta T \propto r^2 \tag{2.6}$$

It is for this reason that most UHPLC columns have a maximum diameter of 2.1 mm (as opposed to the 4.6 or 3.0 mm columns more typically used in HPLC). Thus, when scaling an HPLC method to UHPLC, three column dimensions (column diameter, particle size, and column length) are often varied simultaneously.

2.2.4 Transfer of Isocratic Methods from HPLC to UHPLC

The simplest case for illustrating the way in which a change in column diameter, particle size, and column length necessitates the adjustment of other method parameters is that of an isocratic separation. Here, the number of parameters to be adjusted is limited and are related to the change in the volume of a unit length of column. In general, these adjustments are made to normalize to this change in volume. The injection volume (V_{inj}) must be scaled as the column radius (r) changes:

$$V_{inj} \propto r^2 \tag{2.7}$$

Similarly, the flow rate (F) should be adjusted based on column radius:

$$F \propto r^2 \tag{2.8}$$

A happy consequence of the smaller diameter of UHPLC columns is that they are ideally suited to mass spectrometric (MS) detection, as the total volumetric flow is generally considerably lower than that typical for HPLC methods that use 4.6 mm columns. This lower flow obviates the need for post-column flow splitting. Of course, higher flow rates can be, and often are, used to shorten analysis time (with little penalty in terms of lost plates).

These two adjustments alone are sufficient for the simplest scaling of an isocratic method. A case study described by Yang and Hodges (39) introduces additional considerations. These authors found that the increased resolving power (due to a higher number of theoretical plates, as described earlier) of UHPLC particles allowed them to reduce analysis time by increasing the strength of the eluent (a change in column length was impractical as the original method already used a 50 mm column). Through a combination of increased linear velocity and increased eluent strength, the authors achieved a better than sixfold decrease in analysis time.

Solich et al. (40) describe additional case studies for the conversion of isocratic methods. Here, existing HPLC methods were modified by switching to a shorter, smaller diameter column. The injection volume was scaled as per Eq. (2.7). The flow rate was similarly adjusted as per Eq. (2.8), and then increased to shorten run times. A sevenfold decrease was found in run times. Interestingly, while adequate separation was maintained, the chromatographic profile of the scaled analyses did not match that of the original methods. This discrepancy can be easily attributed to differences in selectivity between the original columns and the UHPLC columns (which were the product of a different manufacturer). This observation is critical, as any modification to a method where the chemistry of the stationary phase is not held constant can be problematic and may be more correctly referred to as method development (rather than method transfer or scaling). Further discussion of this topic will be found later in this chapter.

2.2.5 Cautions on Transferring Isocratic Methods from HPLC to UHPLC

One additional consideration in the scaling of isocratic methods is the increased effect of extra-column band broadening (σ_{ext}^2) as injection volumes decrease. For best results, the ratio of extra-column to on-column broadening (σ_{col}^2) should be kept as small as possible. In practice, this is achieved in UHPLC instruments by minimizing extra-column volume and dispersion through careful attention to flow path geometries. A more detailed discussion of these considerations may be found at the end of this chapter. Longer isocratic retention (increasing $\sigma^2{}_{col}$) will also serve to reduce the $\sigma_{ext}^2/\sigma_{col}^2$ ratio. For this reason, the effects of extra-column band broadening are most clearly seen in the early-eluting peaks in a chromatogram. This

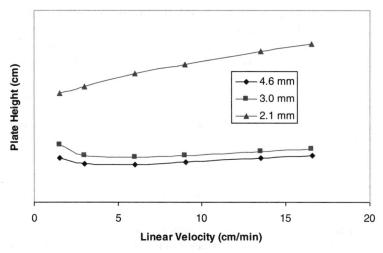

Figure 2.1   Effect of column diameter on the shape of van Deemter curves for an isocratic determination of benzoic acid.

effect has been observed in several case studies (34, 41). Empirical data illustrating the increased effect of extra-column broadening on smaller diameter columns are presented in Figure 2.1. A loss of efficiency is noted as column diameter decreases, particularly from 3 to 2.1 mm. These data are the result of an isocratic determination of benzoic acid, where the only variable was the diameter of the chromatographic column. Flow rate was adjusted as described earlier. Mobile phase, stationary phase, and all other chromatographic parameters were held constant.

This effect was well illustrated in a case study reported by Wu and Clausen (34). Here five alkylbenzenes were separated by an HPLC method. When this method was directly scaled to UHPLC, the separation was adequate, but the column efficiencies were lower than would be predicted. It is likely that the lost efficiency was due to the effects of extra-column band broadening. Pressure drops also did not scale as predicted. This is likely due to physical differences between the HPLC and UHPLC particles. In a similar manner to that described, it was noted that selectivity factors did not remain constant as particle size changed. For these reasons, it is emphasized that the theoretical scaling factors given earlier are approximations, and the observed relationships will likely vary from theory because of additional uncontrolled factors. Again, it is imperative that the stationary phase chemistry remain constant as the method is scaled to an alternate platform.

2.2.6 Transfer of Gradient Methods from HPLC to UHPLC

Gradient elution introduces complexities not observed in isocratic separations. While the mechanism of gradient elution has been thoroughly described in the literature, (42, 43) the application of these theories to the scaling of HPLC method conditions to UHPLC platforms is more recent. Guillarme et al. (28) present an approach to calculating appropriate UHPLC gradient conditions from a source HPLC method. Injection volume and flow rate are treated in the same manner as in isocratic methods. Other parameters are generally scaled on the basis of column volume (V_0).

$$V_0 \propto L \cdot r^2 \tag{2.9}$$

Gradient methods can consist of multiple segments, commonly including an initial isocratic segment. The time (t) for such a segment (whether gradient or isocratic) can most simply be expressed as

$$t \propto \frac{V_0}{F}; \text{ thus } t \propto L \text{ and } t \propto r^2 \tag{2.10}$$

Extra-column broadening (or at least pre-column broadening) has less of an effect on gradient than on isocratic methods because of the focusing effect of gradient elution. In other words, because analytes will be retained at the head of the column until the eluent strength is high enough to cause them to elute, the pre-column volume is less important. However, system-to-system differences in dwell volume can cause several difficulties. System dwell volume (V_d) is the volume between the mixing of the mobile phase components and the column head (which includes the pumps and mixer). As such, it contributes an unintended isocratic segment to the beginning of the method. In the case of HPLC, where the ratio of V_d/V_0 is small, the effect of this isocratic segment is generally negligible. UHPLC columns have a smaller inherent V_0; the resultant higher V_d/V_0 ratio can result in isocratic segments that significantly affect the separation. With the exception of delayed injections, there are few options for modifying a UHPLC method to compensate for this effect. For this reason, the minimization of V_d in UHPLC instruments is critical. Alternatively, HPLC methods can be constructed so as to make the conversion to UHPLC easier. The introduction of a lengthy isocratic segment in an HPLC method can render any additional isocratic component introduced by a conversion to UHPLC insignificant.

2.2.7 Cautions on the Scaling of Gradient Methods from HPLC to UHPLC

Wu and Clausen (34) present several case studies for the scaling of gradient HPLC methods to UHPLC. For one example, a ninefold decrease in run time was achieved by increasing flow rate and decreasing column length and diameter while moving from 5 μm conventional particles to 1.7 μm UHPLC particles. However, while adequate separation was achieved, it was noted that column efficiencies were 10%–20% lower than predicted. While extra-column broadening is less significant in gradient elution, other unmodeled effects exist. The authors again emphasize that theoretical relationships can provide only a starting point for method scaling. Actual method parameters must be adjusted based on observed results, method requirements, and instrumental limitations. Because efficiencies are generally lower than predicted, it may be necessary to use longer columns than theoretically indicated.

2.2.8 Case Studies and Recommendations

A set of experiments was performed to illustrate the preceding effects and recommendations. In each case study, a conventional linear gradient HPLC method was developed on a 150 × 4.6 mm, 3.5 μm Agilent Zorbax SB-C18 column at a flow rate of 1 mL/min. Fast LC methods were then developed on a variety of column technology platforms. UHPLC methods were developed on 2.1 mm diameter, 1.8 μm particle Agilent Zorbax SB-C18 columns with lengths of 50 or 100 mm, by converting them as described and then adjusting to maximize flow rate (and minimize run time), targeting a column pressure of 800–850 bar. Gradient time (and slope) was adjusted from theoretical to match the resolution performance, for the critical pair, of the original method.

In the first case study, acebutolol was analyzed in the presence of acetaminophen, nortriptyline, and qunidine (each spiked at a level of 0.1% relative to acebutolol). Results are listed in Table 2.1. Note that shallower (compared to the base HPLC method) gradient slopes (and thus longer than expected run times) were needed under UHPLC conditions to maintain resolution.

In another case study, ethylparaben was analyzed in the presence of benzoic acid, salicylic acid, and isopropylparaben (each spiked at a level of 0.1% relative to ethylparaben). For this study, data is also presented from a method developed using a column employing superficially porous particles (Zorbax Poroshell SB-C18) on an HPLC with a 600-bar pressure rating. Superficially porous particles have seen increasing use in the development of fast LC methods (44). While a detailed description of this technology is outside the scope of this chapter, a single example is provided for the sake of reference and comparison. A more thorough investigation of the merits of superficially porous particle columns is recommended to the reader. For the experiment at hand, results are seen in Table 2.2. Again, a shallower gradient slope was needed to maintain resolution under UHPLC conditions. No adequate (equivalent to the HPLC method) separation was found using a 50 mm UHPLC column. These examples illustrate the need for further method development after the application of the theoretical scaling factors.

Table 2.1 Results of the Transfer of an HPLC Method for Determination of Trace Acetaminophen, Nortriptyline, and Qunidine in an Acebutolol Sample to a UHPLC Column

Run	Run time (min)	Max Pressure (bar)	Slope (%B/column volume)	Time Saved	Solvent Saved
1 SB-C18, 150 × 4.6 mm, 3.5 μm, run on Agilent 1100	25	100	0.08	0%	0%
2 SB-C18, 100 × 2.1 mm, 1.8 μm, run on Agilent 1290	15	800	0.02	39%	57%
3 SB-C18, 50 × 2.1 mm, 1.8 μm, run on Agilent 1290	6	850	0.02	78%	71%

Table 2.2   Results of the Transfer of an HPLC Method for Determination of Trace Benzoic Acid, Salicylic Acid and Isopropylparaben in an Ethylparaben Sample to UHPLC and to SPP Columns

Run	Run time (min)	Max Pressure (bar)	Slope (%B/column volume)	Time Saved	Solvent Saved
1 SB-C18, 150 × 4.6 mm, 3.5 μm, run on Agilent 1100	14	129	0.17	0%	0%
2 Poroshell SB-C18, 150 × 3 mm, 2.7 μm, run on Agilent 1200 SL	7	510	0.18	49%	57%
3 SB-C18, 100 × 2.1 mm, 1.8 μm, run on Agilent 1290	8	850	0.06	44%	58%

The method developed using the Poroshell column was equivalent to the HPLC method, with a similar gradient slope. Due its higher linear velocity, the run time was decreased twofold.

2.2.9   Final Discussion of Practical Challenges

In 2007, Dong (41) described a set of practical challenges that imperil the scaling of HPLC methods to UHPLC, particularly in a regulated environment. These include the following:

1. The inherently lower injection volumes associated with UHPLC can be prone to injection-to-injection peak area imprecision. Injection precision of <0.5% RSD (typically required of pharmaceutical analyses) are readily achievable for injection volumes of >5 μL. This poses a challenge for method translation from HPLC. Taking into account the scaling factor for conversion of a method from a 4.6 to 2.1 mm diameter column, any method with an injection volume of <24 μL will scale to one where $V_{inj} < 5$ μL.

2. Dong also reported decreased detector sensitivity (signal to noise) due to higher baseline noise than was seen in HPLC applications. This was traced to poor mixing of aqueous and organic phases in the 100 μL static mixer of the UHPLC employed in the study. Larger volume mixers alleviate this issue, but contribute additional system dwell volume, with all the associated drawbacks listed earlier.

2.2.10   Conclusions on the Transfer of Methods from HPLC to UHPLC

The theoretical relationships that govern the scaling of methods from HPLC to UHPLC conditions are well established and provide a good framework for the adjustment of other chromatographic parameters as stationary phase particle size is decreased. However, these predictions are necessarily imperfect because other unmodeled effects exist. For this reason, actual method parameters must be adjusted based on observed results, method requirements, and instrumental limitations. In a regulated (e.g., in a pharmaceutical environment where GMP [good manufacturing practices] regulations apply), this implies that unless a direct scaling is found to produce equivalent chromatography, the translation of an HPLC method to UHPLC conditions may more properly be regarded as the development of a new method. Accordingly, method validation may be appropriate.

2.3 TRANSFERRING UHPLC METHODS TO HPLC PLATFORMS

The primary challenge to scaling liquid chromatographic methods between HPLC and UHPLC is accounting for the varying chemistries and particles within the stationary phase. The scaling calculations assume use of the "same" stationary phase for all columns used in the investigation. Such a hypothesis relies on a manufacturer's ability to produce analytical columns with the same packing and coating efficiencies from 1.5–5 μm particles. As such, methods chosen for transferring from UHPLC to HPLC should use columns with the same stationary phase platform from UHPLC to HPLC. Columns meeting these criteria are commercially available from several manufacturers.

Websites for many column manufacturers offer scaling calculations for their columns. They were all designed with the conversion from HPLC to UHPLC in mind. Typically, the critical parameters in scaling from UHPLC to HPLC are not only the column length and particle size, but also the injection volume (27–31). Nueue et al. (30) recently showed that many separations can be interconverted between 5 μm, 15 cm columns to 1.7 μm, 5 cm columns by making appropriate injection volume adjustments and proposed that U.S. Pharmacopeia (USP) accept the equivalence of these methods. However, the authors here propose that dramatic changes should be made only if the chromatographic profile and resolution of the critical pair is maintained (45).

The primary equations used by the scaling calculators are as follows:
Injection Volume:

$$I_2 = I_1 \cdot \{(dc_2^2 \cdot L_2)/(dc_1^2 \cdot L_1)\} \tag{2.11}$$

Where:

I = the injection volume for method 1 and 2
dc = the column diameters used in method 1 and 2
L = the column length in method 1 and 2 columns.

Flow Rate:

$$F_2 = F_1 \cdot \{(dc_2^2 \cdot dp_1)/(dc_1^2 \cdot dp_2)\} \tag{2.12}$$

Where:

F = the flow rate for method 1 and 2
dc = the column diameters used in method 1 and 2
dp = the particle sizes in method 1 and 2 columns.

Gradient Time:

$$T_2 = T_1 \cdot \{(F_1 \cdot dc_2^2 \cdot L_2)/(F_2 \cdot dc_1^2 \cdot L_1)\} \tag{2.13}$$

Where:

T = the gradient time for method 1 and 2.

Most commercial column manufacturers have these types of calculations available on their website or available for download. The use of method translation calculators was recently reviewed by Majors (26). For example, the proprietary algorithm Grace Discovery Sciences (Deerfield, IL) would predict that a typical 4-min UHPLC method on a 50×2.1 mm, 1.7 μm UHPLC column operating 1 mL/min and 1 μL injection should operate as a 27-min method at 3 mL/min and 4 μL on a 150×3.0 mm, 3 μm HPLC column. As stated earlier in this chapter, the scaling calculations are best used as projection to the range, which should be initially tested followed by further optimization.

At first glance the UHPLC conditions used should convert to HPLC for all commercial systems that can generate flow at 3 mL/min and inject at 4 μL. However, this is deceiving. The critical parameter in this separation is not the ability of the HPLC to handle the projected method conditions; it is whether these conditions yield a pressure under the pressure maximum for the HPLC. Most traditional HPLC platforms tolerate pressure of up to 400 bar. The pressure generated at 3 ml/min is dependent on the mobile phase being used. A low viscosity mobile phase (e.g., one with high organic content or one at high temperatures) may result in column back pressures under 400 bar, but those routinely used in many methods likely may not. Thus, the UHPLC method that is to be converted to a HPLC platform must first be optimized to run at conditions that project the new method pressure will remain under 400 bar. This often entails slowing down the optimized UHPLC method.

2.3.1 UHPLC to Traditional HPLC Columns

As shown by Webster and Elliott (31), a direct GMP method transfer of chromatographic profiles should not be expected because the directly scaled condition may not yield an equivalent chromatographic profile. Of course if the profile was fully maintained, the only advantage of using UHPLC would be run time. However, we know that UHPLC provides efficiency advantages as well. This is illustrated in Figure 2.2 when an optimized UHPLC profile is scaled to a 150×3.0 mm column

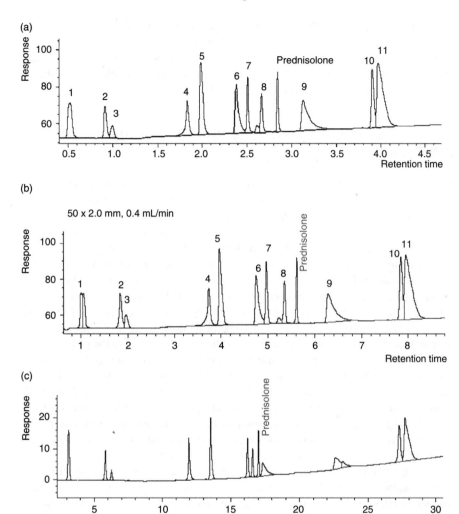

Figure 2.2   Representative UHPLC profile. Analytes: (1) acetaminophen, (2) phenol, (3) benzyl alcohol, (4) salicylic acid, (5) pindolol, (6) quinidine, (7) acebutolol, (8) 3-methyl-4-nitrobenzoic acid, (9) triprolidine, (10) carvedilol, (11) nortripyline. Figures used with permission from reference 31. (a) Original UHPLC separation; (b) Reduced pressure UHPLC separation; (c) Scaled profile on 150 × 3.0, 3 μm column.

format. The resulting profile on an Agilent 1200 is illustrated. This is not the same profile. As a general rule, the change in the profile will be more dramatic as system void volume increases.

Method translation from UHPLC to HPLC for simple separations should be achievable. It has been proposed that optimizing the scaling to include gradient

delays and an optimized flow cell volume can improve the fidelity of the transferred chromatographic profile. While compensating for these parameters certainly
improves the predicted set of conditions and the similarity of the resulting profile, it
does not result in an identical chromatographic profile (31). Flow cell volumes should
primarily affect sensitivity, not profile selectivity.

2.3.2 UHPLC to Superficially Porous Particle HPLC Columns

Recently, several chromatography columns packed with SPP stationary phases have
become commercially available. With the advent of these phases, chromatographic
efficiencies associated with UHPLC conditions can now be achieved on many traditional HPLC platforms. Because of this capability, translating UHPLC methods to
traditional HPLC platforms can better make use of the SPP technology.

Generally, an approximately 2.6 μm SPP particle can be packed in both UHPLC
and traditional HPLC column dimensions (up to 150 mm in length). Figure 2.3
illustrates that a separation produced using 1.8 μm UHPLC particles can have a very
similar profile to that produced using 2.6 μm SPP particles in a column with the same
dimensions. Note, from Eq. (2.12), increasing the particle size should require the need
to scale to a higher flow rate. The peak order change is attributed to the phase change
between manufacturers. Figure 2.4 illustrates running these columns at the same flow
rate. However, the pressure on the SPP particle column was well under 400 bar.
Interestingly, the profile is essentially the same outside the observed peak reversal.
Thus, by simply changing to an SPP stationary phase format the conversion from
UHPLC to traditional HPLC was accomplished. However, this cannot be considered
a direct GMP method translation as the peak order was not maintained and was in
reality the development of a new method for the same separation.

Using the flow rate of the UHPLC method on the 2.6 μm SPP column results
in a separation that can be run with a 400 bar traditional HPLC platform, in a
manner similar to that discussed earlier for the 3 and 5 μm particle columns.
Figure 2.4 illustrates the chromatographic profile of a UHPLC method using a 1.8 μm
C18 stationary phase, the method scaled to a 2.6 μm SPP stationary phase with the
same dimensions, and the method on a SPP column in a 100 × 3.0 mm format.
Thus, transferring a UHPLC method to a traditional HPLC platform was more readily accomplished using a SPP stationary phase format. If the SPP stationary phase
format had been used for the original UHPLC method, scaling to a 400 bar instrument
may have been more readily achieved. The authors propose that this approach may
be of value and should be considered going forward. As discussed throughout this
chapter, such a design comes closest to the ideal of using the same stationary phase
throughout the method conversion. Adapting a method from UHPLC to SPP HPLC
columns seems to work better in regards to maintaining the chromatographic profile
and run conditions than when switching to traditional 3 and 5 μm particle columns.
Further investigation of how best to scale from UHPLC to the SPP platform needs to
be done.

(a)

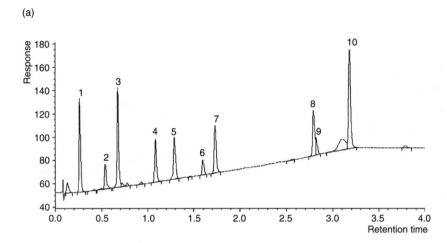

(b)

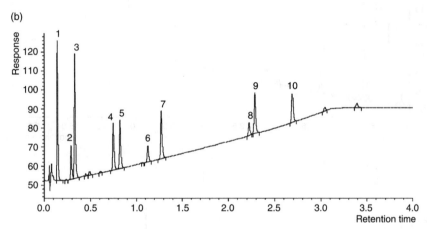

Figure 2.3   Antioxidant UHPLC profile on SPP phase. Analytes: (1) propyl gallate, (2) TBHQ, (3) ethoxyquin, (4) NGDA, (5) BHA, (6) ionox-100, (7) octyl gallate, (8) layryl gallate, (9) BHT, (10) ascorbyl palmitate. (a) Mackerey-Nagel 1.8 μm, C18 (50 × 0.20 mm, 1.4 mL/min, $P_{init} = 996$ bar, LC: Agilent 1290); (b) Phenomenex 2.6 μm, Kinetex XB-C18, 50 × 2.0 mm, 2.1 mL/min, $P_{init} = 781$ bar, LC: Agilent 1290.

2.3.3 UHPLC to HPLC Method Transfer

When working with UHPLC separations for molecules having similar chemistries, the analytical separation should model and scale well. Direct scaling from UHPLC to HPLC should not be assumed to work without confirming that the chromatographic profile and resolution of critical pairs are maintained. Preliminary studies show that direct scaling from UHPLC to HPLC platforms should preferentially use SPP

(a)

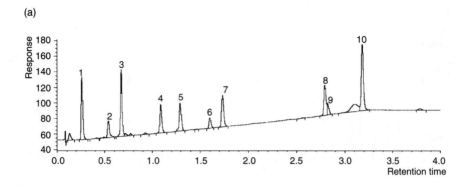

(b)

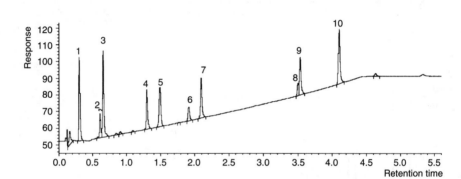

(c)

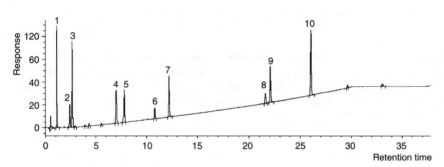

Figure 2.4   Antioxidant UHPLC scaled to SPP phase for higher pressure HPLC platform. Analytes:
(1) propyl gallate, (2) TBHQ, (3) ethoxyquin, (4) NGDA, (5) BHA, (6) Ionox-100, (7) octyl gallate,
(8) layryl gallate, (9) BHT, (10) ascorbyl palmitate. (a) Mackerey-Nagel 1.8 μm, C18 (50 × 0.20 mm,
1.4 mL/min, P_{init} = 996 bar, LC: Agilent 1290); (b) Phenomenex 2.6 μm, Kinetex XB-C18 (50 ×
2.0 mm, 1.0 mL/min, P_{init} = 389 bar, LC: Agilent 1290); (c) Phenomenex 2.6 μm, Kinetex XB-C18
(100 × 3.0 mm, 1.0 mL/min. P_{init} = 343, LC: Agilent 1200).

columns to maintain the chromatographic profile. Starting with the existing UHPLC separation, the analyst should consider the following:

1. Use a common stationary phase format, when available.

2. Scale the UHPLC separation to an approximately 400–600 bar maximum pressure, depending on the HPLC being transferred to.

3. Look to see if a direct SPP column in a phase similar to the UHPLC phase is available. If so, no adjustment of column dimensions may be required.

4. If not, scale the UHPLC method to a longer HPLC column using a suitable scaling calculator or the relationships described earlier.

5. Note the gradient, flow, and injection volume changes needed for the HPLC systems.

6. Confirm the new method on an HPLC system maintains pressure under the system maximum.

7. Confirm performance of the new method, particularly that the profile, system figures of merit, and critical pair selectivity all have been maintained.

2.4 TRANSFERRING LC METHODS TO OTHER LABS

In this current global marketplace, transferring methods from the lab where they were developed to other labs, be it in the same company or to contract facilities, is a necessity. In the pharmaceutical industry, where method transfers are commonplace, method transfers can be done via the following (46, 47):

1. Comparative Testing
   - Comparative testing is the most common form of method transfer. It involves two (or more) laboratories testing identical, well-defined samples. Success of the transfer is based on a preapproved protocol that includes the acceptance criteria, including acceptable variability.

2. Covalidation Between Laboratories
   - Any laboratory that performs the validation of an analytical method is qualified to run the procedure. Therefore, if the receiving laboratory is involved in the validation, once the validation is completed, it will be considered qualified. In a manner similar to that used in comparative testing, the acceptance criteria are predefined. By including one or more laboratories in the validation, the method has a more rigorous assessment of reproducibility.

3. Method Validation/Revalidation
   - By repeating some or all of the validation experiments, a laboratory could be considered qualified to run a method, if it met all the original validation acceptance criteria. Often only the key parameters of a method would need to be revalidated to qualify a laboratory, for example, for an impurity method, the limit of detection is key.

4. Transfer Waiver

- In some cases the method transfer can be waived. Such cases include the transfer of a compendial method, a transfer in which the method to be transferred is similar to one already in use by the receiving laboratory, or one where the scientist that developed, validated, and/or runs the method on a regular basis transfers to the receiving lab. If the method is to be used with a new product and the new product is similar in composition to an existing product, then a waiver may also be applied.

As discussed, UHPLC adds complications to the process, especially in the two most common cases where (1) the receiving lab does not have identical UHPLCs as the transferring lab or (2) the receiving lab does not have UHPLCs.

2.4.1 UHPLC Instrument Issues Encountered During Method Transfers

2.4.1.1 System Volume

System volume is one of the parameters that instrument manufacturers have striven to reduce when the engineering of UHPLC systems began. Some reductions were simple. In others, for instance, in autosamplers, the volume reduction was more complicated. For this reason, total system volumes are not uniform across all available instrumentation. Some manufacturers strictly define their system volume, and the qualification of the instruments for use in regulated environments hinges on this volume being fixed. Other manufacturers offer more flexible options. Virtually all UHPLC systems are modular and can have different components (column switcher vs. single column, binary pump vs. quaternary pump, multiple detectors), and for some manufacturers, two systems may have the same components but be plumbed differently. All of these configurations have an effect on the system volume. Because some of the benefits of UHPLC are the speed of analysis, sharpness of the peaks, and retention of resolution, extra dead volume is undesirable, as it reduces these benefits. To ease method transfer problems, one must account for these types of differences.

One solution for dealing with different system volumes is to calculate and report retention times corrected for dead volume. Method documentation should discuss the calculation of either the dead volume of a system or the retention time of an unretained peak. The retention times and/or retention time windows can then be reported as "corrected retention times" as calculated by Eq. (2.14):

$$t'_R = t_R - t_M \qquad (2.14)$$

Where:

t'_R = corrected retention time,
t_R = retention time of peak/compound of interest
t_M = retention time of unretained peak.

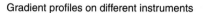

Gradient profiles on different instruments

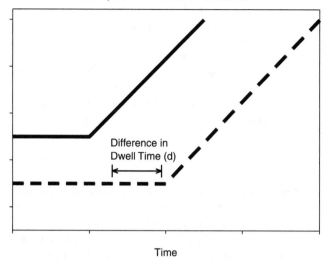

Difference in
Dwell Time (d)

Time

Figure 2.5   Example of systems with differing dwell times.

By taking into account the system volume, errors in peak, or component identification in method transfer, exercises should be reduced.

A second way to account for differing system volumes, especially for gradient
runs, is to correct the gradient for dwell time. One must determine the dwell time, in
minutes, for the system being used. This is often done using a zero-volume connector
and solvents to visualize the gradient (35). To correct for dwell time the method must
have an isocratic hold that is greater than the dwell time of any instrument that might
be used to run the method. For most UHPLC methods, an isocratic hold of 1 min
should suffice; however, this should be a consideration while developing methods.
As is seen in Figure 2.5, it is possible for the same programmed gradient to produce
different gradient profiles based on system geometry. To correct for this, one must
adjust the original isocratic section of the method for the difference in dwell time (d).
Table 2.3 shows a correction for a theoretical gradient, taking into consideration the
difference in dwell time.

Table 2.3   Correcting a Gradient
Program for Varying Dwell Times

Time	%B
0.0	5
100–*d*	5
30.0	95

Another way manufacturers have tried to reduce system volume is to use semi-micro and micro flow cells. Sometimes these smaller flow cells reduce their volume by reducing their diameter, sometimes by reducing their path length, and sometimes by doing both. Therefore, when transferring methods, especially when analyzing components that are at low concentrations, sensitivity checks should be made to determine whether the geometry of flow cells in the sending and receiving labs are comparable.

Finally, it is important to understand the instrumental system and columns well enough to ensure that adequate inter-run equilibration time is allowed.

2.4.1.2 Mixers

Mixers are one of the instrument components that can add to the total system volume discussed. Choice of mixer and even how you make your mobile phases are important, especially when using trifluoroacetic acid (TFA) in the mobile phase.

Briefly, TFA can react with either the residual exposed silica or with the attached phases of reverse phase columns. This causes the TFA concentration in the mobile phase to fluctuate. When the mobile phase content is fluctuating or incompletely mixed, the TFA equilibrium on the column is altered, which can cause amplification of the mixing ripples that are already normally caused by the column. This is further increased in an ultraviolet light (UV) detector, because TFA absorbs more UV light than most other mobile phase components (see Table 2.4 for the UV cutoff of common chromatographic mobile phase solvents and Figure 2.6 for the UV spectrum of TFA).

One way to reduce some of the turbulence of mixing is to premix the mobile phases instead of having just an aqueous phase in line A and an organic phase in line B, that is, premix mobile phase A to be 95:5:0.1 of water: acetonitrile:TFA and a mobile phase B to be 5:95:0.1 of water: acetonitrile:TFA. In this case, the gradient mixing is less dependent on the mixer (although by no means completely independent), and one would run a gradient from 0%–100% A to B, instead of starting at 95:5 and running to 5:95. While this has benefits, some manufacturers discourage these gradients on binary systems that start with only one pump in use. However, even this may not be enough to reduce the rippling. This is when the volume of the mixer will become an important parameter. Different volume mixers

Table 2.4 The UV Cutoff of Common Chromatographic Mobile Phase Solvents

Solvent	UV Cutoff (nm)
Water	190
Acetonitrile	190
Methanol	205
Trifluoroacetic acid	210

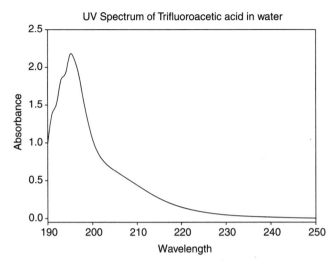

Figure 2.6   Ultraviolet spectrum of trifluoroacetic acid (0.1% TFA in water).

are available depending on the instrument vendor. Some of the current sizes available are 20, 50, 100, 130, and 180 μL. Mixer sizes of at least 100 μL are required for efficient mixing of TFA to avoid baseline ripple. If the larger mixers are required to successfully run the method, the mixer volume must be included in the method, and that same mixer (volume) should be used during validation of the method. This volume also may be a significant portion of the total system volume. Note that this extra volume is not revealed by comparing the elution time of an unretained peak. As with validation, during method transfer activities, the geometry of the mixers used needs to be well understood, and any differences between the sending and receiving laboratories must be thoroughly evaluated and shown to be equivalent, or the gradient chromatography at each site will be quite different. Figure 2.7 shows the same method, using TFA as a modifier, using a system with a 20 μL mixer and on a different system using a 130 μL mixer. Clearly the beginning of the gradient is affected using the smaller mixer.

Finally one must keep in mind that the larger the mixer, the better the mixing, but the further away one is from making the most out of the advantages of UHPLC.

2.4.1.3   Injector and Injection Volume

Similar to HPLCs, UHPLC injection loops vary in size and type and can be readily switched out. As with the mixer, choice of sample loop size depends on balancing the injection volume required and the amount of extra system volume that can be tolerated. The smaller the sample loop, the less loss of sample and the smaller the total flow path. Some instruments include the needle in the flow path, and some instruments

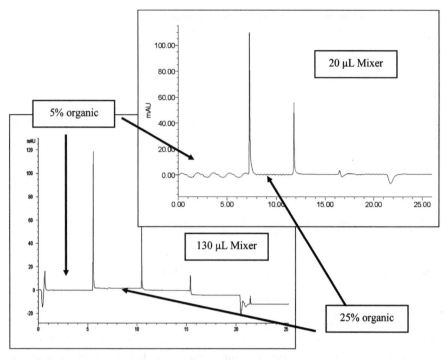

Figure 2.7 The effect of mixer volume on an elution gradient, using TFA as a modifier.

isolate the needle. Here again, the difference in total volume can be troublesome when transferring methods to instruments that have different injector configurations.

Because UHPLC injections tend to be smaller (more in the range of 0.1–10 μL), the tolerances for calibration on the expelled injection volumes have to be much tighter. A tolerance of ±0.3 μL would have very little effect on 25 or 10 μL injections, but would be useless at 0.2 and 0.5 μL injection volumes, which are not uncommon for UHPLC.

A thorough understanding of the injection system and tight calibration tolerances are what is required to avoid method transfer issues due to the injector.

2.4.1.4 System Carryover

Carryover is another problem that is addressed differently between labs and instrument manufacturers. Because by design, the flow path in UHPLCs is shorter than in HPLCs, carryover may be a smaller problem (fewer places to collect), or it may be increased (smaller overall volume, therefore effects of carryover are less dilute and more noticeable). Carryover does drive some manufacturers to continuous improvements. Some nonsample related factors contributing to the differences in carryover are the amount of time the needle is in the flow path and how rigorous the needle wash program is. Most systems include one or even several choices for washing the injection needle; however, if such procedures are governed by site standard operating procedures (SOPs) and are not by the written method, differences in the amount of carryover are possible. Therefore, if carryover has been identified as a problem throughout method development, then the incorporation of blank injections into the sequence and/or explicit instructions on use of a needle wash or needle wash program are required for a successful method transfer.

2.4.2 Back to LC Basics

UHPLCs have many advantages over HPLCs, but HPLCs have had 30 years to improve and evolve, thus many of the systems currently available are so rugged that the analyst no longer worries about some of the more technical issues of chromatography. For instance, gone are the days that one has to degas one's mobile phase or sit and inject every sample with a syringe and a turn of a valve. As UHPLC came online, these issues did not return, however, analysts, once again, must truly understand their chromatography system. As discussed, knowledge of total system volume, type of mixer, flow path of the injector, and needle wash capabilities are things that demand more attention than more recent work with HPLCs requires.

The landscape of analytical chromatography, after a relatively established period of traditional HPLC, is changing rapidly with technological improvements in systems and column technology. It is again an exciting period in analytical chromatography. Technological changes will continue to impact industry in the foreseeable future and necessitate that industry analytical chemists keep pace with the current developments, trends, and the associated regulatory impact of this dynamic time in chromatography.

REFERENCES

1. WANG, X., BARBER, W.E., LONG, W.J. *J. Chromatogr. A.* 2012, 1228: 72–88.
2. MACNAIR, J.E., PATEL, K.D., JORGENSON, J.W. *Anal. Chem.* 1999, 71: 700–708.
3. TOLLEY, L., JORGENSON, J.W., MOSELEY, M.A. *Anal. Chem.* 2001, 73: 2985–2991.
4. NOVAKOVA, L., MATYSOVA, L., SOLICH, P. *Talanta.* 2006, 68: 908–918.
5. WREN, S.A.C., TCHELITCHEFF, P. *J. Pharm Biomed. Anal.* 2006, 40: 571–580.
6. WREN, S.A.C., TCHELITCHEFF, P. *J. Chromatogr A.* 2006, 1119: 140–146.
7. WANG, G., HSIEH, Y., CUI, X., CHENG, K.C., KORFMACHER, W.A. *Rapid Commun. Mass Spectrom.* 2006, 20: 2215–2221.
8. APOLLONIO, L.G., WHITTALL, I.R., PIANCA, D.J., KYD, J.M., MAHER, W.A. *Rapid Commun. Mass Spectrom.* 2006, 20: 2259–2264.
9. KAWANISHI, H., TOYO'OKA, T., ITO, K., MAEDA, M. *J. Chromatogr. A.* 2006, 1132: 148–156.
10. JOHNSON, K.A., PLUMB, R. *J. Pharm. Biomed. Anal.* 2005, 39: 805–810.
11. BARANOWSKA, I., KOWALSKI, B. *Water Air Soil Pollut.* 2010, 211: 417–425.
12. MESSINA, C.J., GRUMBACH, E.S., DIEHL, D.M. *LCGC.* 2007, 25: 1042–1049.
13. BARANOWSKA, I., WILCZEK, A., BARANOWSKI, J. *Anal. Sci.* 2010, 26: 755–759.
14. SUBASRANJAN, A., SRINIVASULU, C., HEMANT, R. *Drug Testing Analysis.* 2: 107–112.
15. MEZCUA, M., AGUERA, A., LLIBERIA, J.L., CORTES, M.A. *J. Chromatogr. A.* 2006, 1109: 222–227.
16. LEANDRO, C.C., HANCOCK, P., FUSSELL, R.J., KEELEY, B.J. *J. Chromatogr. A.* 2006, 1103: 94–101.

17. VENTURA, M., GUILLEN, D., ANAYA, I., BROTO-PUIG, F. *Rapid Commun. Mass Spectrom.* 2006, 20: 3199–3204.
18. CUNLIFFE, J.M., MALONEY, T.D. *J. Sep. Sci.* 2007, 30: 3104–3109.
19. WU, N., LIPPERT, J.A., LEE, M.L. *J. Chromatogr. A.* 2001, 911: 1–12.
20. ANSPACH, J.A., MALONEY, T.D., BRICE, R.W., COLON, L.A. *Anal. Chem.* 2005, 77: 7489–7497.
21. WREN, S.A. *J. Pharm. Biomed. Anal.* 2005, 38: 337–343.
22. SONG, W. *J. Pharm Bio Anal.* 2009, 491–500.
23. BRICE, R.W., ZHANG, X., COLON, L.A. *J. Sep. Sci.* 2009, 32: 2723–2731.
24. CUNLIFFE, J.M., ADAMS-HALL, S.B., MALONEY, T.D. *J. Sep. Sci.* 2007, 30: 1214–1223.
25. SUN, X., WRISLEY, L. *Am. Pharm. Rev.* 2009, Dec: 26–32.
26. MAJORS, R.E. *LC-GC.* 2011, 29: 476–484.
27. GUILLARME, D., NGUYEN, D.T.T., RUDAZ, S., VEUTHEY, J.L. *Eur. J. Pharm. Sci.* 2007, 66: 475–482.
28. GUILLARME, D., NGUYEN, D.T.T., RUDAZ, S., VEUTHEY, J.L. *Eur. J. Pharm. Biopharm.* 2008, 68: 430–440.
29. DITTMANN, M.M. "Approaches Towards Method Compatibility Between HPLC and UHPLC Systems," Paper 1970-6, Pittsburgh Conference & Exposition on Analytical Chemistry and Applied Spectroscopy, March 2010, Orlando.
30. NEUE, U.D., MCCABE, D., RAMESH, V., PAPPA, H., DeMITH, J. *Pharmacopeial Forum.* 2009, 35: 1622–1626.
31. WEBSTER, G.K., ELLIOTT, A. *Am. Pharm. Rev.* 2011, 14: 32–40.
32. VAN DEEMTER, J.J., ZUIDERWEG, F.J., KLINKENBERG, A. *Chem. Eng. Sci.* 1956, 271.
33. GIDDINGS, J.C. *Unified Separation Science.* John Wiley, New York, 1991.
34. WU, N., CLAUSEN, A.M. *J. Sep. Sci.* 2007, 30: 1167–1182.
35. SYNDER, L.R., KIRKLAND, J.J., GLAJCH, J.I. *Practical HPLC method Development.* Wiley, New York, 1997.
36. CHESTNUT, S.M., SALISBURY, J.J. *J. Sep. Sci.* 2007, 30: 1183–1190.
37. McNAIR, J.E., LEWIS, K.C., JORGENSON, J.W. *Anal. Chem.* 1997, 69: 983.
38. JORGENSON, J.W., LUKACS, K.D. *Anal. Chem.* 1981, 53: 1298.
39. YANG, Y., HODGES, C.C. *Separation Science Redefined (Supplement to LC-GC).* 2005: 31–35.
40. NOVAKOVA, L., MATYSOVA, L., SOLICH, P. *Talanta.* 2006, 68: 908–918.
41. DONG, M.W. *LCGC*, 2007, 25: 656–658, 660–666.
42. SYNDER, L.R., DOLAN, J.W., GANT, J.R. *J. Chromatography.* 1979, 165: 3–30.
43. SYNDER, L.R., DOLAN, J.W., GANT, J.R. *J. Chromatography.* 1979, 165: 31–58.
44. MAJORS, R.E. *LCGC.* 2010, 28: 1014–1015, 1017–1018, 1020.
45. WEBSTER, G.K., BASEL, C.L. *LCGC.* 2003, 21: 286–294.
46. AHJUTA, S., SCYPINSKI, S. *Handbook of Modern Pharmaceutical Analysis, Second Edition.* Academic Press (Elsevier), London, UK, 2011.
47. USP35-NF30, General Chapter <1224>: Transfer of Analytical Procedures, 2012.

Chapter 3

Practical Aspects of Ultrahigh Performance Liquid Chromatography

Naijun Wu, Christopher J. Welch, Theresa K. Natishan,
Hong Gao, Tilak Chandrasekaran, and Li Zhang

3.1 INTRODUCTION

Ultrahigh pressure liquid chromatography (UHPLC) was originally developed by academic laboratories in the late 1990s to enable the investigation of ultrahigh pressures (50,000–100,000 psi) on the operation of capillary columns packed with 1.0–1.7 μm particles (1–5). Various commercial UHPLC systems with new hardware have been introduced in the past few years to make such higher pressure (15,000–20,000 psi) systems more widely available to the general community of high performance liquid chromatography (HPLC) users (6). These systems allow sub-2 μm particles to be used at higher linear velocities for improved speed and resolution. UHPLC operation at very high pressures can result in a variety of new problems that were relatively unimportant for previous users of HPLC. For example, frictional heating for 2.1–4.6 mm internal diameter columns (7, 8) can impact both retention and selectivity for analytes, while potentially significantly reducing column efficiency (9, 10). Additionally, the effect of mobile phase compressibility on separation, a phenomenon with little impact on conventional HPLC operation, can become more significant at ultrahigh pressures.

While the response of the HPLC community to the introduction of UHPLC technologies is still evolving, several significant changes to previous conventions are already apparent. For example, the standard analytical column dimensions, which have been more or less fixed at 4.6 mm i.d. × 25 cm length since the early 1970s, have been undergoing rapid change as the increased efficiency afforded by small particle columns has enabled the use of much shorter columns. Similarly, studies suggesting

Ultra-High Performance Liquid Chromatography and Its Applications, First Edition. Edited by Quanyun Alan Xu.
© 2013 John Wiley & Sons, Inc. Published 2013 by John Wiley & Sons, Inc.

that the frictional heating effect can be minimized by reducing the column internal diameter (1, 11) have led to the increased use of smaller internal diameter columns to minimize the temperature gradient across the column. An additional advantage of these smaller diameter columns is a greatly decreased requirement for mobile phase, which has led to an increasing appreciation for these "greener" approaches to LC separation (12, 13).

These trends toward the use of shorter columns with narrower diameter mean that the overall volume of today's UHPLC columns can be much smaller than the HPLC columns of the past. Consequently, the extra-column band-broadening effect on efficiency can become especially significant in UHPLC as the effect is closely related to column dimensions (14, 15). Recently, extra-column band-broadening has been extensively investigated in UHPLC (16–19), conventional HPLC (20), capillary HPLC (21), and modified conventional HPLC (22). Additionally, the effects of extra-column volume on other chromatographic parameters including retention factor, selectivity, and pressure drop also become more significant in UHPLC.

Although commercial UHPLC systems have been available for more than 6 years, it is reasonable to project that UHPLC and HPLC systems will continue to coexist for a period of time, considering the high costs of instrument replacement and the difficulty of effecting change in highly regulated analytical laboratories. Consequently, the transfer of methods between HPLC and UHPLC is important now and will continue to be an important matter across research and development and quality control laboratories for the next few years. It is, therefore, critical to use appropriate parameters for transferred methods to ensure separation performance and method robustness that are comparable to those for the original methods (23–26).

In this chapter, three of the most important practical considerations for practicing chromatographers working with UHPLC are presented. First, the effects of extra-column volume on performance are considered, including impacts on retention factor, pressure drop, column efficiency, and a general discussion on the influence of the dwell volume on gradient separations. Second, the effect of high pressure and frictional heating on separations is reviewed, and strategies to minimize the negative effects of frictional heating effect are addressed. Finally, approaches for method transfer between UHPLC and HPLC systems are reviewed, and case studies for practical method transfer between systems are presented.

3.2   EFFECT OF EXTRA-COLUMN VOLUME ON PERFORMANCE OF SUB-2 μM PARTICLE-PACKED COLUMNS IN UHPLC

In this section, the effect of the extra-column volume on chromatographic performance is discussed for columns and connection tubings of various internal diameters. Relevant theoretical equations are summarized, and the effects of extra-column volume on apparent retention, selectivity, efficiency, and column pressure are discussed. The influence of the dwell volume on separation is also discussed, and recommendations on acceptable ratios of extra-column to column volume dimensions for sub-2 μm particle-packed columns operating under typical UHPLC conditions are presented.

3.2.1 Theoretical Considerations

3.2.1.1 Effect of Extra-Column Volume on Linear Velocity, Pressure Drop, and Retention

For a conventional 4.6 mm i.d. × 25 cm length column, the column void volume is typically about 2.5 mL, while the extra-column volume arising from connecting tubing, etc. is typically about 50 μL, or about 2% of the column void volume. Thus, the amount of time the analytes spend within the connecting tubing is negligible. When a 2.1 mm i.d. × 10 cm length column (for which the column void volume is about 210 μL) is connected to this same system, the 50 μL extra-column volume now constitutes almost 25% of the column void volume. In such a situation, the analytes spend a much more significant amount of time simply being transported to and from the column, resulting in significant degradation of chromatographic efficiency through diffusion and band-broadening. For smaller internal diameter columns (e.g., ≤2.1 mm), a much lower volumetric flow rate is used to obtain the same linear velocity as for a conventional 4.6 mm column. The actual linear velocity should be corrected with the migration time in extra-column volumes, when pressures and efficiencies are compared between columns with different internal diameters:

$$u_{corr} = \frac{L}{(t_0)_{obs} - (t_0)_{ec}} \tag{3.1}$$

where L is the column length, and $(t_0)_{obs}$ and $(t_0)_{ec}$ are the migration times for an unretained compound from a system with and without an LC column, respectively.

The apparent pressure drop, ΔP_{obs}, generated by the column and extra-column tubing is given by:

$$\Delta P_{obs} = \Delta P_c + \Delta P_{ec} = \phi \frac{\eta L}{d_p^2} u + \frac{8 \eta L}{r^2} u \tag{3.2}$$

where ΔP_c is the pressure drop across the column, ΔP_{ec} is the extra-column pressure, ϕ is the column resistance factor, η is the viscosity of the mobile phase, d_p is the particle size of packing material, u is the mobile phase linear velocity, and r is the internal radius of connection tubing (27). Note that the pressure for a packed column is practically independent of the column internal diameter, because the pressure from micro-particles is much greater than that from a typical empty column (1.0–4.6 mm). Equation (3.2) suggests that the pressure from extra-column tubing is proportional to the internal diameter of the tubing. In UHPLC, small internal diameter tubing (≤0.1 mm) is used to connect various components from the pump to the flow cell, so the pressure generated by this tubing is usually no longer negligible. Thus, the actual column pressure ΔP_c should be corrected by subtracting the extra-column pressure ΔP_{ec} from the observed total pressure with the column (ΔP_{obs}).

A retention factor should also be corrected with the migration time arising from the extra-column volume:

$$k_{corr} = \frac{(t_R)_c - (t_0)_c}{(t_0)_c} = \frac{[(t_R)_{obs} - (t_R)_{ec}] - [(t_0)_{obs} - (t_0)_{ec}]}{(t_0)_{obs} - (t_0)_{ec}} \approx \frac{(t_R)_{obs} - (t_0)_{obs}}{(t_0)_{obs} - (t_0)_{ec}} \tag{3.3}$$

where $(t_R)_c$ and $(t_0)_c$ are the retention times contributed solely from a column for the retained and unretained compounds, respectively, $(t_R)_{obs}$ is the observed retention time for a retained analyte when a column is used, and $(t_R)_{ec}$ is the migration time for the retained compound when a zero-dead volume union is used in place of the column. Here it is assumed that $(t_R)_{ec}$ and $(t_0)_{ec}$ values are comparable for a short connection tubing in a UHPLC system. Equation (3.3) suggests that a selectivity factor would not be affected by the extra-column volume, because it is a ratio of two retention factors, and the effect of extra-column volume on t_0 is cancelled by dividing the two retention factors. The relationship of the observed selectivity factor (α_{obs}) and the corrected selectivity factor (α_{corr}) can be expressed as:

$$\alpha_{obs} = \frac{k_{2(obs)}}{k_{1(obs)}} = \frac{(t_R)_{2(obs)} - (t_0)_{2(obs)}}{(t_R)_{1(obs)} - (t_0)_{1(obs)}} = \frac{k_{2(corr)}}{k_{1(corr)}} = \alpha_{corr} \tag{3.4}$$

3.2.1.2 Column Efficiency

In the absence of extra-column band-broadening of any type, the column performance can be described by the van Deemter equation in a reduced form (28, 29):

$$h = A + \frac{B}{v} + Cv \tag{3.5}$$

where h is the reduced plate height ($= H/d_p$), v is the reduced linear velocity ($= u\, d_p/D_m$, D_m is the diffusion coefficient of the analyte in the mobile phase), and A, B, and C are constants that account for contributions to band-broadening from eddy diffusion, longitudinal diffusion, and mass transfer resistance. In practical separations, the overall performance of a chromatographic system is given by the performance of the column itself and by all the other contributions outside the column due to band-broadening. Assuming that all contributions to peak variance are independent and the peak is in a Gaussian shape, the observed total variance σ^2_{obs} of a peak is the sum of these contributions:

$$\sigma^2_{obs} = \sigma^2_c + \sigma^2_{ec} \tag{3.6}$$

where σ_{ec}^2 is the variance from extra-column volume effects. The column efficiency with correction of extra-column band-broadening can be expressed as (30):

$$N_c = 5.54 \frac{[(t_R)_{obs} - (t_0)_{ec}]^2}{(w_{1/2}^2)_{obs} - (w_{1/2}^2)_{ec}} \tag{3.7}$$

where N_c is the theoretical plate number contributed solely by the column, and $(w_{1/2})_{obs}$ and $(w_{1/2})_{ec}$ are the peak widths at half height from an LC system with a column and from an LC system with a zero-dead volume connection union in place of the column, respectively.

3.2.1.3 Extra-Column Band-Broadening

To achieve maximum performance for a column, extra-column band-broadening σ_{ec}^2 in Eq. (3.6) must be minimized. There are basically two types of extra-column contributions. The primary one is volumetric in nature and originates from the injection volume, the detector volume, and the volume of connection tubing between the injector and detector. The other type of band-broadening derives from time-related events, such as the sampling rate and the detector time constant. For modern detectors and data acquisition software, the latter contribution can be negligible when fast data acquisition speed and low detector time constants are used.

The loss in theoretical plate number by extra-column band-broadening effects should not exceed 10% (15). For a specific LC system, the maximum acceptable variance due to extra-column band-broadening can be expressed as:

$$\sigma_{ec(acc)}^2 \leq 0.10\sigma_c^2 \leq 0.10 \frac{\pi^2 L^2 r_c^4 \varepsilon^2 (1+k)^2}{N_c} \tag{3.8}$$

where $\sigma_{ec(acc)}^2$ is the maximally acceptable variance due to extra-column broadening effects, r_c is the column radius, and ε is the porosity of the column. Equation (3.8) suggests that columns with small internal diameters are more susceptible to these extra-column volume effects. Furthermore, sub-2 particle-packed columns are more vulnerable to extra-column band-broadening, where column efficiencies are usually high (sharper peaks) or column lengths (L) are short. Finally, the less retained peaks (lower k) can be negatively impacted more by the extra-column volume than highly retained components.

Primary contributions to extra-column band-broadening in Eq. (3.6) are from the injection volume, connection tubing, and the flow cell. Under given chromatographic conditions, the variance of a sample plug from a connection tubing is described by the Taylor–Aris Eq. (3.1):

$$\sigma_t^2 = \frac{\pi r_t^4 L_t F}{24 D_m} \tag{3.9}$$

where σ_t^2 is the variance originating from the connection tubing in volume units, r is the radius, L is the length of the capillary tubing, and F is the flow rate.

3.2.1.4 Effect of Dwell Volume

The volume between the solvent mixing point and column inlet is defined as dwell volume, and the corresponding time it takes for the selected composition of mobile phase to reach the column is called dwell time. Although the dwell volume is not the part of the extra-column volume, it is discussed in this section, as it impacts chromatographic performance. Under gradient conditions, the dwell volume for an HPLC system can have a significant effect on separation parameters including retention times (t_r) and apparent retention factors k_g'.

$$t_r = t_0(1 + k_g') + t_{dwell} \tag{3.10}$$

where t_{dwell} is the delay time of an analyte due to the dwell volume, and t_0 is the migration time of an unretained analyte. Although the dwell volume is significantly reduced in UHPLC systems, the migration time in the dwell volume becomes more significant when a small internal diameter column is utilized, and thus low flow rates are used to achieve similar linear velocities as a larger internal diameter column.

3.2.2 Extra-Column Volumes and Column Volumes

Extra-column volumes for typical HPLC and UHPLC systems are listed in Table 3.1, with the extra-column volume for a UHPLC being about 4.5 times lower than that for an HPLC. It should be pointed out that experimentally determined extra-column volumes are typically somewhat higher than those calculated in Table 3.1 due to the parabolic flow profile of the mobile phase in the connection tubing. The experimental value of extra-column volume is dependent on mobile phase viscosity, flow rate, mobile phase composition, tubing configuration, etc. (32, 33).

Table 3.1 Typical Extra-Column Volume for a UHPLC System and a Conventional HPLC System

Source	UHPLC§	HPLC¶
Inlet Tubing	6.3 μL (500 × 0.127 mm)	17.7 μL (350 × 0.254 mm)
Outlet Tubing	2.5 μL (200 × 0.127 mm)	10.1 μL (350 × 0.254 mm)
Injection Volume	2.0 μL	10 μL
Needle Seat Capillary	NA	3.7 μL (150 × 0.178 mm)
Flow Cell Volume	0.5 μL	8 μL
Total	11.3 μL	49.6 μL

*Acquity UPLC; ¶Agilent 1100 HPLC.

Table 3.2 Ratios of Extra-Column Volume to Column Void Volume

Column Dimension Length × Internal Diameter (mm)	Column Void Volume* (μL)	Ratio of Extra-column Volume to Column Void Volume	
		UHPLC ($V_{ec} = 11.3$ μL)	HPLC ($V_{ec} = 49.6$ μL)
250 × 4.6	2492	1: 220.5	1:50.2
150 × 4.6	1495	1:132.3	1:30.1
100 × 4.6	996	1:88.2	1:20.0
50 × 4.6	498	1:44.1	1:10.0
50 × 3.0	212	1:18.8	1:4.3
150 × 2.1	312	1:27.6	1:6.3
100 × 2.1	208	1:18.4	1:4.2
50 × 2.1	104	1:9.2	1:2.1
50 × 1.0	24	1:2.1	1:0.5

*Assuming that the column porosity is 0.6. From Wu, N., et al. *J. Sep. Sci.* 2012, *35*, 2018–2025, with permission.

Ratios of extra-column volume to column void volumes are calculated in Table 3.2 for typical UHPLC and HPLC columns and systems. It can be seen that the ratios for typical UHPLC columns and the system (highlighted in gray) are lower than those for typical HPLC columns and the system (underscored), although the extra-column volume is reduced in the UHPLC system. The column of 50 × 4.6 mm can be used for most UHPLC and HPLC systems in terms of the ratio of extra-column volume to column void volume. The extra-column volume for a 50 × 1.0 mm column in UHPLC is approximately 50% of the column void volume, and the resulting efficiency loss would be substantial, especially under isocratic conditions.

3.2.3 Effect of Extra-Column Volume on Performance

3.2.3.1 Retention Time, Retention Factor and Selectivity

The separation time for an analyte is determined by its linear velocity and the column length. Thus, to match linear velocities for columns of different internal diameters, the volumetric flow rate should be adjusted in a proportion to the square of column internal diameter for a given column length, as shown in Table 3.3. As previously noted, the migration time of an analyte in the extra-column volume, mainly in the connection tubing, increases as the column internal diameter decreases (Table 3.3), becoming non-negligible as column diameter drops (e.g., 40% of the apparent retention time in a 1.0 mm i.d. column). Thus, the difference between corrected and observed retention factors increases from 5% to 41% as the column internal diameter decreases from 4.6 to 1.0 mm (presuming the same connecting tubing is used for both columns). It is therefore advisable, when possible, to appropriately scale the volume of connecting tubing as column dimensions are reduced. The data in Table 3.3 also reveal that the relative difference in k' is independent of the retention factor (thus mobile phase

Table 3.3   Flow Rate, Retention (Migration) Times, Retention Factors and Relative Difference Between Corrected and Observed Retention Factors

Column I.D. (mm)	Volumetric Flow Rate (mL/min)	Linear Velocity (mm/s)	$(t_R)_c$ (min)	$(t_0)_c$ (min)	$(t_0)_{ec}$ (min)	k'_{obs}	k'_{corr}	$\%\Delta k'$
4.6	1.4	3.1	1.735	0.282	0.014	5.15	5.42	5%
3.0	0.60	3.1	1.838	0.298	0.032	5.17	5.79	11%
2.1	0.29	3.0	1.919	0.343	0.061	4.59	5.59	18%
1.0	0.067	2.9	2.168	0.528	0.214	3.11	5.22	41%

50%, v/v acetonitrile in water at 25°C, uracil as an unretained analyte and benzophenone as a reteaned analyte. From Wu, N., et al. *J. Sep. Sci.* 2012, *35*, 2018–2025. with permission.

composition) but related to the ratio of extra-column volume to column void volume, with the apparent retention factor and the relative difference in k' increasing with the rising ratio of extra-column volume to column void volume.

Figure 3.1 shows the effect of various mobile phase compositions on the selectivity factors for the separation of phenol/aniline and toluene/benzene pairs on 1.0, 2.1, 3.0, and 4.6 mm i.d. columns operating at a given velocity. The selectivity factors for the four columns are practically the same, even though retention times were not corrected to compensate for the differences in extra-column migration times. This finding is in agreement with Eq. (3.4) that shows that selectivity factors are independent of extra-column volume and column internal diameter.

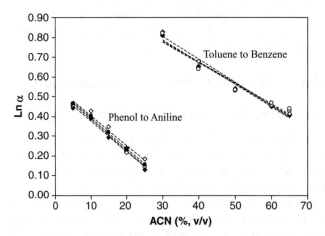

Figure 3.1   Relationship of selectivity factor with the MeCN concentration for four columns with various internal diameters. *Conditions:* Zorbax SB Extend C-18, 1.8 μm particles; 50 mm column length; 0.1 mg/mL solutes in a 50/50 (v/v) water to acetonitrile diluent; acetonitrile in water with various ratios as mobile phases for benzene and toluene and acetonitrile in 10 mM phosphate buffer (pH = 7) with various ratios as mobile phases for phenol and aniline; 0.05 mg/mL uracil as unretained marker; 210 nm UV detection; 25°C. The mobile phases were premixed. From Wu, N., et al. *J. Sep. Sci.* 2012, *35*, 2018–2025, with permission. Key: (●) 4.6 mm i.d.; (○) 3.0 mm i.d.; (◆) 2.1 mm i.d.; and (◇) 1.0 mm i.d.

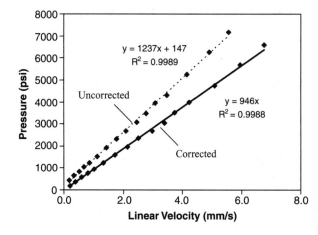

Figure 3.2 Comparison of column pressures between apparent and corrected with the extra-column volume effect. *Conditions:* Premixed 50/50 (v/v) water to acetonitrile as the mobile phase; 0.2 mg/mL uracil as unretained marker. From Wu, N., et al. *J. Sep. Sci.* 2012, *35*, 2018–2025, with permission.

3.2.3.2 Apparent and Corrected Column Pressures

The relationship between the pressure drop and the linear velocity for a 2.1 mm internal diameter column is shown in Figure 3.2. It can be seen that the uncorrected pressure drop is significantly higher than the corrected pressure drop, and that this difference increases with increasing linear velocity or flow rate. This result is in agreement with Eq. (3.3), which shows that a substantial contribution to pressure drop from extra-column tubing can become important at higher linear velocities.

Table 3.4 shows the contribution of extra-column volume to the observed pressure drop for a variety of column sizes operating at the same linear velocity with the same connecting tubing. The smaller i.d. columns require a lower flow rate, which results in a lower extra-column pressure arising from flow through the connecting tubing. Therefore, the relative contribution of the pressure from the extra-column volume sources can be seen to diminish as the column internal diameter decreases, with 31% of the observed pressure attributed to extra-column sources with the 4.6 mm i.d. column, and only 10% of the pressure attributed to extra-column sources for the 1 mm i.d. column.

Table 3.4 The Contribution of Pressure Drop from Extra-Column Volume

Column I.D. (mm)	Flow Rate (mL/min)	ΔP_{obs} (Psi)	ΔP_{ec} (Psi)	ΔP_{corr} (Psi)	$\%\Delta P$ (Psi)
4.6	1.4	3396	1043	2353	31%
3.0	0.6	3382	544	2838	16%
2.1	0.29	3083	382	2701	12%
1.0	0.067	2895	281	2614	10%

Conditions: the same as in Figure 3.

3.2.3.3 Effect of Column Internal Diameter and Extra-Column Volume on Efficiency

Equation (3.6) indicates that the observed or apparent band-broadening arises from two sources: the column itself and the extra-column volume. The relative contribution of these two sources to extra-band-broadening is primarily related to the retention factor of an analyte, the column internal diameter, and the extra-column volume if other factors are kept constant.

3.2.3.3.1 Effect of Retention Factors

The contribution of extra-column band-broadening to column efficiency is more significant for early eluted peaks, because they typically have narrower peak widths. In other words, early eluted analytes are subject to more loss in efficiency than late eluted peaks for a given column at the same chromatographic conditions. Figure 3.3 shows the effect of retention factor on the contribution of extra-column effects for a 2.1 mm internal diameter column using benzyl alcohol, acetophenone, propiophenone, and benzophenone as analytes. Apparent plate heights increase significantly with a decreasing retention factor at high linear velocities. With the correction of extra-column volume effect as described in Eq. (3.7), the plate heights for all the analytes with various retention factors are essentially the same. This confirms that the lower efficiencies for the analyte with lower retention factors are due to the relatively greater contribution of extra-column band-broadening. It should be pointed out that the contribution of extra-column

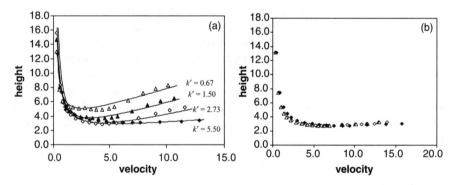

Figure 3.3 Van Deemter curves for various retention factors. (a) Apparent reduced plate height vs. apparent reduced linear velocity. (b) Reduced plate height vs. reduced linear velocity, corrected with the extra-column volume. *Conditions:* Premixed 50/50 (v/v) water to acetonitrile as the mobile phase; 2.1 mm internal diameter column. 0.2 mg/mL each of benzylalcohol (BA), acetophenone (AP), propiophenone (PP), and benzophenone (BP) were used as analytes with different retention factors to plot van Deemter curves for each analyte. Diffusion coefficients for BA, AP, PP, and BP were estimated as 1.08×10^{-5}, 1.02×10^{-5}, 9.36×10^{-6}, and 8.10×10^{-6} cm^2/s, respectively, using the Walke-Chang equation and empirical models for the acetonitrile water system described in Li, J.; Carr, P.W.; Anal. Chem. 1997, 69, 2530–2536. 0.2 mg/mL uracil as unretained marker. From Wu, N., et al. *J. Sep. Sci.* 2012, *35*, 2018–2025, with permission.

volume to band-broadening is decreased when gradient conditions are used (34) or when large molecule analytes (with slower rates of diffusion) are studied.

3.2.3.3.2 Effect of Column Internal Diameter

Theoretically, column efficiency is independent of the column internal diameter for a packed column, as none of the A, B, and C terms in the van Deemter Eq. (3.5) are dependent upon the column internal diameter. However, it can be seen from Eq. (3.6) that the observed peak variance is the sum of the peak variance and the variance from the extra-column volume, which is proportional to the 4th power of column internal diameter, as indicated in Eq. (3.8). As a result, the apparent or observed column efficiency is closely related to the column internal diameter, owing to this important contribution of extra-column volume.

Figure 3.4 shows typical migration times and band-broadening due to extra-column volume for three columns with different internal diameters all operating at the same linear velocity. Although the extra-column volume of a UHPLC system is minimized compared to a conventional LC system, extra-column band-broadening can still become significant for the smaller internal diameter columns. To achieve a 2 mm/s linear velocity for the three columns, volumetric flow rates of 1.0, 0.21, and 0.048 mL/min were used for the 4.6, 2.1, and 1.0 mm internal diameter columns, respectively. Figure 3.4 shows that the peak widths at half height attributable to extra-column band-broadening from the UHPLC system were 0.36, 1.6, and 4.1 seconds, respectively, a dramatic illustration of the important impact that extra-column volume can have on band-broadening, especially with narrow i.d. columns.

Figure 3.5 illustrates the efficiency loss for 1.0, 2.1, 3.0, and 4.6 mm i.d. columns using four analytes with different retention factors. The loss in efficiency is affected significantly by the column internal diameter and the retention factor of analytes, as predicted by Eq. (3.8). The columns of 2.1–4.6 mm i.d. provide more than 80%

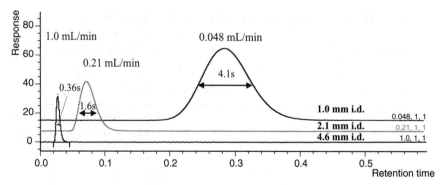

Figure 3.4 Effect of extra-column volume on migration times and band-broadening for a series of columns of different internal diameter operating at the same linear velocity. *Conditions:* A zero-dead volume union was used in place of a column to connect the injector to the detector; premixed 50/50 (v/v) water to acetonitrile as the mobile phase; 0.05 mg/mL uracil as an analyte. From Wu, N., et al. *J. Sep. Sci.* 2012, *35*, 2018–2025, with permission.

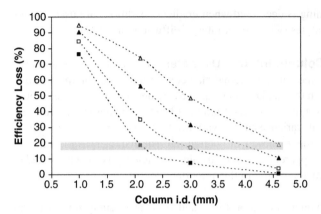

Figure 3.5   Efficiency loss for columns of various internal diameters. *Conditions:* Premixed 50/50 (v/v) water to acetonitrile as the mobile phase; 0.2 mg/mL each of analytes and uracil as unretained marker; 50 mm column length. The flow rate for various columns are shown in Table 3.3. Injection volumes for 1.0, 2.1, 3.0, and 4.6 mm i.d. columns are 0.23, 1.0, 2.0, and 4.8 μL, respectively. From Wu, N.; Bradley, A.C.; Welch, C.J.; Zhang, L.; *J. Sep. Sci.* 2012, *35*, 2018–2025, with permission. Key: (Δ) $k' = 0.67$, benzylalcohol; (▲) $k' = 1.50$, acetophenone; (□) $k' = 2.73$, propiophenone; and (■) $k' = 5.50$, benzophenone.

efficiency for the analytes with retention factors greater than 5. In any event, the use of a 1.0-mm internal diameter column is not practical because the relative contribution of the extra-column volume is greater than 70% even for a highly retained compound. Longer column lengths or gradient conditions can minimize the effect of extra-column band-broadening on efficiency for such a micro-bore column (34). Alternatively, a specially designed micro-LC system should be used for a column with an internal diameter of 1.0 mm or less.

3.2.3.3.3   Effect of Extra-Column and Injection Volumes

Four inlet tubings with different internal diameters are used in a UHPLC to study the effect of extra-column volume on efficiency. As shown in Table 3.5, the total extra-column

Table 3.5   Ratios of Extra-Column Volume to Column Void Volume

Internal Diameter of Inlet Tubing (mm)	Total Extra-column Volume (μL)*	Ratio of Extra-column Volume to Column Void Volume Ratio for a 50 × 2.1 mm Column (104 μL)**
0.064	6.6 (1.6)	1:15.7
0.127	11.3 (6.3)	1:9.2
0.254	30.3 (25.3)	1:3.4
0.508	106.3 (101.3)	1:1.0

*The extra-column volume other than inlet tubing is 5 μL for UPLC (See Table 1).
**Assuming that the column porosity is 0.6.

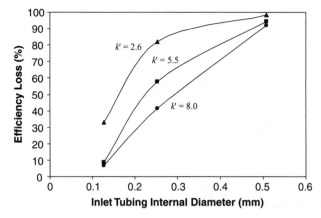

Figure 3.6   Efficiency loss for inlet tubings of various internal diameters. *Conditions:* Premixed 50/50 (v/v) water to acetonitrile as the mobile phase; 0.2 mg/mL each of propiophenone, benzophenone, pentanophenone, and uracil as unretained marker; 50 × 2.1 mm column; 0.3 mL/min flow rate; 210 nm; 25°C. Stainless steel tubings with various internal diameters were used as inlet connection tubing for the column. From Wu, N., et al. *J. Sep. Sci.* 2012, *35*, 2018–2025, with permission.

volume and thus the ratio of extra-column volume to column void volume increases substantially for a 50 × 2.1 mm column as the internal diameter of inlet tubing increases.

The loss in column efficiency versus inlet tubings of various internal diameters is shown in Figure 3.6. It can be seen that the column efficiency decreases dramatically with an increasing extra-column volume. Additionally, the early eluted peaks are more susceptible to extra-column volume, and the loss in efficiency for those peaks is more significant. Finally, the chromatograms provided by 0.254 and 0.508 mm inlet tubings with a length of 50 cm were too deteriorated to be used in practice.

The effect of injection volume on efficiency loss is shown in Figure 3.8. The column efficiency decreases more significantly from 1.0–5.0 µL and levels off from 5.0–20.0 µL. The early eluted peaks are also more susceptible to injection volume. Compared to the volume of inlet tubing, the effect of injection volume on efficiency is relatively less significant. The peak variance from the injection volume is approximately $V_{inj}^2/12$ while it is V_{tub}^2 for the inlet tubing volume. The results in Figure 3.7 also suggest that injection volumes should not be greater than 5 µL to avoid the significant efficiency loss for intermediately retained small molecules in UHPLC.

3.2.3.3.4   Effect of Dwell Volume on Separation

As indicated in Eq. (3.10), the retention time and separation can be impacted by the dwell volume of an HPLC system. The delay time of an analyte due to the dwell volume is inversely proportional to the flow rate used for the separation for a given system, as expressed $t_{dwell} = \frac{V_{dwell}}{F}$, where V_{dwell} is the dwell volume for a UHPLC or an HPLC system. Figure 3.8 shows gradient separations of phenones using UHPLC and HPLC with a 50 × 2.1 mm column. The first peak elution time is 2.2 min on UHPLC and 4.9 min in HPLC. The

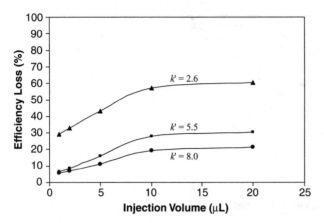

Figure 3.7   Efficiency loss for injection volumes. *Conditions:* Premixed 50/50 (v/v) water to acetonitrile as the mobile phase; 0.2 mg/mL each of propiophenone, benzophenone, pentanophenone, and uracil as unretained marker; 50 × 2.1 mm column; 0.3 mL/min flow rate; 210 nm; 25°C. A stainless steel tubing of 500 × 127 mm was used as inlet connection tubing for the column. From Wu, N., et al. *J. Sep. Sci.* 2012, *35*, 2018–2025, with permission.

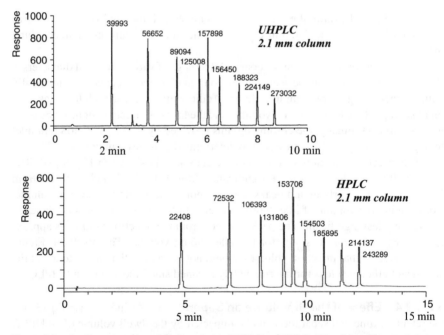

Figure 3.8   Gradient separation of phenones using UHPLC and HPLC with a 2.1 × 50 mm column. *Conditions:* 50 mm columns; 0.31 mL/min flow rate; 0.2 mg/mL of each of phenones; acetonitrile with 0.1% phosphoric acid in water as the mobile phase. Acetonitrile was increased from 10%–90% in 8 min and held at 90% for 5 min; 25°C; 210 nm. The numbers on top of the peaks are the apparent plate numbers.

delay time from the dwell volume is about 3 min more on HPLC than on UHPLC. The dwell volume for UHPLC is 120 μL compared to 1000 μL for HPLC. The flow is 0.31 mL/min and thus the 3 min delay volume in HPLC is ascribed to its bigger dwell volume (880 μL more than UHPLC). It can be seen that the apparent efficiency for HPLC are also lower than that for UHPLC due to the larger extra-column volume of the HPLC system. Thus, the delay time should be considered during a method transfer between UHPLC and HPLC. The result also suggests that a column with 2.1 mm or smaller internal diameter is not suitable for a conventional HPLC system with a large dwell volume and an extra-column volume.

Results in this section suggest that a reduction in column dimension be limited in UPLC due to the increased contribution of extra-column band-broadening to column efficiency for miniaturized columns. According to Eq. (3.8) and Tables 3.2 and 3.5, it is recommended that the maximum ratio of extra-column volume to column void volume be 1:10 for a UHPLC or an HPLC system to achieve acceptable performance for small molecules. For example, columns with the minimum 2.1 mm internal diameter (for ≥ 50 mm length) should be used for a UHPLC system with an extra-column volume of <20 μL to achieve acceptable column efficiency; likewise columns with the minimum 3.0 mm internal diameter (≥ 100 mm) should be used for a conventional HPLC system with an extra-column volume of <50 μL. As suggested in Eq. (3.9), extra-column band-broadening can be effectively reduced by using short and small radius connection tubings as well as small injection volumes and flow cells. It should be pointed out, however, that too small internal diameter tubings (such as <25 μm) would significantly increase the system back-pressure.

3.3 COLUMN PRESSURE AND FRICTIONAL HEATING

3.3.1 Column Pressure vs. Particle Size

It is natural for scientists becoming involved with UHPLC to wonder about the physical and chemical consequences of these ultrahigh pressures. The higher operating pressure of UHPLC allows chromatographers to take advantage of smaller particle columns operating at high linear velocities, leading to the improved separations and reduction in analysis time that are so valuable to practicing chromatographers. However, are there any negative consequences of the ultrahigh pressures of UHPLC?

The pressure drop across a column packed with particles is given by Eq. (3.11):

$$\Delta P = \frac{\phi \eta L u}{d_p^2} \tag{3.11}$$

The definitions of these variables in Eq. (3.11) are the same as in Eq. (3.2). From Eq. (3.11) it can be seen that the pressure drop across the column is inversely proportional to square of the particle diameter. In other words, halving the diameter of the particles packed within the column will result in a fourfold increase in pressure drop.

The optimum linear velocity is also dependent on the inverse of particle diameter:

$$u_{opt} = \frac{3D_m}{d_P} \qquad (3.12)$$

where u_{opt} is the optimum linear velocity, D_m is the diffusion coefficient, and d_p is the particle size. Combining Eqs. (3.11) and (3.12), the relationship of pressure drop and particle size can be expressed as:

$$\Delta P_{opt} \alpha \frac{1}{d_p^3} \qquad (3.13)$$

The pressure at a given optimum linear velocity is inversely proportional to the cube of particle diameter, assuming constant column length and mobile phase viscosity (1, 35).

For a given theoretical plate number (N), the column length (L) is proportional to the particle size: $L = N{\cdot}h{\cdot}d_p$. Packing of smaller particles provides higher column efficiency, allowing a shorter column length to be used for a given separation. Replacing L with $N{\cdot}h{\cdot}d_p$ in Eq. (3.12), the pressure required to achieve the optimum linear velocity is inversely proportional to the square of the particle diameter when the required column efficiency and other parameters are kept constant:

$$\Delta P_{opt} \propto \frac{1}{d_p^2} \qquad (3.14)$$

This ability of UHPLC technologies to deliver highly efficient separations with short columns packed with sub-2 μm particles has led to a paradigm shift, precipitating a widespread abandonment of longer column lengths that were typically used in conventional HPLC (36). At the same time, the high pressures encountered with newer small particle chromatography have taken researchers into a new realm of analytical chromatography.

3.3.2 Phenomenon of Frictional Heating

One consequence of the high pressures used in UHPLC is the phenomenon of frictional heating—a phenomenon with little impact on conventional HPLC for typical flow rate, but which can have dramatic negative consequence on UHPLC chromatographic performance if not adequately addressed.

As the mobile phase moves through a column packed with fine particles, this not only causes a pressure drop across the column but can also lead to a considerable rise in temperature of the mobile phase owing to friction. The heat generated is dissipated through the walls of the column, resulting in both axial and radial temperature

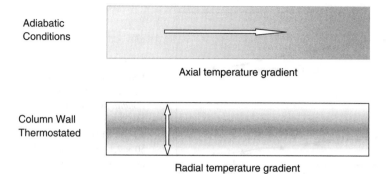

Axial temperature gradient

Radial temperature gradient

Figure 3.9 Schematic representation of temperature gradient inside a column. From Gritti, F., Guiochon, G. *J. Chromatogr. A.* 2009, 1216: 1353–1362, with permission.

gradients. In axial temperature gradients, the temperature changes from the inlet to the outlet of the column, and in radial temperature gradient, the temperature changes from the center to the walls of the column. The degree of these gradients is dependent on the dimensions and thermal insulation characteristics of the column.

As shown in Figure 3.9 (37), the axial or longitudinal temperature gradient is largest when the column is under an adiabatic condition, owing to the fact that no heat transfer from the column to its surrounding takes place. Similarly, the radial temperature gradient is largest when the column wall temperature is kept constant, that is, in a thermostated system where the heat transfer is not efficient. For these reasons, columns in UHPLC systems are typically operated with thermal insulation in an attempt to reduce axial and radial temperature gradients.

3.3.3 Effect of Frictional Heating on Retention Factor

Column temperature is an important factor in separation of analytes in liquid chromatography. Column temperature can affect retention, selectivity, peak shape, and column pressure. UHPLC separations are frequently carried out at elevated column temperature, as this helps reduce the back pressure that results from operating columns packed with smaller particles at high linear velocities.

Increasing column temperature generally reduces retention and selectivity in reversed-phase liquid chromatography, depending on enthalpy and entropy values of analytes (38). Figure 3.10 shows the effect of pressure-induced frictional heating on retention in UHPLC with a still air column heater. In this experiment, chromatography was carried out at a relatively low back pressure (1000 psi), then at a greatly increased back pressure (10,500 psi) with increased flow rates. The retention factors for two test analytes used in the study decreased by 12% at the higher pressure, a result that can be attributed to the increase in temperature within the column due to frictional heating. It is worth noting that these results were observed even when the column heater was under near adiabatic conditions.

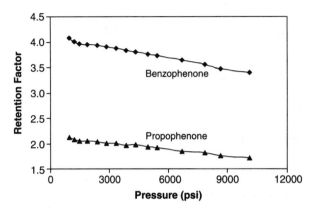

Figure 3.10   Effect of pressure on retention factor. *Conditions:* 50/50 (v/v) MeCN/H$_2$O; 25°C; 210 nm; 1.7 μm BEH Shield C18; 50 × 2.1 mm column; 1 μL injection; 0.2 mg/mL analyte in the mobile phase; Acquity UPLC system.

3.3.4 Factors Leading to Frictional Heating

As we have seen from the previous example, even when column heaters are set to a constant temperature over an extended period of time we can still experience a significant decrease in retention factor due to frictional heating. For a selected temperature, the column experiences longitudinal and radial temperature gradient due to the incoming solvent, which is at a different temperature compared to the oven temperature, leading to changes in retention factor and selectivity.

With a thermally equilibrated column (Figure 3.11a) (39), the temperature at the center and walls of the column will be the same; affording a typical peak shape.

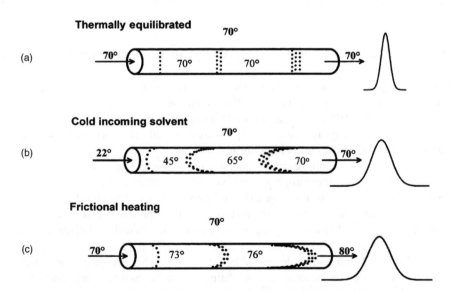

Figure 3.11   Band-broadening due to thermal effects: (a) thermally equilibrated, (b) negative radial temperature gradient with significant axial gradient, (c) positive radial temperature gradient with minor axial gradient. From Wolcott, R.G., et al. *J. Chromatogr. A.* 2000, 869: 211–230, with permission.

Figure 3.11b shows that when the incoming solvent is colder than the column, a radial temperature gradient will be established, owing to the higher temperature at the column walls relative to the column center. This gradient is most pronounced at the head of the column, with the difference in temperature between the center of the column and the outer wall becoming less pronounced further along the column. This radial temperature gradient leads to a coning effect as seen in Figure 3.11b, which has an overall negative effect on chromatographic peak shape. In addition, the solvent viscosity decreases from the center to the walls, creating a band migration, which is faster at walls as opposed to the center, also resulting in band-broadening or peak distortion.

A similar phenomenon arises from the frictional heating resulting from the flow of solvent through the column, leading to an in increase in column temperature from inlet to outlet as seen in Figure 3.11c. Again, this temperature remains higher at the center of the column, as opposed to the walls, causing a faster band migration at the center and resulting in a coning effect opposite to what is seen in Figure 3.11b. Consequently, if a packed column has poor heat dissipation properties, a temperature gradient can easily be produced within the column, resulting in decreased efficiency and performance. At constant wall temperature the radial temperature gradient from the center of column to the walls of column is given by Eq. (3.15):

$$\Delta T_R = \frac{qR_c^2}{4\lambda} \tag{3.15}$$

where ΔT_R is radial temperature gradient, q is the heat flux, R_c is column radius, and λ is radial heat conductivity.

As pointed out earlier, viscous heat dissipation leads to the formation of a longitudinal temperature gradient under adiabatic conditions. An adiabatic condition may be experienced using column heaters, where still air is used to regulate column temperature. The heat transferred between still air and column wall is poor and results in longitudinal temperature gradient at higher pressure. The magnitude of the longitudinal gradient temperature can be calculated using Eq. (3.16):

$$\Delta T_L = \frac{\Delta P}{C_P} \tag{3.16}$$

where C_P is the specific heat capacity of the solvent (in J/m^3).

The amount of heat generated (E) in frictional heating is a product of pressure and flow rate as seen in Eq. (3.17):

$$E = F\Delta P \tag{3.17}$$

From Eq. (3.21) we know the pressure at a given linear velocity is inversely proportional to the cube of particle diameter. Hence combining Eqs. (13) and (17), the relationship between energy and particle size can be expressed as for a fixed column length:

$$E \propto \frac{1}{d_p^3} \tag{3.18}$$

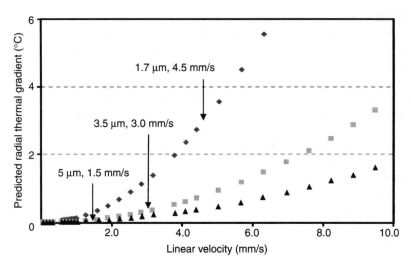

Figure 3.12   Radial thermal gradient vs. linear velocity for hexylbenzene on particles that are 1.7 μm (red diamonds), 3.5 μm (green squares), and 5.0 μm (blue triangles) in diameter. *Conditions:* Column dimensions, 2.1 × 50 mm; mobile phase: acetonitrile:water 7:3 (v/v); temperature, 25°C. From Mazzeo, J.R., et al. *Anal. Chem.* 2005, 77: 460A–467A, with permission.

From Eq. (3.18), the amount of heat generation is inversely proportional to the cube of particle size at the respective optimum linear velocity. For example, the predicted radial temperature gradients are 0.1, 0.35, and 3.6°C for 5.0, 3.5, and 1.5 μm particles, respectively, as shown in Figure 3.12 (6). Therefore, frictional heating is of special concern when using small particle columns.

The temperature gradient also affects the analyte diffusivity and retention on the column. Thus, both diffusivity and retention factors are a function of pressure and temperature. Frictional heating can cause a nonuniform increase or radial gradient in temperature inside a column, which can have a detrimental effect on the separation, resulting in band-broadening and poor peak shape (40, 41).

3.3.5 How to Resolve Thermal Gradient Issues

There are ways to minimize the formation of temperature gradients in UHPLC, which can directly influence the chromatographic efficiency, mobile phase viscosity, analyte retention factor, and diffusion coefficient.

The first alternative is to work at lower pressure ranges. Equation (3.17) suggests that the amount of heat generated is directly proportional to pressure. Most chromatographers use UHPLC to perform faster separations and will understandably be reluctant to operate at lower pressures, slower flow rates, longer analysis times, and lower efficiency. Reducing the column diameter also helps minimize temperature gradients, as narrow bore columns tend to have better heat dissipation properties than wider bore columns. Using a lower eluent temperature than that of the column is

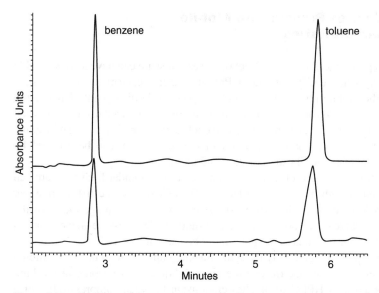

Figure 3.13   Comparison between peak shapes obtained for benzene and toluene using the Acquity column heater (top trace) and water bath (bottom trace). *Conditions:* Acquity BEH C18 Column (100 × 2.1 mm i.d., 1.7μ) at 40°C; 0.7-mL/min. From Jandera, P., Churacek, J. *J. Chromatogr. Library*. 1985, 31: 190–223, with permission.

another potential way to minimize radial temperature gradients and loss of column efficiency; however, this practice is difficult to control in routine work.

As seen in Figure 3.13 (42), the separation efficiency for a column thermostated with a typical still-air column heater (top chromatogram) is higher than that of a column heated with a water bath (bottom chromatogram). A typical still-air heater provides a near adiabatic condition for a column, allowing the radial temperature gradient that impacted column efficiency to be minimized. When the column was thermostated by a water bath, the radial temperature gradient was more significant, as water was a much better conductor than the air, and the temperature inside the column is always higher than the water temperature due to the frictional heating inside the column.

In UHPLC operation, frictional heating is typically dealt with by operating the column at a thermostated temperature, often above room temperature. Column operation at elevated temperature also helps reduce solvent viscosity, and subsequently, pressure. One should ensure the column chamber is insulated properly, as oven insulation is critical for heat flow from oven to column. It is necessary to allow sufficient time to equilibrate the column before making an injection.

Another significant factor that impacts column efficiency is the temperature mismatch of an eluent entering a column. This effect can be minimized by a preheating coil or a micro-thermostat; however, it should be noted that such heating coils or tubing can lead to an increased extra-column volume and thus band-broadening, as previously discussed.

3.3.6 Effect of Pressure on Mobile Phase Characteristics

Although less pronounced than the effect on gases, pressure can have a considerable effect on the physical chemistry of liquids. Pressure can influence many of the physical attributes of the mobile phase such as melting point or boiling point, density, and viscosity. In addition, the attributes of a compound in solution can also be somewhat influenced by pressure, for example, diffusion coefficient. The distribution of a solute between solid stationary phase and liquid mobile phase can also be affected by pressure, for example, equilibrium constant and phase ratio.

The effect of pressure on the melting point of a liquid under high pressure can be extreme, and at high pressure afforded by UHPLC, we run the risk of freezing chromatographic eluents, leading to overpressure and pump damage (43). On the other hand, the decrease in boiling point of liquids under pressure means that an extremely volatile mobile phase can potentially be used in UHPLC.

Although liquids are generally thought of as incompressible in the conventional HPLC, non-negligible volumetric compression of liquids can be seen at the high pressure afforded by UHPLC (43). The relationship between compressibility of a liquid and pressure is given by Eq. (3.19):

$$\beta_T = -\frac{1}{V}\left[\frac{dV}{dP}\right]_T \tag{3.19}$$

where β is the compressibility, V is the volume, P is the pressure, and T is the temperature.

Table 3.6 shows the effect of pressure on temperature and pressure on water volume. In going from 0 to 1000 atm (a typical UHPLC pressure drop), the volume of water is reduced by 4% at 40°C. The effect on chromatography due to compression of mobile phase can be seen in gradient devices using two or more pumps to mix components at high pressure. It is important for the pumps to deliver the exact volume accurately; otherwise the flow rate and thus retention time could be affected during a

Table 3.6 Volume of Water Compressed in Relation to Pressure and Temperature [1]

Pressure (atm)	Temperature (°C)						
	−10	0	10	20	40	60	80
1	1.0017	1.0000	1.0001	1.0016	1.0076	1.0168	1.0287
500	0.9788	0.9767	0.9778	0.9804	0.9867	0.9967	1.0071
1000	0.9581	0.9566	0.9591	0.9619	0.9689	0.9780	0.9884
1500	0.9399	0.9394	0.9424	0.9456	0.9529	0.9617	0.9717
2000	0.9223	0.9241	0.9277	0.9312	0.9386	0.9472	0.9568
2500	0.9083	0.9112	0.9147	0.9183	0.9257	0.9343	0.9437
3000	0.8962	0.8993	0.9028	0.9065	0.9139	0.9225	0.9315

gradient run. Most UHPLC instruments are equipped with either an electronic sensor or a flow rate feedback system, which corrects for these variations. Without these sensors, the reproducibility of the pump can be affected, and this will in turn influence the quantitation of retention time of an analyte (44).

Viscosity for a liquid is normally treated as being independent of pressure; however, the change in viscosity under ultrahigh pressures is no longer negligible. Viscosity in simple terms is the force experienced by liquid under shear stress. Fluid viscosity must be considered when an appropriate flow rate is chosen. Liquids such as water and hydrocarbons follow a Newtonian behavior in which viscosity remains constant with shear rate of agitation, that is, the pump speed is directly proportional to the flow rate. The pressure drop due to viscosity of the mobile phase can be a significant contributor to system back pressure and can affect the retention time and volume. This effect is rather insignificant in the conventional HPLC. However, the influence of pressure becomes more important in UHPLC systems (>9000 psi).

Diffusion of solutes within the column is critically important in liquid chromatography. Under typical HPLC conditions, viscosity of a mobile phase and the diffusion coefficient of a solute do not vary greatly with pressure. The Stokes-Einstein equation of diffusivity of solvent shows that the relationship of an analyte diffusion coefficient (D) is inversely proportional to the viscosity of mobile phase (η). As seen in Eq. (3.20):

$$D = \frac{kT}{6\eta r} \tag{3.20}$$

where k is the Boltzman constant, T is temperature, η is viscosity, and r is radius of the solute. It has been reported that the viscosity of a mixture of acetonitrile-water increases with rising pressure, and accordingly the diffusion coefficient also decreases with an increase in pressure. (45). Therefore, it is important to correct for the diffusion coefficient as a function of pressure; otherwise data generated for column performance using the van Deemter equation is inaccurate.

3.3.7 Effect of Pressure on Retention Factor with Minimal Frictional Heating

As discussed, mobile phase properties can be affected by high pressures. Additionally the solvation state of analytes and stationary phases can be changed under ultrahigh pressures. These changes in the properties of solutes and mobile and stationary phases at ultra-high pressures can result in changes in analyte molecular volume, potentially leading to changes in retention factors for solutes. The change in k' as a function of pressure at constant temperature can be expressed by Eq. (3.21) (46):

$$\ln\left(\frac{k'}{k'_0}\right) = -\frac{\Delta V}{RT} \cdot P + \ln\left(\frac{\beta}{\beta_0}\right) \tag{3.21}$$

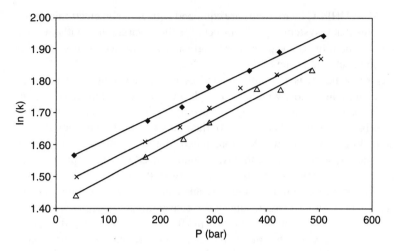

Figure 3.14   Plot of ln$k$ vs. P for large polar compounds. *Conditions:* Xbridge C18 BEH; mobile phase 25% acetonitrile, 75% water at 0.2 mL/min; UV detection at 254 nm. From Ringling, J., et al. *Am. Pharm. Rev.* 2008, 11: 24–32, with permission. **Key:** hydrocortisone (diamonds); prednisone (crosses); and prednisolobe (triangles)

where k_0' and β_0 are the retention factor and the phase ratio under reference conditions (which are taken as atmospheric pressure), R is the gas constant, T is the absolute temperature, and ΔV is the change in molar volume associated with the solute's transition between the mobile and the stationary phases: $\Delta V = V_{stat.} - V_{mob.}$. To study the effect of the pressure on retention factor, the frictional heating effect should be minimized to keep the column temperature constant. Two cases are discussed here. The first case is where a 50×2.1 mm column packed with 5 μm particles was used to minimize friction between mobile phase and stationary phase, and the column pressure was increased by using a capillary restrictor at the column outlet.

The linear increase in lnk' with an increasing pressure is demonstrated in Figure 3.14. Retention factors of hydrocortisone, prednisone, and prednisolobe increased by approximately 50% when the pressure was increased from 40 to 500 bar (or 60 to 7500 psi). The frictional heating effect is minimal for this study because the pressure from the 5 μm particles was much lower than that from 1.7 μm particles (see Figure 3.12). The high pressure mechanism is different from that shown in Figure 3.11. It was also reported that the increase in retention with pressure is much higher for larger molecules compared to smaller ones, which is due to larger changes in molar volume from mobile phase to stationary phase as a result of losing a part of their hydration layers when entering a hydrophobic stationary phase (45).

In addition to the molecular size, changes in k' for ionized compounds with pressure also depend on the mobile phase pH and on the analyte pK_a. As a result, selectivity between compounds of various pK_a values can be achieved merely by increasing the pressure of the separation. Figure 3.15 (46) shows the separation of a mixture of various ionizable compounds pyridine (pK_a 5.27), 2,6-lutidine (pK_a 6.75),

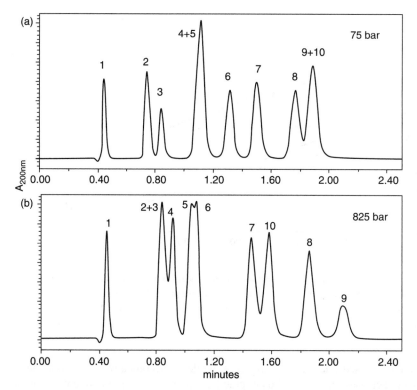

Figure 3.15   Chromatograms of a mixture of ionizable compounds at different average column pressures: (a) 75 bar (b) 825 bar obtained without and with a 25 cm restriction capillary, respectively. *Conditions:* XBridge C_{18} BEH; mobile phase acetonitrile 0.025 M potassium phosphate pH 5.8 (15:85, v/v); flow 0.3 mL/min; UV detection 200 nm. Peaks: (1) thiourea; (2) 2-methylbenzylamine; (3) pyridine; (4) 2,6-lutidine; (5) 2-picoline; (6) 2,4-lutidine; (7) 3-picoline; (8) aniline; (9) benzyl alcohol; (10) 3,4-lutidine. From Ringling, J., et al. *Am. Pharm. Rev.* 2008, 11: 24–32, with permission.

2-picoline (pK_a 5.97), 2,4-lutidine (pK_a 6.74), 3-picoline (pK_a 5.52), aniline (pK_a 4.62), 2-methylbenzylamine (pK_a 9.50), and 3,4-lutidine (pK_a 6.5) together with a neutral compound (benzyl alcohol) using acetonitrile-0.025 M phosphate buffer pH 5.8 (15:85, v/v) at 75 bar (1125 psi) without a restriction capillary and at 825 bar (12,375 psi) with a 25 cm restriction capillary. These different changes in retention associated with the increased pressure caused the changes in selectivity and even a reversal in the elution order of peaks 8 and 10 (aniline/3,4-lutidine). Pressure is thus potentially another parameter that could be used to adjust the selectivity of separations (46). Furthermore, the effect of pressure on selectivity should be taken into consideration for method transfer between, for example, a 5 μm column and a sub-2 μm particle column as separations may not be exactly the same on columns of different particle size due to their different operating pressures, even if the stationary phase chemistry is identical. However, an increase in pressure typically increases

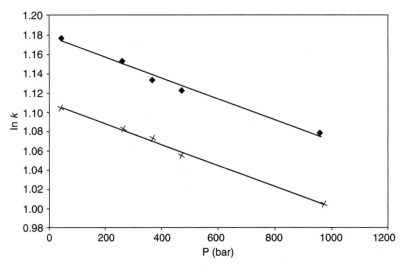

Figure 3.16   Plot of ln$k$ vs. P at 0.6 mL/min for amitriptyline (diamonds) and nortriptyline (crosses). *Conditions:* Atlantis Silica, 5 μm, 50 × 2.1 mm column; mobile phase: 10% water, acetonitrile 90% with 5 mm ammonium formate, pH 3. From Ringling, J., et al. *Am. Pharm. Rev.* 2008, 11: 24–32, with permission

retention in reversed-phase chromatography, while the accompanying frictional heating effect decreases retention. Thus in practice when comparing the retention factor and selectivity of columns packed with particles with different sizes, the pressure and pressure-induced thermo effects may partially cancel out.

Changes in k' for polar compounds with pressure can also be affected by the mode of liquid chromatography. For example, the opposite effect was noticed in hydrophilic interaction liquid chromatography (HILIC)—a near normal phase mode of LC, that is, retention factor decreased with an increase in pressure as illustrated in Figure 3.16. This can be explained by the fact that the HILIC stationary phase was more hydrated than the mobile phase, and thus the solubility for the polar solutes increased with an increasing pressure as the protonated anilines at pH 3 were preferably retained in the more polar stationary phase.

In another case where capillary columns are used, the frictional heating effect is negligible and the change in retention factor in UHPLC can be primarily attributed to the increased pressure. Figure 3.17 shows that the retention factor for 4-methylcatechol is increased by approximately 50% as pressure increases from 1000 to 7000 bar (or 15,000 to 105,000 psi) for capillary columns of various internal diameters, packed with 1.0 μm nonporous particles (35). This increase can be mainly ascribed to the compressibility of the mobile phase, the change in molecular volume, and the potential solvating state of the stationary phase. It should be pointed out that in this study k' varies in a linear fashion with column pressure while lnk' shows a linear relationship with pressure in a broader pressure range using capillary UHPLC, as demonstrated in Figure 3.14. Additionally, the frictional heating effect is negligible for these capillary columns compared to conventional 2.1–4.6 mm columns, where

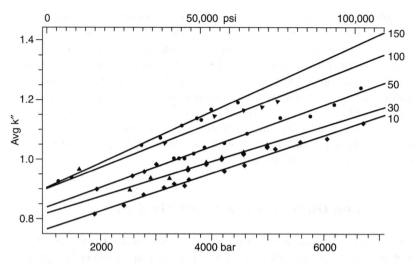

Figure 3.17   The relationship of retention factor vs. pressure for capillary columns of various internal diameters. *Conditions:* 1.0-μm nonporous silica C_{18} bonded particles; 4-methylcatechol as the analyte. Each fit is labeled with its corresponding diameter. From Colon, L.A., et al. *Analyst.* 2004, 129: 503–504, with permission.

the opposite trend was observed in reversed-phase UHPLC, as shown in Figure 3.10. It was also observed that increasing column diameter results in an increase in retention factor at any given pressure. As the column diameter decreases, the packing of particles is less ordered and rather more randomly packed, suggesting that the larger i.d. columns can pack more densely, leading to increased k' values.

In summary, it is important to note that there are various effects of frictional heating on column performance including decreased retention factors and deteriorated column efficiency. Fortunately, the effect on column efficiency is minimal in the current operating pressure range (<1200 bar or 18,000 psi) for a still-air column heater, which is widely used in UHPLC. The effect of ultrahigh pressure on retention factor is more complex, even under the conditions where frictional heating is negligible. The retention factor can be increased, unchanged, or decreased, depending on mobile phase composition, molecular size, polarity of analytes and stationary phases, etc. However, the effect of frictional heating and ultrahigh pressure on retention factor and potentially selectivity should be considered when a method for HPLC is converted to that for UHPLC. Also it is important to address these effects that will eventually play more significant roles once the uncharted operating pressure range (>18,000 psi) is reached.

3.4 METHOD TRANSFER BETWEEN UHPLC AND HPLC AND OTHER TIPS

Although scientists in research and development areas have started to replace HPLC with UHPLC systems, it will take some time for the entire industry, especially quality control laboratories, to fully upgrade their instruments. Therefore, smooth and efficient method transfer between the two platforms is very important within this transition period. Conversion of conventional HPLC methods to fast UHPLC methods has been demonstrated in the literature (24,25, 47–49). In practical applications, there are also a few occasions where a method developed for UHPLC must be converted to an HPLC method, especially when a method is to be utilized in a quality control laboratory where UHPLC is not yet available. In both scenarios, the goal for a method transfer is to achieve similar resolution, retention factor, sensitivity, and robustness. In this section, the basic parameters that need to be considered for a successful method transfer between the two platforms will be discussed.

3.4.1 Column Dimension and Particle Size

The first consideration to make a method equivalent between HPLC and UHPLC is to use columns that have comparable efficiency on each system. As shown in Eq. (3.22), column efficiency (N) is determined by the ratio of column length (L) to particle size (d_P):

$$N \propto \frac{L}{d_P} \tag{3.22}$$

Currently, the most commonly used HPLC columns are 15 cm long with a particle size of 3 or 3.5 μm. To achieve the same efficiency on UHPLC columns packed with sub-2 μm particles, a relatively shorter column, for example, 10 or 7.5 cm, can be used. The packing uniformity of the sub-2 μm particles is not as good as the conventional columns with 3 or 3.5 μm particle sizes, and thus the real efficiency is less than the theoretical values. It is therefore prudent to use slightly longer UHPLC columns than what is calculated using the preceding equation. To maintain the same column efficiency during method transfer between the two platforms, it is also important to keep the ratio of extra-column volume to column volume (V_{ex}/V) comparable. UHPLC columns usually have smaller dimensions; therefore, V_{ex} needs to be minimized by using narrower connecting tubing and a smaller flow cell. In practice, one can choose the right column internal diameter during transfer between HPLC and UHPLC methods to maintain similar V_{ex}/V.

More recently, fused core columns from Supelco and core-shell columns from Phenomenex and Agilent have gained much interest in the field of fast chromatography. The particles used in these columns are made of a sub-1.2 to -2 μm solid core and a 0.35–0.5 μm porous outer shell, with an overall particle size below 3 μm, providing a compromise between high efficiency and modest operating pressures. Consequently, one can run these columns on conventional HPLC instruments to obtain very rapid separations. In UHPLC, however, smaller internal diameter columns can be used to reduce solvent consumption. Users can refer to several papers for advantages and applications of these two types of columns (22, 50–52).

Table 3.7   HPLC and UHPLC Columns Provided by Major Manufacturers

Manufacture	Brand	HPLC ($\geq 3\ \mu m$)	HPLC/UHPLC (2–$3\ \mu m$)	UHPLC ($\leq 2\ \mu m$)
ACE	ACE	3	-	-
Agilent	Zorbax	3.5	-	1.8
Phenomenex	Synergi	-	2.5	-
Phenomenex	Luna	3	2.5	2
Phenomenex	Kinetex	-	2.6	1.7
Restek	Pinnacie DB	-	-	1.9
Shimadzu, XR	Shimadzu	-	2.2	-
Supelco	Ascentis Express	-	2.7	-
Thermo	Hypersil	3	-	1.9
Waters	Atlantis	3.5	-	1.8
Waters	Xbridge	3.5	2.5	1.7
Waters	Sunfire	3.5	2.5	-
Waters	HSS	3.5	-	1.8
YMC	Pro C18	3	-	2

Note: Hyphen (-) indicates that the column is unavailable.

3.4.2 Column Chemistry and Mobile Phase

The selectivity of a chromatographic method is determined by the column stationary phase chemistry and the mobile phase composition; consequently, it is advisable to keep the mobile phase and column chemistry the same when the method is transferred between HPLC and UHPLC. This means that in labs where exchange of methods between instrument types is anticipated, chromatographers are advised to choose column brands where both UHPLC and HPLC columns with the same stationary phase chemistry are available. Table 3.7 lists the column brands currently provided by the major manufactures. It can be seen that not all brands carry both columns packed with the conventional 3–5 μm and the sub-2 μm particle sizes. Within each brand, not all stationary phases are available in both platforms. In such situations, column equivalence assessment needs to be performed to find the best alternative column by using tools such as the reversed-phase column selectivity charts available from column vendors.

3.4.3 Flow Rate and Gradient Profile

After the column and mobile phase are selected, the next step in method transfer is to adjust the flow rate and gradient profile to keep the same retention factor for the analytes of interest. The conversion between HPLC and UHPLC for isocratic methods has been reviewed before (24, 25). Here, we will focus on gradient elution, which is more commonly used. The gradient retention factor (k^*) is determined by gradient time (t_G), flow rate (F), gradient step or the difference in organic composition from the initial gradient to the end time ($\Delta \phi$), and the column void volume (V_m), as shown in Eq. (3.23).

$$k^* = \frac{t_G \times F}{\Delta \phi \times V_m} \tag{3.23}$$

To arrive at the same retention factors, the same gradient steps are typically employed, with appropriate adjustments of gradient time and flow rate. The first parameter to be considered is flow rate. The optimum linear velocity is inversely proportional to particle size; therefore, the flow rate scaling factor between two methods can be expressed as shown in Eq. (3.24):

$$\frac{F_U}{F_H} = \frac{u_U \times d_{C,U}^2}{u_H \times d_{C,H}^2} = \frac{d_{P,H} \times d_{C,U}^2}{d_{P,U} \times d_{C,H}^2} \tag{3.24}$$

where F is the flow rate, d_c is the column internal diameter, u is the linear velocity, d_P is the particle size, and subscripts H and U denote HPLC and UHPLC, respectively. Once flow rate is determined, the gradient time scaling factor can be derived by combining Eqs. (3.23) and (3.24), and the result is shown in Eq. (3.25):

$$\frac{t_{G,U}}{t_{G,H}} = \frac{L_U \times d_{C,U}^2 \times F_H}{L_H \times d_{C,H}^2 \times F_U} \tag{3.25}$$

where L is the column length. Figure 3.18 shows the method transfer between an HPLC method and a UHPLC method with gradient profiles and flow rates determined using the preceding scaling approach. The elution profile is very similar and the average resolution is comparable.

The scaling approach described can be considered as a good starting point for some method transfers. In certain circumstances where there is no concern for closely eluting peaks, such as in-process control or potency-only methods, the initially converted UHPLC method can be further optimized for even shorter run times by increasing the flow rate and/or solvent strength. For UHPLC columns packed with small particles, van Deemter curves are relatively flat; therefore, it is possible to operate above the optimum linear velocity (within the system pressure limit) without a significant loss in efficiency. Figure 3.19 compares two chromatograms obtained on the same column and with the same conditions except the flow rate. With flow rate increased from 0.5 to 1.0 mL/min, the analysis time was decreased from 20 to 10 min, and no significant change in separation profile was observed.

When a flow rate reaches the system pressure limit, the run time can be shortened by increasing gradient steepness $\Delta \phi / t_G$ or switching to mobile phases with a greater eluting strength. Here, the loss in resolution can be offset by gain in efficiency on the smaller particles, and the overall resolution can still be comparable between the two methods.

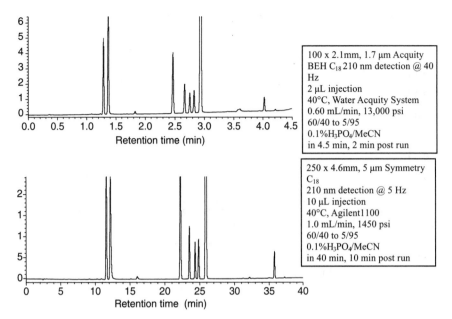

Figure 3.18   Chromatograms of an impurity mixture analyzed using UHPLC and HPLC columns.

3.4.4 Injection Volume

As shown in Figure 3.8, an increased injection volume can have a negative effect on the column efficiency. Therefore, the injection volume should be appropriately scaled down when an HPLC method is transferred to a UHPLC method or vice versa to achieve the same sensitivity and avoid overloading or detector saturation and extra-column band-broadening. The scaling factor is mainly based on column dimension, as shown in Eq. (3.26). However, lower injection volumes than calculated can often be used on UHPLC to achieve the same sensitivity due to enhanced peak heights from use of the high-resolution columns and low carryover from the injector:

$$\frac{V_{inj,U}}{V_{inj,H}} = \frac{L_U \times d_{C,U}^2}{L_H \times d_{C,H}^2} \tag{3.26}$$

Here V_{inj} is the injection volume, and other parameters are defined in Eq. (3.24).

3.4.5 Dwell Volume

Within the dwell time, the sample is subject to an extra isocratic or gradient step that is not reflected in the gradient profile, as shown in Eq. (3.10). A conventional HPLC system generally has a large dwell volume, that is, 500–1000 μl, whereas on the UHPLC systems, the dwell volume is decreased to about 100–200 μl. Basic gradient

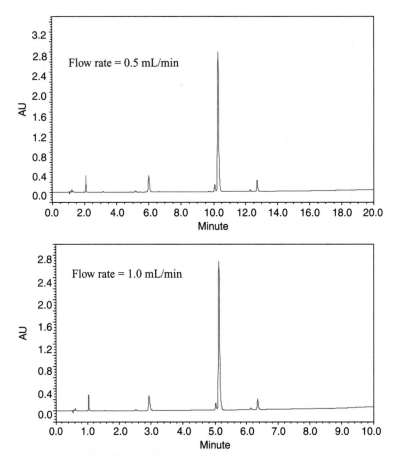

Figure 3.19 Chromatograms of an impurity mixture analyzed on Thermo Hypersil PFP column (100 × 3.0 mm i.d., 1.9 μm) with 0.1% H$_3$PO$_4$/MeCN/IPA gradient.

conversions based on the scaling approach discussed above are valid only when the dwell time is similar and/or analytes have high apparent retention factors. When the dwell time between systems is significantly different, retention time variations for early eluters and resolution change between critical pairs are expected, as shown in Figure 3.8. To maintain the same relative retention factor and resolution, the ratio of dwell time to column void time (t_d/t_0) or dwell volume to column void volume (V_{dwell}/V_0) needs to be kept relatively constant; therefore, the gradient profile needs to be adjusted accordingly.

Figure 3.20 shows the separation of a mixture on four HSS C18 columns with different particle sizes and dimensions (53). The column packed with sub-2 μm particles was run on Waters Acuity UHPLC, and the other three conventional columns were run on Waters Alliance HPLC systems. The difference in V_{dwell}/V_o between each set up was compensated by changing the length of initial gradient hold, and

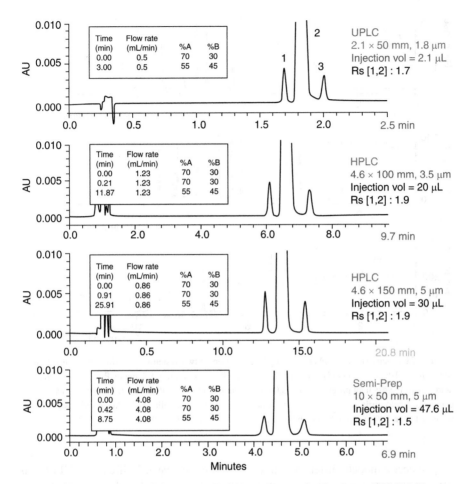

Figure 3.20 Gradient adjustment to account for difference in t_{dwell}/t_0. *Conditions:* HSS C18 SB columns; column temperature, 30°C; UV detection, 295 nm. Peaks: (1) related compound B; (2) paroxetine; (3) related compound F. From Ruta, J., et al. *J. Sep. Sci.* 2010, 33: 2465–2477, with permission.

therefore, the chromatographic resolution and selectivity were maintained across all chromatograms.

Users sometimes need to run the exact same method on a conventional HPLC system as well as a UHPLC system. In this case, the dwell volume difference must be taken into consideration during the method transfer. As shown in Figure 3.21, when the same gradient was used on both systems, the retention time for the first peak was delayed about 1.3 min due to the 0.88 mL greater dwell volume of the HPLC system. When a 1.3 min hold was applied at the beginning of the gradient for the UHPLC gradient profile, the retention times for the peaks are much more comparable.

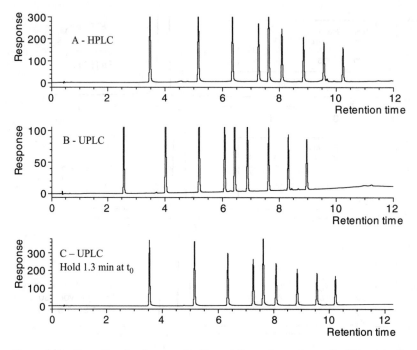

Figure 3.21   Separation of a pharmaceutical formulation on two systems with different dwell volume: (A) Agilent 1100, (B) Waters UPLC with the same gradient program, (C) Waters UPLC with 1.3 min hold at t_0 of the gradient program. 50×3.0 mm column, 0.6 mL/min flow rate. 0.2 mg/mL of each of phenones. Acetonitrile with 0.1% phosphoric acid in water as the mobile phase. Acetonitrile was increased from 10%–90% in 8 min and held at 90% for 5 min, 25°C, 210 nm.

To accommodate differences in dwell volume between UHPLC and HPLC, an appropriate isocratic hold time can be adjusted accordingly when the method is run on different instruments with different dwell volumes. Several vendors also offer simple software programs to determine appropriate changes needed to produce similar separations using different dwell volumes. Employing dwell volume adjustments to the method would not require revalidation by the United States Pharmacopeia section on chromatography (USP<621>).

3.4.6 UV Detection

Variable UV-visible or photodiode array detectors are the most commonly used detectors for UHPLC and HPLC. During method transfer most detection parameters should be kept the same, but some settings need to be adjusted to accommodate the difference in peak width. To ensure reproducible integration, the corresponding chromatography software needs at least 10–15 points across a peak. With the much narrower peak widths obtained by UHPLC, higher sampling speeds and shorter filter times are required. For an HPLC system, the sampling rate is usually several Hz, whereas filter time is typically set to a couple of seconds. In a UHPLC system, 20–40 Hz is typically used as the sampling rate, with a time constant about 0.1–0.2 sec, depending on the peak width. In addition, a smaller flow cell is required on UHPLC to minimize the extra column volume, as described in Table 3.1.

3.4.7 Other Tips

3.4.7.1 Instrumentation Differences

Differences in commercially available instruments need to be considered when performing a method transfer from HPLC to UHPLC. UHPLC equipment from different vendors may contain different configurations such as pressure limit, system delay volume, and injector design. This may lead to method robustness issues in which methods developed on one type of UHPLC may require some modifications for another model. Chromatographers must consider such differences during method development to avoid future method transfer issues. For example, quaternary pump systems are commercially available for most conventional HPLC systems but are currently limited to only a few UHPLC systems. This factor should be considered before pursuing the development of methods employing tertiary or quaternary gradient solvent systems.

3.4.7.2 Pressure Limit

Pressure limits of current UHPLC instruments from different vendors range from 9000 to 19,000 psi. Additionally, the range of flow rates capable of delivering back pressures greater than 5000 psi can vary from vendor to vendor. The use of smaller particle columns provides a wider range of optimal linear flow rates. As a result, flow rate adjustments may be needed between different UHPLC instruments. Consequently, a method developed on a UHPLC that provides greater than 10,000 psi back pressure can be modified by decreasing the flow rate to one that is suitable for UHPLC systems with lower back pressure limits. Although the run times will need to be increased proportionally to the decreased flow rate, the resolution between critical pairs of impurities should be maintained.

3.4.7.3 Mobile Phase Mixer Volume

When using low UV wavelengths with mobile phase additives, inefficient mixing can provide periodic baseline noise during gradient elution, making it difficult to identify impurity peaks and limiting method sensitivity. This effect can be more pronounced on UHPLC systems that typically incorporate smaller volume mixers in an effort to decrease instrument dwell volume. Consequently, mixing efficiency should be considered when performing the method transfer from HPLC to UHPLC, as larger volume mixers or new mixer designs may be needed to achieve the desired method sensitivity. The effect of mixer volume is shown in Figure 3.22 for the separation of a

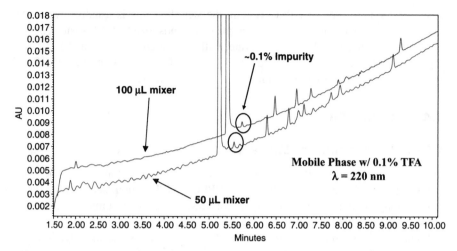

Figure 3.22   Chromatograms of an impurity mixture analyzed on a UHPLC system utilizing two different mixture volumes. From Abrahim, A., et al. *J. Pharm. Biomed. Anal.* 2010, 51: 131–137, with permission.

pharmaceutical using 0.1% TFA(aq):acetonitrile mobile phases. The larger mixer volume effectively reduced the UV baseline noise in the UHPLC method (26). Recently several new low-volume mixers have been introduced that are either based on multi-layer microfluidics technology or microreactor design to deliver efficient, low-volume mixing (35–50 μL); the mixers have been found to deliver stable baselines (55).

3.4.7.4   Column Temperature

The column temperature should be kept the same for transfer of the method from HPLC to UHPLC. However, for columns packed with sub-2 μm particles, it has been found that both longitudinal and radial temperature gradients exist within the column due to the frictional heating between the mobile phase and particles as discussed in Section 3.3. Although the radial temperature gradient is generally not a concern for columns in a forced longitudinal direction, a significant temperature increase can occur along the longitudinal direction up to 20°C higher in the column outlet than the inlet (42, 54, 55). The longitudinal temperature gradient can cause a decrease in retention and may also change selectivity if the separation is sensitive to temperature. The column temperature may need to be decreased in UHPLC to provide equivalent separations to HPLC. This effect is shown in Figure 3.23 in a separation where the flow rate and temperature was adjusted to maintain the separation.

3.4.7.5   Filtration of Mobile Phases and Sample Solution

As the particle size for UHPLC columns is significantly smaller than that for HPLC, the pore size of frits used for holding the particles in the columns is usually smaller.

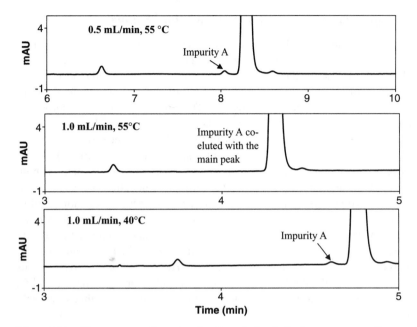

Figure 3.23   Chromatograms illustrating the effect of a longitudinal temperature gradient may have on retention. At a 0.5 mL/min flow rate, the impurity in front of the main peak was well separated using a column temperature of 55°C. However, increasing the flow rate to 1.0 mL/min caused the impurity to co-elute with the main peak. Adequate separation was achieved by reducing the column temperature to ~40°C. From Abrahim, A., et al. *J. Pharm. Biomed. Anal.* 2010, 51: 131–137, with permission.

For example, 0.5 μm frits are usually used for 1.7–1.8 μm particles, whereas 2 μm frits are used for 3–5 μm particles. There is a greater possibility of having clogging issues for the columns packed with sub-2 um particles, especially when dirty samples are analyzed. UHPLC solvents and sample solutions must be carefully filtered using a 0.2 μm membrane prior to analysis to ensure that there is no plugging of the instrument tubing or column, although this is a good practice for all LC systems.

3.4.7.6 Equilibration Time

The equilibration time for an LC system is primarily related to the dwell volume, column volume, and flow rate under gradient conditions. As both small internal diameter columns and correspondingly reduced flow rates are used in UHPLC, the typical equilibration time for the columns should be similar for both UHPLC and HPLC, although larger columns are used in HPLC. The equilibration time needed for the dwell volume can be calculated based on flow rate. The equilibration time for column chemistries and pressures for gradient conditions needs to be carefully evaluated as well. To reduce waiting time, pumps have been designed without pulse dampers or with lower-volume dampers. Pulse dampers are components that smooth out pump pulses to create a uniform flow.

3.4.7.7   Regulatory Considerations and Method Equivalency

The strategy for utilization of UHPLC methods in regulatory filings is still in development in the pharmaceutical industry. At this time, it may be practical to provide equivalent UHPLC and HPLC methods in the New Drug Application (NDA)/Marketing Authorization Application (MAA) regulatory filings to allow maximum flexibility of either method to be used for analytical testing. This approach would be appropriate during the transition of HPLC to UHPLC laboratories globally. This will require both methods to be validated as usual, but in the end, it will save time and allow the UHPLC method to be used for the drug substance in the future during the product's lifecycle. Full method validation may not be necessary to complete on both methods, but at a minimum, the sensitivity confirmation, injection precision, result equivalence, and robustness is performed in the UHPLC method validation. Additionally, with pharmaceutical development organizations often leveraging external suppliers to manufacture intermediates or drug substance and to conduct analytical characterization, it is necessary to ensure that the method can be run in multiple laboratories and to understand that having HPLC and UHPLC methods is advantageous and maximizes method flexibility. Until such time that all laboratories have UHPLC, it is the prudent approach to have equivalent HPLC and UHPLC methods available for analysis.

3.5   CONCLUDING REMARKS

The recent introduction of UHPLC has been considered a breakthrough by the chromatography community as it is capable of performing much faster and more efficient separation than conventional HPLC. However, operation at very high pressures can result in a variety of new problems that were relatively unimportant for users of conventional HPLC. The extra-column volume effect can become more important because column dimensions have been significantly reduced in UHPLC. The extra-column volume in UHPLC can lead to significant changes in apparent chromatographic parameters including increased pressure drop, linear velocity, and retention factor when small i.d. columns are used. The extra-column band-broadening effect becomes increasingly significant as the ratio of column to extra-column volumes decreases. Thus, appropriate column dimension and injection volume should be used to minimize this effect.

High pressures utilized in UHPLC can result in various effects that were not significant previously in conventional HPLC. Frictional heating caused by high pressures on 2.1–4.6 mm i.d. columns packed with sub-2 μm particles can lead to reduced retention factors and a change in selectivity of analytes. Additionally, the column efficiency can be significantly deteriorated by frictional heating if it is not appropriately addressed. Finally, high pressures can also lead to changes in retention factor and selectivity, depending on the properties of the solutes, the mobile phase, and the stationary phase. Although these effects can be complex as they often occur simultaneously, benefits of UHPLC in speed and efficiency are much greater than these downside effects.

The smooth and efficient method transfer between the two platforms will remain important within the transition period between UHPLC and HPLC. The column dimension, particle size, stationary phase chemistry, mobile phase flow rate and gradient profile, dwell volume, injection volume, and data acquisition are key factors to be carefully evaluated to ensure improved or maintained performance of transferred methods. Other factors including regulatory requirements should also be considered for a successful method transfer to quality control labs. The extra-column volume effects such as increased retention factors and decreased efficiency and high pressure effects (e.g., frictional heating and changes in retention factors) should be considered when various methods are compared and transferred.

The relatively significant effects caused by high pressure and extra-column volume in current UHPLC provide an opportunity for instrument manufacturers to further improve their systems. Along with these improvements, it would be desirable for increased operation pressures (>18,000 psi) for the next generation of UHPLC systems.

REFERENCES

1. MacNair, J.E., Lewis, K.C., Jorgenson, J.W. *Anal. Chem.* 1997, 69: 983–989.
2. MacNair, J.E., Patel, K.D., Jorgenson, J.W. *Anal. Chem.* 1999, 71: 700–708.
3. Lippert, A.J., Xin, B., Wu, N., Lee, M.L. *J. Microcol.* 1999, 11: 631–643.
4. Wu, N., Collins, D.C., Lippert, A.J., Xiang, Y., Lee, M.L. *J. Microcol.* 2000, 12: 462–469.
5. Wu, N., Lippert, A.J., Lee, M.L. *J. Chromatogr.* 2001, 911: 1–12.
6. Mazzeo, J.R., Neue, U.D., Kele, M., Plumb, R.S. *Anal. Chem.* 2005, 77: 460A–467A.
7. Halasz, I., Endele, R., Asshauer, J. *J. Chromatogr.* 1975, 112: 37–60.
8. Lin, H.-J., Horvath, C.S. *Chem. Eng. Sci.* 1981, 36: 47–55.
9. Gritti, F., Guiochon, G. *J. Chromatogr. A.* 2007, 1138: 141–157.
10. Gritti, F., Martin, M., Guiochon, G. *Anal. Chem.* 2009, 81: 3365–3384.
11. Colon, L.A., Cintron, J.M., Anspach, J.A., Fermier, A.M., Swinney, K.A. *Analyst.* 2004, 129: 503–504.
12. Welch, C.J., Wu, N., Biba, M., Hartman, R., Brkovic, T., Gong, X., Helmy, R., Schafer, W., Cuff, J.F., Pirzada, Z., Zhou, L. *Trends Anal. Chem.* 2010, 77: 667–680.
13. Chen, S., Kord, A. *J. Chromatogr. A.* 2009, 1216: 6204–6209.
14. Yeung, E.S. in Novotny, M., Ishii, D. (eds.). *Microcolumn Separations*, Amsterdam: Elsevier, 1985, pp. 117–144.
15. Krejci, M. *Trace Analysis with Microcolumn Liquid Chromatography.* New York: Marcel Dekker, 1992, pp. 35–37.
16. Fountain, K.J., Neue, U.D., Grumbach, E.S., Diehl, D.M. *J. Chromatogr. A.* 2009, 1216: 5979–5988.
17. Usher, K.M., Simmons, C.R., Dorsey, J.G. *J. Chromatogr. A.* 2008, 1200: 122–128.
18. Cabooter, D., Billen, J., Terryn, H., Lynen, F., Sandra, P., Desmet, G. *J. Chromatogr. A.* 2008, 1204: 1–10.
19. Cabooter, D., deVilliers, A., Clicq, D., Szucs, R., Sandra, P., Desmet, G. *J. Chromatogr. A.* 2007, 1147: 183–191.
20. Grittia, F., Felinger, A., Guiochon, G. *J. Chromatogr. A.* 2006, 1136: 57–72.
21. Prüß, A., Kempter, C., Gysler, J., Jira, T. *J. Chromatogr. A.* 2003, 1016: 129–141.

22. ALEXANDER, A.J., WAEGHE, T.J., HIMES, K.W., TOMASELLA, F.P., HOOKER, T.F. *J. Chromatogr. A.* 2011, 1218: 5456–5469.
23. WU, N., CLAUSEN, A., WRIGHT, L., VOGAL, K., BERNARDONI, F. *Am. Pharm. Rev.* 2008, 11: 24–33.
24. GUILLARME, D., NGUYEN, D.T.T., RUDAZ, S., VEUTHEY, J.L. *Eur. J. Pharm. Biopharm.* 2007, 66: 475–482.
25. GUILLARME, D., NGUYEN, D.T.T., RUDAZ, S, VEUTHEY, J.L. *Eur. J. Pharm. Biopharm.* 2008, 68: 430–440.
26. YIN, Z., FOUNTAIN, K.J., McCABE, D., DIEHL, D.M. *LC-GC Europe.* 2009, 22: 33–34.
27. GIDDINGS, J.C. *Unified Separation Science.*John Wiley & Sons, New York, 1991, p. 65–66.
28. VAN DEEMTER, J.J., ZUIDERWEG, F.J., KINKENBERG, A. *Chem. Eng. Sci.* 1956, 5: 271–289.
29. SNYDER, L.R., KIRKLAND, J.J. *Introduction to Modern Liquid Chromatography.* 1979, John Wiley & Sons, New York, Chapter 2.
30. YAN, B., CARR, P.W. et al. *Anal. Chem.* 2000, 72: 1253–1262.
31. FREEBAIRN, K.W., KNOX, J.H. *Chromatographia.* 1984, 19: 37–47.
32. GRITTIA, F., GUIOCHON, G. *J. Chromatogr. A.* 2010, 1217: 7677–7689.
33. McCALLEY, D.V. *J. Chromatogr. A.* 2010, 1217: 4561–4567.
34. LESTREMAU, F., WU, D., SZÜCS, R. *J. Chromatogr. A.* 2010, 1217: 4925–4933.
35. PATEL, K.D., JERKOVICH, A.D., LINK, J.C., JORGENSON, J.W. *Anal. Chem.* 2004, 76: 5777–5786.
36. COLON, L.A., CINTRON, J.M., ANSPACH, J.A., FERMIER, A.M., SWINNEY, K.A. *Analyst.* 2004, 129: 503–504.
37. GRITTI, F., GUIOCHON, G. *J. Chromatogr. A.* 2009, 1216: 1353–1362.
38. LI, J. *Anal. Chim. Acta.* 1998, 369: 21–37.
39. WOLCOTT, R.G., DOLAN, J.W., SNYDER, L.R., BAKALYAR, S.R., ARNOLD, M.A., NICHOLS, J.A. *J. Chromatogr. A.* 2000, 869: 211–230.
40. WU, N., CLAUSEN, A.M. *J. Sep. Sci.* 2007, 30: 1167–1182.
41. GRITTI, F., MARTIN, M., GUIOCHON, G. *Anal. Chem.* 2009, 81: 3365–3384.
42. DE VILLIERS, A., LAUER, H., SZUCS, R., GOODALL, S., SANDRA, P. *J. Chromatogr. A.* 2006, 1113: 84–91.
43. MARTIN, M., GUIOCHON, G. *J. Chromatogr. A.* 2005, 1090: 16–38.
44. JANDERA, P., CHURACEK, J. *J. Chromatogr. Library.* 1985, 31: 190–223.
45. KAISER, T.J., THOMPSON, J.W., MELLORS, J.S., JORGENSON, J.W. *Anal. Chem.* 2009, 81: 2860–2868.
46. FALLAS, M.M., NEUE, U.D., HADLEY, M.R., McCALLEY, D.V. *J. Chromatogr. A.* 2010, 1217: 276–284.
47. GRUMBACH, E., WHEAT, T., McCABE, D., DIEHL, D., MAZZEO, J. *LC-GC North America.* 2006, 24: 80–81.
48. YANG, Y., HODGES, C.C. *LC-GC North America.* 2005, 23: 31–35.
49. RINGLING, J., WOOD, C., BORJAS, R., FOTI, C. *Am. Pharm. Rev.* 2008, 11: 24–32.
50. TYLOVA, T., KAMENIK, Z., FLIEGER, M., OLSOVSKA, J. *Chromatographia.* 2011, 74: 19–27.
51. RUTA, J., GUILLARME, D., RUDAZ, S., VEUTHEY, J.-L. *J. Sep. Sci.* 2010, 33: 2465–2477.
52. ABRAHIM, A., AL-SAYAH, M., SKRDLA, P., BEREZNITSKI, Y., CHEN, Y., WU, N. *J. Pharm. Biomed. Anal.* 2010, 51: 131–137.
53. KLEINTOP, B.L., WANG, Q. *Am. Pharm. Rev.* 2010, 13(4): 67–72.
54. GRITTI, F., GUIOCHON, G. *Anal. Chem.* 2008, 80: 6488–6499.

Chapter 4

Coupling UHPLC with MS: The Needs, Challenges, and Applications

Julie Schappler, Serge Rudaz, Jean-Luc Veuthey, and
Davy Guillarme

4.1 INTRODUCTION

Currently, analysts have to deal with extremely complex samples, which necessitate ever more powerful analytical methods. Even though liquid chromatography (LC) coupled to mass spectrometry (MS) is considered to be the gold standard for many applications, the resolving power, throughput, sensitivity, and selectivity of this approach can still be inadequate (1).

Classic LC experiments are performed with a 150 × 4.6 mm C18 column packed with 5 μm particles at a temperature of 30°C and a flow rate of 1 mL/min using an HPLC instrument that can withstand pressures of up to 400 bar. Under such conditions, dozens of compounds can be separated in a time frame of 20–40 min, which includes column equilibration. Over the last few years, various strategies have been used to improve performance in terms of throughput and/or resolution. These approaches are (i) monolithic supports, which consist of a bimodal structure made from mesopores and macropores (2,3); (ii) multiplexing (i.e., parallel analysis of various samples) (4); (iii) elevated mobile phase temperatures (up to 200°C in some cases) (5–7); (iv) columns packed with sub-3 μm shell particles (also known as fused-core, core-shell, or shell particle technology) (8–10); and (v) the combination of columns packed with porous sub-2 μm particles with chromatographic systems able to work at pressures equal or greater than 1000 bar (11,12), which was introduced in 2004 as ultra-high pressure liquid chromatography (UHPLC). So far, this last technique seems to be the most promising and has been used to separate numerous compounds in 2–4 min (fast or ultra-fast analysis) with 50-mm-long columns. UHPLC

Ultra-High Performance Liquid Chromatography and Its Applications, First Edition. Edited by Quanyun Alan Xu.
© 2013 John Wiley & Sons, Inc. Published 2013 by John Wiley & Sons, Inc.

has also been used to increase the number of compounds that can be separated two- to threefold by using longer analysis times and columns (13–15).

At the same time, MS has rapidly evolved, particularly since the introduction and development of soft atmospheric pressure ionization (API) sources for coupling with fluid-based separation techniques. Electrospray ionization (ESI), developed by Fenn, is certainly the most widely used ionization source for LC-MS and became commercially available at the end of the 1980s (16, 17). ESI possesses some obvious advantages, particularly for the ionization of biomolecules. In the early 1990s, mass spectrometers were primarily based on quadrupole and ion trap analyzers that required highly technical skills to operate because of their complexity. However, since the end of the twentieth century, two different families of analyzers have been primarily employed, depending on the type of analysis to be performed. On the one hand, quadrupole-based instruments remain the reference for targeted analysis (i.e., the determination of known substances (18, 19)) and are user-friendly, sensitive (improvement of the ionization yield and ion transmission from the source to the detector), and able to perform quantitative determination (20). On the other hand, time-of-flight (TOF/MS) instruments are widely utilized in the field of untargeted analysis (i.e., the determination of unknown substances) (21).

Because LC and MS have both evolved rapidly over the last ten years, the majority of recent mass spectrometers is compatible with fast chromatography, as demonstrated in Figure 4.1, which makes MS(/MS) the detector of choice for many UHPLC applications. In this chapter, the compatibility between fast-LC and MS

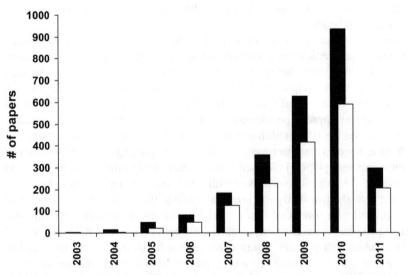

Figure 4.1 Number of papers published each year in the field of UHPLC and UHPLC-MS since 2003. Black bars were obtained using the key words *UPLC* and *UHPLC*; white bars were obtained using an additional filter (key word *MS*). *Source:* Scifinder Scholar 2007 search of the Chemical Abstracts database from 2003–2009. Date of information gathering: August 2011.

will be discussed, with an emphasis on the needs and challenges to obtaining efficient coupling (e.g., flow rate compatibility, acquisition rate, band-broadening issue, etc.). Then, applications of UHPLC-MS will be presented and classified into various sections, including bioanalytical assays, drug metabolism studies, multiresidue screenings, metabolomics, and the analysis of proteins.

4.2   TECHNICAL REQUIREMENTS FOR THE COUPLING OF UHPLC WITH MS

4.2.1   Mobile Phase Flow Rate Compatibility

An ESI source operating at atmospheric pressure is generally employed for the coupling of LC to MS. The role of this source is (i) to evaporate the mobile phase while eliminating the large amount of eluent and maintaining the vacuum level required within the mass spectrometer and (ii) to ionize the compounds in the gas phase (22).

It has been demonstrated that pure electrospray processes (formation of a Taylor cone and Coulomb fission) are limited to flow rates below 10 μL/min (23). Beyond this value, the electrolytic droplet charging process that occurs at the capillary tip is inefficient, which leads to a poor ionization yield and low sensitivity. However, in a conventional LC-MS with a column i.d. of 2.1 mm, the flow rate is between 100 and 300 μL/min. To avoid, or at least limit, the need to split the mobile phase, pneumatic assistance (i.e., addition of nitrogen gas at high flow) is employed by modern ESI sources to assist in the nebulization process and thereby improve the compatibility and stability of the high flow rate spray (24). Under these conditions, the optimal flow rate of ESI is 100–500 μL/min, even though higher flow rates have been reported (25). When using UHPLC, the choice of column i.d. is more restricted primarily because of (i) the frictional heating effects that can occur at ultra-high pressures in columns with a large i.d. (26, 27) and (ii) the rate of solvent consumption (28). For these reasons, 2.1 mm i.d. columns represent the best compromise for limiting the frictional heating for pressures up to 1000 bar. Additionally, the new generation of LC systems remains compatible with such columns in terms of both gradient delay volume and extra-column band-broadening (29, 30). The mobile phase flow rate in UHPLC should theoretically increase threefold as the particle size decreases from 5 to 1.7 μm (i.e., the mobile phase linear velocity is inversely proportional to the particle size) (31). Therefore, it is common to perform UHPLC with mobile phase flow rates ranging from 500 to 1000 μL/min, depending on the size of the investigated compounds and the acceptable efficiency loss (optimal linear velocity determined by the van Deemter equation) (32).

However, to limit the sensitivity loss that can occur in MS using a mobile phase flow rate above the optimal value for pneumatically assisted ESI, it is necessary to improve the interface and make it more readily compatible with elevated flow rates. In this context, the two most prominent manufacturers of UHPLC-MS systems have launched new source designs that can operate at elevated flow rates.

One solution, designed by Waters (Milford, MA, USA), is based on the well-known Z-spray technique in which the probe is perpendicular to the sampling cone; this design also places the second extraction cone perpendicular to the ion beam (33). With this source geometry, elimination of the mobile phase is accomplished by adding a heated nebulizing gas at up to 1200 L/hr and 650°C. Using a higher mobile phase flow rate requires using a higher nebulizing gas flow rate and temperature. The source block is also heated to 150°C to assist in the desolvation process. Under these conditions, the ESI source is compatible with mobile phase flow rates of up to 1000 μL/min with only a moderate impact on the sensitivity. This type of source design was employed in UHPLC-MS, and the sensitivity was evaluated for various compounds, including both acidic and basic drugs, at flow rates ranging from 300 to 1000 μL/min (34). Between these flow rates, the sensitivity was slightly reduced (less than 15% at 1000 μL/min); however, the maximal sensitivity was, surprisingly, attained at a flow rate of 600 μL/min for basic compounds (two- and threefold compared to 300 and 1000 μL/min, respectively). This behavior was related to neither the ionization yield nor the adduct distribution, but most likely to the chromatographic step separation based on the dependence of the peak width (and thus peak height) on flow rate (van Deemter curves) in the gradient mode (34).

Another approach, commercialized under the trademark Jet Stream Thermal Gradient Focusing Technology, has been proposed by Agilent (Waldbronn, Germany). This technique uses a heated sheath gas (11 mL/min at 350°C), in addition to the nebulizing gas used earlier, that confines the nebulized spray and desolvates the ions more effectively by concentrating them in a thermal confinement zone (35), which improves the sensitivity. A five- to tenfold gain for flow rates ranging from 0.25–2 mL/min was reported. It is noteworthy that the operating parameters related to the heated sheath gas (i.e., flow rate and temperature) are only slightly dependent on the mobile phase flow rate, which reduces the need to re-optimize the ESI parameters when the flow rate changes. In conclusion, to reach high MS sensitivities at the flow rates optimal for UHPLC, a highly efficient desolvation must be performed. Therefore, the instrument should withstand high gas flow rates and temperatures. The risk of thermal degradation of the compounds of interest is limited because the travel time over the heated zone is less than hundreds of milliseconds.

Finally, when no solution can be obtained through the modification of the ion source, it is possible to employ either a post-column flow split with 2.1 mm i.d. UHPLC columns or a smaller i.d. (such as a 1 mm) column at a more reasonable flow rate. However, in both cases it is required to evaluate the compatibility of the system (e.g., the effect of the gradient viscosity on the split ratio) with these strategies in terms of additional band-broadening (36).

4.2.2 Acquisition Rate and Data Quality

Another issue related to the coupling of UHPLC to MS is the very small peak widths generated by UHPLC. In conventional LC, the average baseline width is 10–20 seconds (s), which is reduced by 2–5 s in UHPLC. For ultra-fast analyses (less than

1 min) performed at elevated mobile phase flow rates (i.e., 1 mL/min), the peak widths are theoretically as low as 1 s (for fast gradients of 1 min). At least fifteen acquisition points per peak are recommended for quantitative analysis, which could be critical depending on the acquisition speed of the MS device.

4.2.2.1 *Operation in SIM, SRM or Scan Mode*

With single quadrupole instruments, the single ion monitoring (SIM) mode is generally used. In this operation mode, the quadrupole parameters (amplitude of the DC and RF voltages) are set for the filtration or selection of only one specific *m/z*. This mode provides higher sensitivity than scan mode in which only a short period of time is spent on each ion. The period of time to collect data at a particular mass is called the dwell time. If several masses have to be detected, then the instrument works sequentially; starting with the lowest mass to be analyzed, the instrument collects data for the duration of the selected dwell time and then progresses to the next mass. For older MS devices, the dwell time usually ranges from tens to hundreds of milliseconds, which induces a relatively low rate of data acquisition. With modern quadrupole-based instruments, the dwell time is reduced to 5 ms for most MS devices and even 1 ms for the most powerful ones. The inter-channel delay (i.e., the time required to switch from one *m/z* to another one) has also been drastically reduced, to as little as 1 ms in some cases, without any cross-contamination of the *m/z* channels.

Selected reaction monitoring (SRM) is generally the mode of choice for powerful data acquisition when using triple quadrupole (QqQ) instruments and consists of two mass filtration stages. In the first stage, the ion of interest (the precursor) is preselected by the first quadrupole. With the help of an inert gas (e.g., argon), a collision-induced dissociation (CID) process occurs in the second chamber, which leads to fragmentation. Specific fragments are monitored in the third quadrupole to increase the selectivity and thus sensitivity of the MS method. Similarly, the required dwell time (i.e., the time spent acquiring the specific SRM transition during each cycle) has been drastically reduced by most manufacturers over the last few years. An example of an MS/MS instrument operation is presented in Figure 4.2a. The acquisition rate for the determination of ten different compounds with a 10 ms dwell time is 6.7 points per second (pts/s). Nowadays, the QqQs from various manufacturers can accommodate dwell times as little as 5 ms or even only 1 ms while maintaining the detection sensitivity.

Finally, one of the most important advances has been made in TOF and QqTOF technologies, which are well adapted to quickly record and store data over a broad mass range without compromising sensitivity. With the last generation of TOF/MS instruments, high mass resolution (e.g., greater than 10,000 full width at half maximum [FWHM]) can usually be attained at speeds of 20 spectra per second and up to 40 spectra/s for a few instruments. More recently, the TripleTOF instrument was launched; this instrument is a QqTOF able to acquire data with a resolution of 40,000 FWHM at 100 spectra/s with a high degree of sensitivity. Another technology featuring a very fast TOF instrument has also been released that uses the patented technology Folded Flight Path (FFP) to provide full-range mass spectra at speeds of

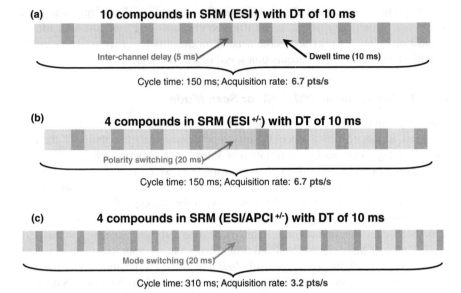

(a) **10 compounds in SRM (ESI⁺) with DT of 10 ms**

Inter-channel delay (5 ms) Dwell time (10 ms)

Cycle time: 150 ms; Acquisition rate: 6.7 pts/s

(b) **4 compounds in SRM (ESI⁺/⁻) with DT of 10 ms**

Polarity switching (20 ms)

Cycle time: 150 ms; Acquisition rate: 6.7 pts/s

(c) **4 compounds in SRM (ESI/APCI⁺/⁻) with DT of 10 ms**

Mode switching (20 ms)

Cycle time: 310 ms; Acquisition rate: 3.2 pts/s

Figure 4.2 Schematic representation of the different steps of the MS/MS process for a mixture of several compounds when acquiring data in (a) ESI^+, (b) $ESI^{+/-}$, and (c) $ESI/APCI^{+/-}$. Adapted from Schappler, J., et al. *Talanta* 2009, with permission.

up to 200 spectra/s and resolutions of up to 100,000 with mass accuracies of less than 1 ppm.

Because quadrupole-based MS instruments collect data sequentially, their ability to decrease SIM and SRM dwell times to 1 ms remains of interest for the monitoring of a significant number of compounds. However, TOF-based instruments acquire data over a wide mass range, and an acquisition rate of 10–20 spectra/s is sufficient for UHPLC experiments because their peak widths at baseline would never be less than 1 s. Currently, the only interest in faster TOF/MS devices concerns coupling to fast or ultra-fast gas chromatography (GC), which has peak widths at the baseline as small as dozens of milliseconds.

4.2.2.2 Polarity Switching and Multimode Ionization

A recent trend in MS devices is to attempt to perform detection in both the positive and negative ionization modes within the same run (37, 38). For this purpose, it has become possible to alternate the polarity (+/−) at a speed between 15 and 20 ms for the most recently developed instruments. This polarity switching is very fast and can be employed to increase productivity, particularly when dealing with the simultaneous analysis of both acidic and basic compounds. However, because the MS instrument works sequentially and the peak widths in UHPLC are very small, the use of this feature is not recommended as it would compromise the quantitative performance because the number of data points across the peak would be too small.

As illustrated in Figure 4.2b, if the data are collected for four different compounds, the cycle time for a 10 ms dwell time, 5 ms inter-channel delay, and 20 ms polarity switch would be 150 ms (corresponding to only 6.7 points/s). This would make it impossible to quantify more than four compounds with an average UHPLC peak width of 2 s. Conversely, it can be useful to select the best ionization mode for each of the individual compounds prior to carrying out experiments so the polarity mode that provides the highest sensitivity is used. The sampling rate would not be compromised in this case, and only a limited sensitivity decrease (20%) has been reported (34).

Another option offered by the most recent generation of instruments is the capability to work with multimode ionization sources able to simultaneously carry out both ESI and atmospheric pressure chemical ionization (APCI) experiments (39). The time required to switch from ESI to APCI mode has been reduced to only 20 ms for most instruments (34). As previously mentioned, even if dual sources can be useful to increase screening productivity, they always compromise the sampling rate and sensitivity relative to a single ionization mode, as shown in Figure 4.2c. Finally, the APCI mode of these dual sources is clearly not equal to a dedicated APCI source, and most apolar compounds appear not to be as efficiently ionized.

To conclude, although it is possible to collect data for $ESI^+/ESI^-/APCI^+/APCI^-$ within the same run to obtain the maximum amount of information possible, doing so would significantly decrease the sensitivity relative to analyses performed in a dedicated mode. Furthermore, because of the very thin peaks generated by UHPLC, it is not advisable to use such conditions because they severely compromise the sampling (34).

4.2.3 Band-Broadening in UHPLC-MS Conditions

It has been demonstrated that MS instruments could represent a non-negligible source of extra-column band-broadening in UHPLC compared to UV detectors for reasons other than the acquisition rate (40). Indeed, the column volume in UHPLC is drastically less than in HPLC (approximately 100–200 μL compared to 1.5–2.5 mL). For this reason, the sources of band-broadening (injection, tubing, and UV detection) should be minimized relative to conventional LC (29, 30). For instance, it was found that TOF/MS was a non-negligible source of broadening (40). A peak capacity reduction of 15%–30% was observed in gradient mode in MS when compared to UV detection because of the ionization chamber volume, transfer capillary volume, and electronic signal treatment. Such behavior is demonstrated in Figure 4.3, which compares the band-broadening of UHPLC-UV to UHPLC-TOF/MS (40). This study used model compounds to measure the peak capacities of both UV and TOF/MS detectors for gradients of 20 and 60 min, respectively, using a 150 × 2.1 mm, 1.7 μm column with a mobile phase flow rate of 400 μL/min. Roughly speaking, the peak capacity decreased by approximately 15% for the 60 min (5%–95% acetonitrile [ACN]) gradient and 30% for the 20 min (5%–95% ACN) gradient. The extra-column dispersion was measured at 10 $μL^2$ for the UHPLC-UV and approximately 70 $μL^2$ for the UHPLC-TOF/MS, which caused an important peak capacity loss, particularly

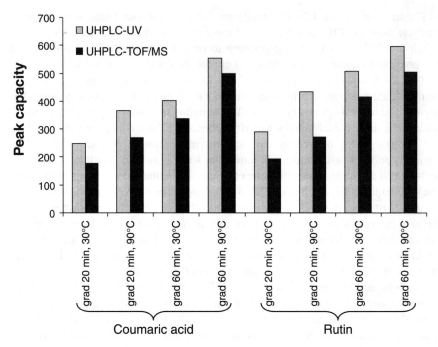

Figure 4.3   Comparison of peak capacities in UHPLC-UV and UHPLC-TOF/MS configurations calculated for coumaric acid (S = 6, MW: 164 g/mol) and rutin (S = 12, MW: 610 g/mol). Adapted from Grata, E., et al. *J. Chromatogr. A.* 2009, with permission.

in the fastest gradient. In this study, only long UHPLC columns (i.e., L_{col} > 150 mm) and gradient times (t_{grad} > 20 min) were taken into consideration; therefore, the loss in peak capacity was quite reasonable. However, if short (50–100 mm) or thin (1 mm i.d.) columns were used, the performance loss could be more pronounced.

Finally, the peak width could also be strongly dependent on the acquisition rate of the detector. In this respect, polarity switching, which represents an interesting method for gathering more information from a single run, further contributes to the apparent peak broadening, and an additional 30% peak capacity loss was measured for a UHPLC-TOF/MS device using the polarity switching feature (40). Therefore, the positive and negative ion modes should be employed independently to maximize the peak capacity and to separate closely related compounds.

4.3   UHPLC-MS FOR BIOANALYTICAL ASSAYS

Bioanalysis samples (saliva, urine, whole-blood, plasma, serum, cerebrospinal fluid, tissue, etc.) may contain water, ionic salts, proteins, lipids, anticoagulants, stabilizers, and other compounds that can interact on either a temporary or permanent basis with the stationary phases. These interactions decrease the column lifetime and interfere with ionization (i.e., matrix effect), thus hampering the quantitative determination of compounds in such matrices. Therefore, appropriate sample pretreatment and preparation procedures, such as dilution, filtration, protein precipitation, solid phase extraction (SPE), and liquid-liquid extraction (LLE), have been implemented to avoid the degradation of the analytical performance. The choice of which procedure to use largely depends on the desired selectivity and sensitivity as well as any throughput constraints (41). In addition, there is a constant demand for lower-concentration methods to quantify a given drug molecule or candidate, the primary objective of which is developing a bioanalytical method that meets the selectivity, accuracy (trueness and precision), and linearity requirements set by authorities (42).

4.3.1 Selectivity Issues and Matrix Effects

An important issue regarding selectivity concerns the assessment and quantitation of the matrix effects. Because UHPLC yields peaks with bases as narrow as 1 s, an overall enhancement of the chromatographic resolution is obtained. The potential co-elution and ion suppression are thus reduced, which enhances the sensitivity and reliability of the MS. However, while UHPLC improves the separation throughput and resolution, practical issues with using MS may arise because acquiring sufficient data points (>15 points per peak) is essential to ensure reliable quantitation. As previously mentioned (Section 4.2.2), instruments with high acquisition rates and low dwell times are, therefore, preferentially selected for quantitative determination.

An internal standard (IS) is required to reduce the impact of the system variability on method performance, and its selection is important for quantitative bioanalysis. Structural analogues and stable isotopically labeled (SIL) compounds remain the gold standards of ISs; however, structural analogues differ from the compounds of interest and may have different ionization behaviors than the analytes. Even with a closely eluting IS and analyte, they reach the ionization source at different retention times, and short-term variations in the ionization process may be of concern, particularly for ESI.

In contrast to structural analogues, ionization changes can be efficiently corrected when using SIL compounds, which possess similar ionization responses and fragmentation patterns. Therefore, a deuterated IS can be used to correct both the overall method variability (e.g., sample preparation and chromatographic process) and matrix effects because the amount of suppression from interferents should be similar. However, the total concentration of the analyte and SIL compound should be below the ionization process saturation point. It is noteworthy that, because of the high resolution afforded by UHPLC, SIL compounds with a slight difference in polarity can be separated from their non-deuterated analogues, which circumvents the ionization variability correction. Additionally, for numerous candidates (e.g., original compounds, metabolites, natural products, etc.), SIL compounds are not commonly available and can be very expensive. For this reason, some researchers produce their own deuterated internal standards by means of protium-deuterium exchange, as was the case for a quantitation study of resveratrol that was performed using UHPLC-QqTOF/MS (43). As the ionization experienced severe matrix effects and an appropriate IS was not available, a stable isotopologue of resveratrol was produced in-house to enable the correction of compound discrimination during the extraction, cleanup, chromatographic separation, and ESI-MS detection. A comparison to the data obtained through quantitation using a structural analogue as the IS confirmed the accuracy of the developed method.

An alternative way to compensate for the effect of the co-extracted matrix components is through the so-called ECHO approach. This strategy consists of using the non-labeled target compound as the IS by injecting it into the LC-MS system after a short time period as an "echo" of the analyte (44). So far, this strategy has not been applied to UHPLC-MS because the injection time is quite long (i.e., 30–45 s) relative to the analysis time for UHPLC (i.e., 2–3 min).

4.3.2 Time Delivery Constraints

Achieving the desired sensitivity and resolution with a short analysis time is still challenging. When high-throughput bioanalyses are performed, the sample preparation may become the limiting step in terms of the total analysis time. Indeed, there is a large contrast between ultra-fast chromatographic analysis and conventional sample preparation techniques, which remain highly labor intensive and time consuming.

Regular LLE procedures are mostly implemented prior to UHPLC and present several benefits in terms of sample cleanup and enrichment. This technique is well adapted to LC-MS analysis because proteins and salts are extensively excluded, which minimizes any matrix effect and/or ion suppression and allows for very low detection limits with good recovery and precision (45). However, LLE usually includes long handling steps, which are not in line with UHPLC throughput.

Other extracting procedures, such as SPE, can be used and feature high selectivities when used with an appropriate retention mechanism and high sensitivities because of the small recovered volumes. In addition, less organic solvent is needed than for LLE, and adverse effects such as foaming are avoided. For these reasons, SPE is currently the leading sample preparation method used in routine bioanalytical assays, even though it may be time consuming and relatively expensive and can suffer from poor batch-to-batch reproducibility. To cope with the requirements for UHPLC, in-line automated or 96-well plate SPE methods are being developed, but only a few studies have been reported so far (46–49).

The selectivity and sensitivity afforded by extraction procedures, such as SPE and LLE, have become less advantageous since highly sensitive and selective MS analyzers, such as TOF/MS and QqTOF/MS, have been coupled to UHPLC. A simple dilution of the sample prior to analysis (dilute-and-shoot approach) can be sufficient to circumvent compound discrimination while maintaining very short delivery times. Such an approach can be considered a general strategy to perform a single-run analysis for a wide range of compounds in a complex matrix. Another advantage is the absence of drug degradation that may occur during sample preparation or treatment.

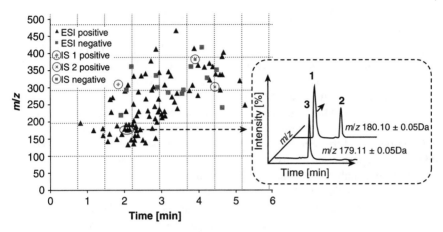

Figure 4.4   Analyte separation (n = 103) of *m/z* as a function of t_R. Data from ESI positive and negative modes are plotted together. The three internal standards are circled with a continuous line. A zone (dashed line) is magnified to show the advantage of coupling UHPLC to a QqTOF/MS mass spectrometer. In the magnified zone, the compounds methylephedrine (1), MDA (2), and nikethamide (3) are separated as a function of time, intensity, and *m/z*. Adapted from Badoud, F., et al. *J. Chromatogr. A.* 2009, with permission.

Consequently, underestimation of drug concentration is avoided. However, the dilute-and-shoot procedure is nonselective and does not remove the interferents responsible for the matrix effects, which prevents quantitative determination. A recently developed example of such a generic approach involves injecting a relatively well-adapted matrix after a simple dilution (Figure 4.4) and features good sensitivity with an analysis time of less than 9 min (including the equilibration periods) for the rapid screening of 103 drugs in urine (50).

The large quantities of proteins present in plasma and serum samples remain a major problem for separation techniques coupled to MS. The fastest way to remove most proteins in biological fluids, such as plasmas or serums, consists of precipitating them by adding an organic solvent, salt, or acid. This procedure is rapid and easy to perform and can be applied to a wide range of samples, which results in a good compromise between throughput and suitable sample cleanup. For example, a method using UHPLC-MS/MS was developed for the analysis of several antifungals and their metabolites in plasma for therapeutic drug monitoring (51). This method featured high throughput due to a simple protein precipitation step, and the analytical run time was less than 7 min with reduced matrix effects, enabling the accurate quantitation of each compound in the plasma.

Blood matrices are a field of great development concern regarding sample collection and handling. Within this field, there is growing interest in the analysis of dried blood spots (DBS) and dried matrix spots (DMS). The use of such supports presents numerous advantages over traditional wet plasma sampling techniques; they are less invasive and use less blood, which enables juvenile toxicology and pediatric studies as well as requiring fewer animals for drug testing. Reduced shipping and storage costs, analyte stability, and improved safety are other promising features of the DBS and DMS techniques. So far, only a few studies have been published on DBS using UHPLC-MS/MS (52, 53). Once again, there is a disparity between the time needed for sample preparation and the fast analysis exhibited by UHPLC because the generic manual extraction technique used to analyze DBS/DMS samples involves many steps (punching a disk from the center of the support, transferring the disk to a tube, adding an extraction solvent containing the IS, shaking the sample for approximately 2 h, centrifuging it, and transferring the supernatant to a fresh tube prior to analysis). In addition, the reported sensitivity was lower and the ion suppression was higher than for wet plasma analyses. Another issue concerns what effects hematocrit has on the physical characteristics of DBS samples and analyte quantitation. Indeed, as the area of the DBS sample decreases, with a corresponding increase in the hematocrit levels (viscosity effect), a bias can be observed that is particularly pronounced at hematocrit levels outside the normal values. Finally, the authorities have not yet described this approach in official guidelines for bioanalytical assays. These are the primary reasons why this technique is only slowly being integrated into bioanalytical laboratories that routinely use protein precipitation in a high-sample-throughput environment. Potential solutions would be to include quality control samples with different hematocrit levels, while direct in-line analytical techniques could be used to completely eliminate the need for manual extraction steps (54). So far, these techniques have not been coupled to UHPLC-MS.

4.4   DRUG METABOLISM STUDIES USING UHPLC-MS

In drug metabolism studies, numerous metabolites have to be detected and identified in a single run. Investigation into the metabolites' properties, such as metabolic stability, interaction and inhibition of cytochrome P450 (CYP) isozymes, reactivity and toxicity, is of the utmost importance. At first, long chromatographic separations were performed to avoid co-elution and took into account such issues as insufficient resolution to differentiate structural isomers and sensitivity loss due to ion suppression. Significant advances in the development of high-throughput and high-resolution analytical methods have been achieved because of the large number of compounds that must be screened in early metabolism assays, the complexity of the samples, and the diversity of the metabolites produced. UHPLC is perfectly adapted to fulfill both of these tasks, whereas the type of MS instrument used depends on the application. QqQ instruments with high acquisition rates are preferred for ultra-fast metabolite stability and inhibition assays (necessitating quantitation), while high-resolution mass spectrometers, such as TOF and QqTOF, are mostly used for metabolite identification. A few metabolism studies have also reported the use of ion trap and Orbitrap instruments. Although these devices are interesting for the elucidation of biotransformation sites thanks to their MS^n capabilities, they suffer from relatively low acquisition rates and are barely compatible with the thin peaks produced by UHPLC.

4.4.1 High-Throughput Metabolite Assays

The two primary approaches proposed for high-throughput analysis are UHPLC-MS and multiplexed UHPLC-MS systems.

The power of UHPLC-MS/MS is illustrated in Figure 4.5, which shows the analysis of two drugs and their respective main metabolites (55). A high selectivity was attained in under 1 min with compounds completely resolved by either UHPLC or MS for analytes with the same MS/MS transition (e.g., tramadol and O-desmethylvenlafaxine, N,O-didesmethylvenlafaxine and M2).

In the abovementioned study, only one isozyme (i.e., CYP 2D6) was involved, whereas simultaneous measurement of multiple enzyme activities is highly informative and should be achievable in a generic workflow (56). This can be done by using a drug cocktail that allows the comprehensive analysis of the metabolic pathways obtained from a single, multiplexed experimental workflow. To increase the desired throughput even further, these assays are performed using ultra-fast analytical techniques. For example, a UHPLC-MS/MS method was developed to evaluate the six CYP probe drugs in a "Pittsburgh cocktail" and their relevant metabolites, which enabled the effects of the CYP enzymes to be characterized for both normal subjects and patients with a variety of diseases (57). The probes were quantified by stable isotope dilution analysis (see Section 4.3.1) from plasma and urine samples, and the method was applied successfully in a clinical phenotyping study of liver disease. SRM for each analyte was performed in duplicate on a QqQ mass spectrometer; one run operated in the positive ionization mode and the other in the negative mode. Both runs used the same chromatographic method, and the entire procedure lasted less than 8 min. The sample preparation, a simple protein precipitation, was adapted to the UHPLC time scale. The major objective of the abovementioned work was to select another drug cocktail that could be analyzed in a single run with all probes being chromatographically resolved using only one ionization mode. Such an advance would further increase the throughput by twofold. Although feasible for a wide range of drugs, the new selection would require extensive clinical validation.

As demonstrated by the above examples, the conventional run times of UHPLC have been reduced to a few minutes, whereas it takes only a few seconds for the mass spectrometer to record the chromatographic peak of interest, which results in it spending most of its time idling and waiting for the next sample. Consequently, the instrument time is not efficiently used, which reduces the sample throughput. To optimize the MS acquisition period, multiplexed or parallel UHPLC-MS could be utilized. Such an approach has not yet been accomplished with UHPLC, but examples with high-throughput LC-MS/MS using core-shell technology already exist (58).

The time available for MS acquisition can also be used to increase the amount of generated data by means of information-dependent acquisition (IDA). The recently introduced TripleTOF system, a hybrid quadrupole TOF/MS platform, features such capabilities. This system also possesses high mass accuracy (<2 ppm) and resolution (up to 40,000 FWHM) as well as rapid acquisition rate (i.e., 100 Hz, highly compatible with UHPLC) afforded by TOF/MS technology, which, when combined with its high sensitivity (i.e., low attomolar range), produces quantitation capabilities similar

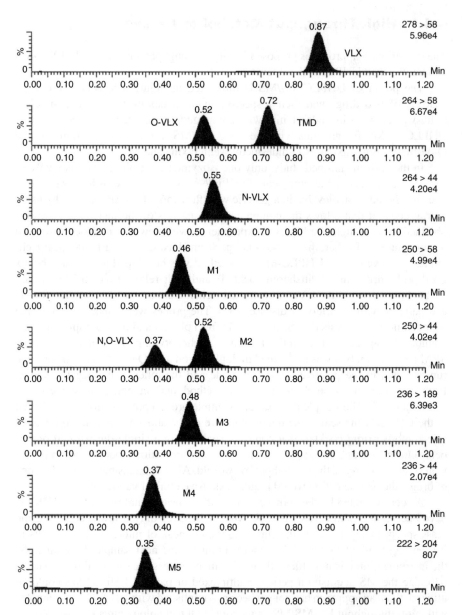

Figure 4.5 Sub-1 min UHPLC-MS/MS analysis of venlafaxine (VLX), tramadol (TMD), and their major metabolites (O-desmethylvenlafaxine, N-desmethylvenlafaxine, and N,O-didesmethylvenlafaxine for VLX, and M1, M2, M3, M4, and M5 for TMD). Adapted from Nicoli, R., et al. *J. Med. Chem.* 2009, with permission.

to a QqQ. These characteristics permit simultaneous qualitative and quantitative workflows (QUAL-QUAN approach) and integrate comprehensive exploration and rapid profiling (QUAL features) with high-resolution quantitation (QUAN features) in the same run (59).

Another approach, called rapid fire, was recently proposed for TOF/MS pharmacokinetics studies (60). In contrast to multiplexing, rapid fire uses only one pump, and the column and autosampler are embedded in the system. All of the injection and switching valves, as well as the chromatographic column, are incorporated in the autosampler arm, which drastically reduces the cycle time. Figure 4.6 shows a typical multi-injection chromatogram obtained for multiple samples using only one data file. This approach minimizes both the time required for communication between the autosampler and mass spectrometer and the time needed to open a new data file. These features allow a sample cycle time of approximately 7 s to be achieved, and an entire 384-well plate of samples can be analyzed in approximately 50 min (compared to 12 h for conventional UHPLC-MS/MS with a sample cycle time of 2 min). The only observed drawback was the carryover, which could bias the absolute quantitative analysis.

Because the analytical run times, which include both the chromatographic separation and MS detection/acquisition, are now very fast and highly informative, and sample preparation has been reduced to a minimum, a new bottleneck has emerged. Data processing has become the limiting step in terms of total analysis time. For instance, the processing of an entire 384-well plate can take several hours unless automated algorithms are used. This bottleneck represents a new challenge to manufacturers who must now include software fast enough to keep pace with the MS devices.

4.4.2 High-Resolution Metabolite Identification

Drug metabolism experiments must detect and identify unknowns, such as highly active or toxic metabolites, as early as possible in the drug discovery process to circumvent compound attrition in late-phase development. The launch of new MS with increased sensitivity has facilitated the detection of minor metabolites and increased the number of metabolites that must be resolved and identified. Because these new and unexpected metabolites, which include various positional isomers, can be detected from the microsomal incubation of a single parent drug, high-chromatographic resolution and mass accuracy are required for fragmentation patterns. UHPLC coupled with high-resolution analyzers are particularly appropriate for both tasks in a high-throughput environment.

The coupling of high-temperature UHPLC (HT-UHPLC) to QqTOF/MS has demonstrated the capability to separate and identify drug metabolites present in urine following oral administration. An example is given of the identification of the numerous metabolites of ibuprofen, which was used as a model compound (61). HT-UHPLC operated under high-resolution conditions ($L_{col} = 150$ mm, $t_{grad} > 20$ min, $T = 65°C$), provided a high peak capacity (approximately 350), and generated

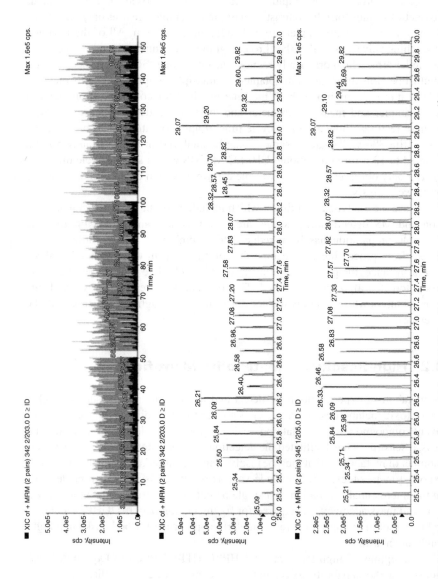

Figure 4.6 Multiple-injection chromatogram from rapid-fire MS/MS for 1'-hydroxymidazolam and its internal standard, $^{13}C_3$-hydroxy-midazolam. Adapted from Brown, A., et al. *Rapid Commun. Mass Sp.* 2010, with permission.

high-quality metabolic data with the detection of both phase I and phase II metabolites. To further improve the structural characterization of metabolites obtained by QqTOF/MS, MS^E experiments were performed and exhaustive data on both the precursor and product ions were obtained simultaneously in a single run. Briefly, MS^E involves the parallel, alternating acquisition of both low- and high-collision energy functions and makes it possible to acquire a full QqTOF/MS scan in the first channel and a QqTOF/MS/MS spectrum in the second channel of the same analytical run. The identification of nine glucuronides of ibuprofen, a side-chain oxidized carboxylic acid acyl glucuronide, and various hydroxylated metabolites was accomplished thanks to both accurate mass determination and the presence of ion fragments common to ibuprofen. This study emphasized the improved resolution and sensitivity afforded by UHPLC-QqTOF/MS. These improved features are explained by a combination of the reduction in both peak width (and the consequent increase in analyte concentration) and ion suppression (occurring with the co-elution of metabolites and endogenous compounds).

The detection and quantitation of adducts produced by the in vitro formation of reactive metabolites was also demonstrated by UHPLC coupled to inductively coupled plasma (ICP) MS and QqTOF/MS. The identification of reactive metabolites is of great interest as they may be involved in the idiosyncratic drug reactions that lead to many drugs being withdrawn from the market. In a recent study, adducts formed from clozapine in the human liver by microsomes supplemented with glutathione (GSH, used as a nucleophile to trap reactive electrophilic metabolites) were determined by UHPLC-MS (62). The adducts were quantified by sulfur-specific ICP-MS and further identified by QqTOF/MS. The use of ICP-MS offers a novel, rapid, and sensitive way to determine the quantity of GSH adducts with reactive drug metabolites formed in the liver. When used in tandem with QqTOF/MS, this technique features the possibility to assign a mass to the sulfur response, which allows for a tentative deduction of the GSH adduct structure.

As mentioned, the improved technology available for metabolism studies has amplified the amount of data generated. Therefore, a great effort has been made to develop the statistical strategies used to analyze the parallel data sets acquired from the sample. For example, statistical heterospectroscopy (SHY) operates through the analysis of the intrinsic covariance between signal intensities in the molecular fingerprints of identical and related molecules measured by multiple spectroscopic techniques, such as ^{1}H-NMR and UHPLC-QqTOF/MS^E. This approach emphasizes UHPLC-MS as a key technique for both metabolism (xenometabolite identification) and metabolomic (endogenous metabolite identification, see Section 4.6) assays. It was recently demonstrated that SHY is well adapted for epidemiological studies because of its potential for extracting structural information on components of urine samples from a normal, uncontrolled, and unselected human population (63). SHY uses direct cross-correlation of the spectral parameters—that is, the chemical shifts from NMR and m/z data from MS combined with the fragment analysis from the MS^E scans—to not only detect the numerous endogenous urinary metabolites, but also to identify the xenobiotic metabolites that result from drug use. In addition, SHY is a data-driven technique that gives rise to unexpected results, such as discovery of the previously unreported metabolites of several drugs.

4.5 MULTI-ANALYTE SCREENING WITH UHPLC-MS

The aim of multi-analyte screening is to rapidly assess the presence or absence of contaminants within a complex sample. Therefore, the developed method should be able to determine as many compounds as possible within a single analytical run. In this context, GC-MS is often employed, but LC-MS remains the gold standard as it is far more generic and does not require a pre-derivatization step to render the compound volatile. In addition, LC-MS is the method generally recommended by official regulatory texts that establish the maximum residue limits (MRL) and the number of required identification points (IP). Multi-analyte screenings are generally applied to a wide variety of compounds and matrices, which include doping agents and veterinary drugs in biological matrices; drugs, pesticides, and herbicides in environmental matrices; and veterinary drugs, drugs, and pesticides in food samples. In most cases, multi-analyte determination consists of a multistep strategy that requires a fast, initial, generic screening. A time-consuming confirmation procedure is subsequently carried out to confirm the presence of contaminants within the investigated sample.

Over the last few years, the multi-analyte methodology has improved drastically. This improvement was driven by the expanding need to get more and better information more quickly in an environment where cost is a major concern. For these reasons, the replacement of traditional HPLC by UHPLC has been evaluated in numerous studies to both improve analysis throughput and reduce the response time (50, 64–73). Another trend is the replacement of QqQ analyzers with time-of-flight devices, which are able to provide a similar sensitivity and dynamic range but with higher mass resolution (>10,000 FWHM) and routine mass accuracy (<10 ppm).

4.5.1 Sample Preparation Procedures

One of the main problems with multi-analyte determination is the increased number of target compounds that have to be investigated (between dozens and several hundreds) (65–69) that often possess very different physicochemical properties. For these reasons, it is necessary to implement a generic procedure for the extraction of multiple components from a complex matrix, which usually involves a compromise in the selection of the experimental conditions used to attain a sufficient recovery of the analytes. Depending on the type of the investigated matrices, various sample preparations can be utilized prior to UHPLC-MS.

For biological fluids, SPE has been widely employed for the extraction of doping agents and veterinary drugs from urine and plasma. However, a number of authors have reported the use of simple sample treatments such as plasma protein precipitation or urine dilute-and-shoot (64,65). These procedures were only used during the initial screening step and not the confirmation step. In addition, TOF/MS devices were generally employed to compensate for the lack of selectivity and sensitivity afforded by these sample preparation procedures.

The use of SPE to analyze environmental matrices, such as waste and surface water, was reported exclusively with hydrophilic-lipophilic balanced reversed-phase (e.g., Oasis HLB) or polymeric mixed-mode sorbents (e.g., Oasis MCX) (66–68). These supports were selected because it is possible to simultaneously extract acidic, neutral, and basic residues with a single cartridge. In addition, it is possible to pre-concentrate the sample considerably to allow for the detection of very low concentrations of analytes. For example, extraction recoveries after SPE using an HLB support were found satisfactory as they generally varied between 70% and 110% for numerous analytes, such as herbicides (68) and pesticides (67), from both waste and surface water. However, the results achieved for pharmaceuticals on both HLB and MCX were less acceptable, particularly because of the strong variability in the physicochemical properties of acidic and basic drugs (66, 75).

Finally, the extraction of pesticides, herbicides, or drugs from food samples such as fruits, vegetables, or wine has often been accomplished by SPE. However, a few papers also reported the use of a QuEChERS-based (Quick, Easy, Cheap, Effective, Rugged, and Safe) sample preparation procedure (76) prior to UHPLC-MS/MS to provide a less time-consuming, harmful, and costly method that ensures that suitable recoveries, precision, and ruggedness are obtained. Recoveries for more than ninety compounds, including pesticides, herbicides, and mycotoxins, from various organic food samples were in the range of 70%–120%, which demonstrates the huge potential of this method (77, 78). Five extraction methods, including two different LLE-based methods, two different QuEChERS procedures, and a single SPE method, were compared prior to analysis by UHPLC-MS/MS for the determination of mycotoxins and pesticides in milk samples. The conclusion of this study was that, while all of the evaluated methods could simultaneously extract both mycotoxins and pesticides, SPE was the only one that could simultaneously extract forty-two pesticides and six mycotoxins with good recovery values (i.e., 60%–120%) at low concentrations (78).

In the literature, 90% of the SPEs were carried out in the conventional cartridge format. As mentioned (Section 4.3.2), this format is clearly incompatible with the high throughput afforded by the UHPLC-MS strategy. Currently, only a few attempts to scale down the time required for sample preparation and to perform SPE in the 96-well plate format have been reported (48).

4.5.2 UHPLC Conditions

To avoid peak co-elution that leads to matrix effects, a multi-analyte screening by conventional LC can take a relatively long time. For this reason, UHPLC was employed to increase the throughput while maintaining sufficient chromatographic resolution to minimize the co-elution of compounds with similar m/z ratios and fragmentation pathways as well as to decrease the ionization alterations.

In UHPLC, a generic gradient is generally used to cover the widest polarity range possible. To improve the MS sensitivity and limit the number of unwanted interactions between analytes and residual silanols on the surface of the stationary phase, 0.1% formic acid is added to the mobile phase. Only a few authors have reported the use of buffer solutions to control the pH to obtain a more robust selectivity. A C18 chemistry-based stationary phase of 50–100 mm is generally combined with a gradient time of 5–20 min, including re-equilibration. It is important to keep in mind that the longest columns often do not provide the highest peak capacity under UHPLC gradient conditions because of the increased back pressure generated (79, 80). For this reason, a 100 mm column can be used with gradients of at least 7 min. Finally, the internal diameter of the UHPLC column is usually 2.1 mm because a larger i.d. would detrimentally affect the frictional heating and solvent/sample consumption. On the contrary, a lower i.d. is critical to maintain compatibility with the existing instrumentation (extra-column band-broadening, gradient delay volume). Only Kasprzyk-Hordern et al. (66, 74) have reported the successful screening of more than fifty pharmaceuticals in wastewater on a 1 mm i.d. column. As expected, the peaks were distorted and broadened in the chromatograms due to the contributions of the external volume, which is particularly critical for MS detection.

The reduction in the analysis time observed between HPLC and UHPLC for multi-analyte determination is generally between three- and fivefold, which is lower than expected based on theory (i.e., ninefold for a geometric method transfer between a 150 × 4.6 mm, 5 μm, and a 50 × 2.1 mm, 1.7 μm column). Indeed, the throughput of multiresidue screening is important, but there is also a strong need to improve the reliability of the data by increasing the chromatographic resolution.

4.5.3   Triple Quadrupole vs. Time-of-Flight Analyzers

In the case of multi-analyte determination, the number of investigated analytes is important, and there is a need for fast MS analyzers. The two types of analyzers employed most frequently in combination with UHPLC were QqQ operating in SRM mode and TOF/MS devices. These two instruments possess several inherent advantages and drawbacks that will be critically discussed in this section. A comparison of LC-MS approaches for the determination of pesticide residues in food matrices based on QqQ, TOF, and QqTOF analyzers was accomplished by Grimalt et al. (81). Figure 4.7 illustrates the capabilities of the three devices, each possessing both advantages and limitations, to perform such screening.

QqQ instruments are well suited for targeted analysis, as it is possible to attain sufficient selectivity and sensitivity. However, their primary limitation is the trade-off between the speed and sensitivity of the instrument; although it was shown that modern MS/MS devices are able to obtain high signal-to-noise ratios with short SRM dwell times (as little as 1 ms in some cases). To attain the best compromise between the sensitivity and acquisition rate for the multiresidue determination of a large number of compounds (>100), the MS/MS method is generally segmented into various time-schedule windows that contain different SRM channels and time intervals. For example, an SRM window method with a dwell time of 15 ms was used to analyze thirty-four pesticides in wastewater (67) and fifty-three pesticides in fruits and vegetables (73), whereas it was reduced to only 5 ms for the determination of 130 unknown substances in urine (64). The main limitations of this procedure arise from the elution-time variability. Indeed, when investigating several hundred compounds, the time windows are very small, and the method could require adjustment when the analytes leave their originally defined window. An alternative strategy is

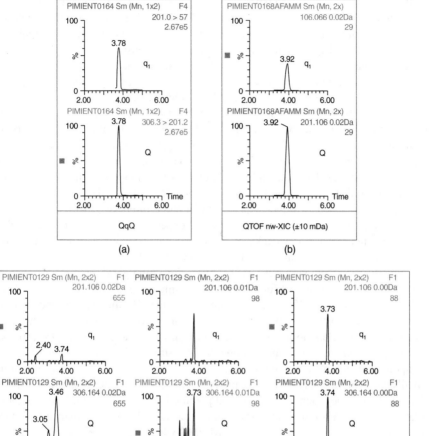

Figure 4.7   Chromatograms for a pepper matrix-matched to a buprofezin standard at 10 μg/L.
(a) LC-ESI-QqQ, (b) LC-ESI-QqTOF/MS (XIC ± 10 mDa), and (c) LC-ESI-TOF/MS (XIC ± 10 mDa),
(XIC ± 5 mDa), and (XIC ± 2.5 mDa). Adapted from Grimalt, S., et al. *J. Mass Spectrom.* 2010, with
permission.

to perform several successive injections of the unknown sample with only a limited
number of SRM transitions; this maintains a high sensitivity level and sufficient
acquisition rate but is time consuming. This method was recently applied to the
determination of forty-eight drugs in wastewater (82) and ninety pesticides in fruit
juices (71).

Time-of-flight instruments are an attractive alternative because they are adapted
for both targeted and nontargeted determination. Indeed, the sensitivity attained by

UHPLC-TOF/MS is almost comparable to that of a QqQ instrument of the same generation, as demonstrated in a recent study dealing with the determination of eleven pesticides in nine different food matrices (an average LOQ of 100 fg for QqQ vs. 300 fg for TOF/MS) (81). In addition to improved sensitivity, TOF/MS provides improved mass resolution ($>$10,000 FWHM) and full spectrum acquisition over a broad mass range with accurate mass values ($<$5 ppm). For multi-analyte screening, the scan time is generally set to 0.2–0.3 s, which corresponds to 3–5 spectra per second (65, 69), and a mass window of 10–20 atomic mass unit (amu) is selected to extract the target compounds from the full spectrum with sufficient selectivity. In addition to this high acquisition rate, TOF/MS is able to (1) assist in the structural elucidation of nontargeted compounds based on accurate mass measurements and isotopic patterns; (2) follow an unlimited number of compounds with no impact from the initial acquisition rate; and (3) offer the ability to store the data and perform any retrospective analysis. However, TOF/MS instruments often have a worse quantitative performance than QqQ devices and are more expensive. An exhaustive discussion of the advantages and drawbacks of these two analyzers coupled to UHPLC for multi-analyte determination can be found elsewhere (72).

Finally, the combination of UHPLC strategies with hybrid instruments, such as QqTOF/MS, presents an obvious advantage for multiresidue screening. In contrast to both QqQ and TOF/MS devices, the number of identification points, and thus the confidence level of the results, is significantly enhanced in QqTOF/MS as shown in recent papers by Barcelo and co-workers (75, 83) and Badoud et al. (48, 50) (see Figure 4.4). This technique also allows for the structural elucidation of unknown compounds in complex matrices because of the valuable fragmentation information. Finally, an Orbitrap analyzer was coupled to UHPLC for the determination of various drugs in fish, honey, porcine kidney, and porcine liver samples (84). It was demonstrated that high-resolution mass spectrometry is suitable for such analyses because its selectivity exceeds that of classical MS/MS when the data are recorded with a resolution of at least 50,000 FWHM and a correspondingly sized time window, thus decreasing the number of false positives.

ESI remains the ionization source of choice for more than 95% of all studies. A study involving the determination of seventeen illicit drugs in oral fluids by UHPLC-MS/MS was performed using ESI, APCI, and atmospheric pressure photoionization (APPI) (85). Values for the limit of quantitation ranged between 0.02 and 2 ng/mL for all of the compounds with each of the three different ionization sources, which showed that APCI and APPI, originally developed for apolar compounds, are also suitable for polar drugs at the sub-ppb level. An advantage of APCI and APPI is the significant reduction in the matrix effects of urine and plasma samples relative to ESI (86, 87).

4.6   UHPLC-MS IN METABOLOMICS

Metabolomics (or metabonomics) is the biological approach to analyzing and modeling the metabolic response to biological interventions and disease processes (88, 89).

This field is expanding rapidly despite the complexity of the approach. Indeed, the number of metabolic intermediates, secondary metabolites, and other signaling molecules contained within a biological sample (90, 91) can be very important and necessitates a multistep process that includes sample collection and preparation, separation, detection, and data mining. This process should also allow for the comprehensive identification and quantitation of all metabolites in a complex biological system (92,93). Therefore, there is a need for analytical systems that provide improved resolution and increased sensitivity for separation and detection, such as that afforded by the UHPLC-TOF/MS and UHPLC-QqTOF/MS platforms that are already in use for numerous metabolomic studies. Until recently, such analytical tools were used to globally fingerprint and/or profile various samples, from biological fluids to plant extracts. In this section, emphasis will be placed on the separation and detection steps; however, it is important to keep in mind that sample collection, sample preparation, and data mining remain of prime importance for successful metabolomics (94,95).

4.6.1 UHPLC-MS for Metabolomics in Human and Animal Tissues

The two types of experiments in metabolomic studies should be differentiated. On the one hand, a metabolomic fingerprinting consists of rapidly assessing the differences between samples using high-throughput analytical tools. Therefore, the reduction in analysis time afforded by the UHPLC vs. HPLC strategy (up to a factor of 10) is beneficial. On the other hand, a high-resolution metabolomic profiling is carried out to confirm the presence of biomarkers from any disease or biological intervention performed on the sample of interest. For this purpose, a longer UHPLC column can be used to increase the amount of information gathered from a single run.

In 2005, Wilson and co-workers first developed a metabolomic strategy that used UHPLC-MS (96–99). Biological fluids from a rat and a mouse were rapidly analyzed on a 50 mm UHPLC column packed with 1.7 μm particles and coupled to TOF/MS detection. Peak widths of approximately 1 s were obtained using a gradient time of only 1 min, which corresponds to a peak capacity of 60. Using the additional information provided by the TOF/MS device, the number of features (i.e., signals observed with a specific m/z and RT that differ from the noise and can be considered as a variable for data treatment) was equal to 1000 and thus was sufficient to differentiate between samples of normal and obese Zucker rodents. It also enabled a rapid discimination between the age, strain, gender, and diurnal variation of rats and mice. As reported, a similar number of features can be observed using conventional HPLC; however, the analysis time is tenfold longer. This discovery was confirmed by Nordstrom et al. (100), who noticed a 20% increase in the number of components detected by UHPLC vs. HPLC during the quantitative analysis of endogenous and exogenous metabolites in human serum.

Other studies performed on animal or human plasma and urine confirmed the potential of UHPLC-MS for nutritional, toxicological, and other metabolomic studies (101–108). In most cases, the analytical conditions that gave a good compromise between throughput and information quality for urinary metabolite profiling were the following: a column length of 50–100 mm, gradient time of approximately 10 min, and mobile phase flow rate of 500 μL/min. Recently, a protocol for the global metabolic profiling of urine was proposed by Want et al. (109). With the exception of the throughput, the UHPLC strategy proved beneficial compared to conventional HPLC, exhibiting improved retention time, reproducibility, and signal-to-noise ratios.

TOF/MS and QqTOF/MS analyzers remain the first choice in terms of detection for gathering the maximal information from a high-throughput run because of their excellent mass resolution and accuracy. It is also of note that an LTQ-Orbitrap coupled to an UHPLC was used for the successful metabolomic profiling of serum (103). In this study, the assessment of the disease biomarkers from complex samples was more easily confirmed because the mass resolution was as high as 30,000 at FWHM and the mass accuracy was as low as 2 ppm.

Based on these results, UHPLC coupled to high-resolution MS seems to be the best choice for metabolomics, and the workflow diagram of a typical MS-based metabolite profiling study is presented in Figure 4.8. However, it is also important to check whether the retention time stability and mass accuracy are sufficient because they can critically influence the number of detected features and the data treatment method. This topic was widely discussed in several recent papers (99, 110). One example reported that the shift in the retention time and the mass shift were only 0.03 min and less than 4 ppm, respectively, after 600 injections of a pooled quality control sample over a 5 day period (102). Similar results were also obtained in another study (101), in which the retention times and mass shifts were estimated to be 0.03 min and less than 5 ppm, respectively. Finally, it was discovered that the first few injections on the UHPLC system were not representative and should be discarded (99). Therefore, based on the assessment of the QC samples, UHPLC-MS provides a powerful and repeatable method for the global metabolomic analysis of biological samples (99, 110).

Finally, two innovative strategies can be applied to further extend the applicability of UHPLC to human metabolomics: the application of very high temperatures in UHPLC and the use of HILIC columns (i.e., hydrophilic interaction liquid chromatography). The impact of the mobile phase temperature on the chromatographic behavior is well known, and an elevated temperature induces a strong decrease in the polarity and viscosity of the eluent. It is therefore possible to (i) attain faster separations by applying a higher mobile phase flow rate without compromising the resolution, (ii) significantly decrease the amount of organic solvent required to elute the compound of interest, and (iii) obtain alternative selectivity for complex samples. The application of high temperatures in UHPLC was recently reported and similar benefits, including a strong reduction in the analysis times and the potential to increase the column length to attain higher resolutions, were observed (111–113). In the field of metabolomics, Wilson and co-workers (99) investigated the use of a temperature gradient (up to 180°C) for the global metabolite profiling of both plasma and urine with pure water as the mobile phase. Two UHPLC columns of 100 mm were connected in series and used in combination with a 21 min gradient. The stability of the stationary phase under such extreme conditions was evaluated,

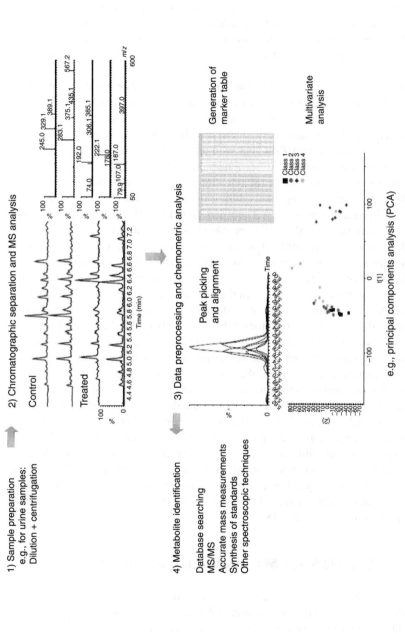

1) Sample preparation
e.g., for urine samples:
Dilution + centrifugation

2) Chromatographic separation and MS analysis

Control

Treated

Time (min)

m/z

3) Data preprocessing and chemometric analysis

Peak picking
and alignment

Generation of
marker table

Multivariate
analysis

e.g., principal components analysis (PCA)

4) Metabolite identification

Database searching
MS/MS
Accurate mass measurements
Synthesis of standards
Other spectroscopic techniques

Figure 4.8 Workflow diagram for a typical MS-based metabolite profiling. Step (1) is the sample preparation followed by MS analysis, which is usually coupled to a LC or GC separation step (Step 2). A key component is the data analysis (Step 3), which can be divided into the data preprocessing and chemometric analysis. This is followed by the identification of important metabolites (Step 4). Adapted from Want, E.J., et al. *Nature Protocols*. 2010, with permission.

and no performance loss was observed after more than 250 injections. The potential of the high-temperature UHPLC strategy was assessed by the number of ions detected when a thermal gradient higher than the ambient temperature was applied (approximately 2000 features). However, the metabolite stability under such extreme conditions was neither evaluated nor discussed in this paper (99). Additionally, the metabolomic profiling of the plasma samples was not possible because the eluotropic strength of water was insufficient, even at 180°C, to elute all of the compounds from the column.

The term *HILIC* was coined by Alpert in 1990 (114). This method was originally developed for the analysis of polar compounds such as carbohydrates or amino acids. However, because the mechanism is a complex combination of partition, adsorption, and ion exchange, HILIC stationary phases are also able to retain most of the compounds usually analyzed by RP-LC but with orthogonal selectivity (115, 116). Gika et al. (97) reported the use of HILIC stationary phases packed with sub-2 μm particles as a complementary technique for the retention of polar compounds and the orthogonal selectivity of other components during urine metabolite profiling. Obviously, significant differences were observed between the HILIC-UHPLC profile (95%–50% ACN in 12 min) and the RP-UHPLC profile (5%–100% ACN in 11 min). The elution order was essentially reversed between the two modes with several significant changes in the selectivity for certain classes of compounds. Another important feature of HILIC is the possibility to enhance its sensitivity through MS detection because of the high proportion of acetonitrile in the mobile phase, which is particularly beneficial for the ESI process (34). Despite these observations, the number of features detected by HILIC was lower than for RP-LC (i.e., 2098 vs. 3284), and the two methods provided very different markers for sample type differentiation. Because of the complementary nature of both of the separation modes, which enables an improvement to the metabolome coverage, the use of HILIC methodology in combination with RP-LC was reported in the protocol for urine global metabolic profiling (109). The repeatability of the retention time (relative standard deviation [RSD] below 2%), signal intensity (RSD below 15%), and mass variability (<0.005 amu) demonstrated high within-run repeatability, even with an extended run time (>200 samples, 60 h run time) (88). The only limitation of the HILIC approach remains the poor retention for most acidic metabolites; however, this problem could, in our opinion, be resolved by (1) working under very acidic conditions to place the acids in their neutral forms (117) or (2) using a different stationary phase chemistry (e.g., silica with polar bonds such as amides or diols) (115, 118).

4.6.2 UHPLC-MS for Metabolomics in Plant Extracts

The chemical diversity and complexity of metabolites present in natural products is extremely important. For this reason, metabolomic approaches are also applied to the field of phytochemistry. Two research groups used UHPLC coupled to TOF/MS and QqTOF/MS detectors for untargeted metabolic profiling. Wolfender and co-workers (119, 120) investigated the signaling molecules from the well-defined biological model plant, *Arabidopsis thaliana*, using an UHPLC-TOF/MS platform, and Jia et al. reported the profiling of several medicinal *Panax* herbs (121–123).

Wolfender and co-workers proposed an original multistep strategy that allows the determination, isolation, and identification of signaling molecules after wounding an *Arabidopsis thaliana* leaf, which mimics an herbivore attack (119, 120). The first step of this process consisted of evaluating the intra-sample variability and forming a suitable pool through rapid UHPLC-TOF/MS fingerprinting on a 50 × 1 mm i.d. column with very high linear velocity and using a gradient time of 7 min. Then, a high-resolution metabolite profiling of the selected pool samples was carried out on long UHPLC columns of 150 to 300 × 2.1 mm i.d. To avoid detrimental changes to the selectivity, the rapid fingerprinting approach was geometrically transferred using the basic rules for transferring methods between conventional LC and UHPLC, and the gradient time was extended to more than 100 min. This profiling step was mandatory to confirm the presence of different stress-related compounds based on unsupervised data-treatment methodologies (94). In addition, the high peak capacity attained by the long columns packed with sub-2 μm particles was essential for the complete deconvolution of the various closely related isomers (119). The third step consisted of isolating the signaling molecules for further characterization. For this purpose, the high-resolution profiling method was adjusted to the semi-preparative scale using long 500 × 19 mm columns packed with 5 μm particles of the same chemistry as the LC-MS-triggered preparative isolation. Finally, after isolating the compounds of interest at the microgram level, an unambiguous structural elucidation of the isolated wound biomarkers (including known signaling molecules, as well as the original compounds oxylipins and jasmonates) (120) was achieved using a capillary-NMR probe, and both 1D and 2D spectra of good quality were obtained. Such analytical strategies can be implemented to screen any other plant extract without re-optimization because they are fully generic.

Another UHPLC-QqTOF/MS metabolomic approach concerns the differentiation of various *Panax* herbs using principal component analysis (PCA) (122). Conventional HPLC with a 250 × 4.6 mm i.d. column and an 80 min gradient time was compared to UHPLC using a 100 × 2.1 mm i.d. column and a 20 min gradient. The number of saponins identified as chemical markers by UHPLC was more than twofold larger than for HPLC. However, this comparison did not take into account that a powerful QqTOF/MS was coupled to the UHPLC while only a single quadrupole MS was coupled to the HPLC. The bioactive compounds responsible for the variations between *Panax* herbs were identified by injecting available reference standards (121, 122), which was possible thanks to the high mass accuracy and resolution of the QqTOF/MS analyzer as well as the good retention time repeatability of the UHPLC method.

4.7   ANALYSIS OF PROTEINS WITH UHPLC-MS

A growing number of biopharmaceutical proteins have been produced in recent years that are already, or soon will be, available. The properties of these proteins mainly depend on their molecular weights, possible conformations, solubilities, stabilities, in vivo lifetimes, posttranslational modifications, and microheterogeneity, which results from modifications during production, extraction, purification, formulation, and storage. The efficacy and safety of recombinant proteins may thus be affected because small differences in their manufacturing process can lead to adverse biological and clinical effects. In addition, considering the number of co- and posttranslational modifications (e.g., enzymatic cleavage and the attachment of lipids or glycans) needed to establish the correct activity of the desired protein, generic versions of biological drugs are difficult to create. Consequently, it can be assumed that it is much more complicated to characterize biomolecules than conventional low-molecular-weight molecules; therefore, dedicated techniques are required. Most of the time, a combination of analytical strategies is necessary to obtain complete information (124). As previously mentioned throughout this chapter, the coupling of UHPLC with MS has provided many benefits in terms of analytical throughput, resolution, and identification. The majority of these studies have focused on small molecule analyses, but UHPLC-MS can provide the same advantages for the analysis of proteins.

In this section, only studies involving UHPLC-MS used to separate and identify intact proteins are mentioned. The analysis of intact proteins has several advantages over the bottom-up approach, where proteins are first digested in solution by a specific protease, such as trypsin, prior to peptide products analysis by LC-MS/MS. One such advantage is that the complete protein sequence is potentially accessible, which enables posttranslational modifications (PTMs) to be located and characterized. The identity of the original protein is determined by deconvolution of the mass spectrum from the multiply charged ions produced by ESI, which generates complicated mass spectra that require the use of high-mass-accuracy detection devices, such as QqTOF/MS.

4.7.1 Issues and Solutions in Intact Protein Analysis

The reversed-phase LC of proteins has, historically, been found to be difficult relative to small molecules because of their adsorption, carryover, lack of retention (pore exclusion), and poor chromatographic performance (peak tailing and broadening, multiple peak formation). These issues arise because (1) the size of proteins causes them to have low diffusion coefficients in the pores of the chromatographic column and (2) the protein interaction with both the mobile and stationary phases (secondary interactions) can affect the conformation and retention of the protein. In addition, the latter is strongly dependent on small changes in the solvent strength, and the percentage windows for organic modifiers are quite narrow for each protein. For these reasons, isocratic conditions are rarely used and gradient elution is generally preferred.

The traditional solution to poor resolution has been to increase the run time, which negatively affects the analytical throughput without completely solving the issues. An alternative solution consists of using columns packed with sub-2 μm particles under UHPLC conditions. Reducing the particle size leads to significant improvements

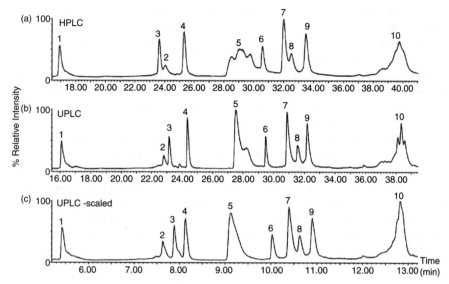

Figure 4.9 Comparison of the analytical performance for a protein standard mixture: (a) HPLC-MS, (b) UHPLC-MS with $F = 75$ μL/min and $t_{gradient} = 60$ min, and (c) gradient-scaled UHPLC-MS with $F = 225$ μL/min and $t_{grad} = 20$ min. Adapted from Everley, R.A., Croley, T.R. *J. Chromatogr. A.* 2008, with permission.

in the kinetic performance, which is particularly useful for large molecules. This effect is based on the van Deemter equation, which shows that the optimal mobile phase flow rate is inversely proportional to the particle diameter, whereas the mass transfer resistance is directly proportional to the square of the particle diameter. Such effects were emphasized in a study where an HT-UHPLC-MS method was developed for the analysis of ten intact proteins ranging in mass from 6–66 kDa (125). As demonstrated in Figure 4.9, the optimized UHPLC-MS method featured enhanced resolution, improved sensitivity, and a threefold increase in throughput compared to the original HPLC-MS method, without compromising the mass spectral data quality.

Adsorption of proteins onto the UHPLC instrument is another issue for intact protein analysis. Inert materials, such as titanium, should be used preferentially to the very hydrophobic polyether ether ketone (PEEK) for connection tubes and injection needles. Recently, manufacturers have taken such adsorption issues into account and produced new UHPLC instruments dedicated to biomolecular analysis.

Manufacturers have also provided a solution for the poor peak shapes that result from secondary interactions by providing silica-based stationary phases with restricted access to any residual silanols (e.g., endcapped, bidentate, hybrid silica, high-density bonding, and embedded polar group stationary phase). In addition, the performance can be improved by raising the temperature of the mobile phase. Because the mobile phase viscosity is reduced at elevated temperatures, analyte diffusion is enhanced, and the kinetic and mass transfer rates are improved, which strongly reduces both peak tailing and broadening. However, the temperature should be set with caution (<90°C) because of the potential thermal degradation of the proteins and limited thermal stability of most stationary phases. Another strategy consists of adding an ion-pairing agent to the mobile phase. Trifluoroacetic acid (TFA) with a concentration of 0.1% is commonly used for protein analysis, as it possesses excellent ion-pairing and solvating characteristics and circumvents both peak tailing and broadening. However, when using TFA in combination with MS detection, ion suppression can occur in both the negative and positive modes. Therefore, formic acid should be used for UHPLC-MS. Finally, because some hybrid silica-based stationary phases are stable over a wide pH range (i.e., up to pH 12), it would be interesting to evaluate how basic pH conditions may further reduce the ionic interactions. So far, this method has only been reported for small molecule analysis (34), but it could certainly be extended to proteins. Because proteins are multicharged molecules and composed of mostly basic, ionizable functional groups, changes to the mobile phase pH will have a pronounced effect on the protein conformation and, thus, their retention behavior and peak shape. It is noteworthy that basic pH may also have an impact on their ionization by ESI-MS.

4.7.2   UHPLC-MS for Intact Protein Analysis

4.7.2.1   RP-UHPLC-MS

Traditional protein liquid chromatography suffers from a number of issues that can be circumvented by using UHPLC with small particles at high flow rates. Furthermore, many traditional techniques (e.g., size-exclusion chromatography and ion-exchange chromatography) are barely compatible with MS detection, whereas the commonly used reversed-phase (RP) UHPLC can be directly coupled to MS, which provides a powerful tool for protein identification, characterization, and eventual quantitation. For these reasons, RP-UHPLC-MS appears to be a promising technology for assessing protein batch purity and highlighting any protein degradation (e.g., truncation, aggregation, oxidation, reduction, glycosylation, isomerization, deamidation, and clipping), misfolding, or PEGylation in a high-throughput environment. So far, few papers have reported the use of RP-UHPLC-MS technologies for this purpose; however, due to recent advances in RP-columns packed with sub-2 μm particles (various types of sorbents are now available, including C4 packing with large pores, e.g., 300 Å) and the straightforward coupling to MS, its usefulness for intact protein analysis will become increasingly important over the next few years.

A good example is given by the use of UHPLC-MS to ensure the product integrity and stability of monoclonal antibody (mAb) therapeutics at various stages in their development and production. Conventionally, enzymatically derived peptides of mAb are generated in solution with a dedicated sample preparation procedure, separated by ion-pairing RP-LC under acidic conditions, and detected by ESI-MS/MS using a bottom-up approach (126). The features of MS/MS include the sequencing of co-eluting peptides by isolating a specific m/z window that contains an isotopic cluster from only one of the peptides. Even though the separation of all peptides is not absolutely required, this strategy is still limited by the duty cycle and sensitivity of the MS instruments, as well as the ion suppression effects inherent to ESI. In addition, because changes in the product are revealed by changes in the retention times and by MS detection, good retention time repeatability is required (typically RSD below 0.1%). Conventionally, the analysis of tryptic digests by UHPLC-MS/MS is carried out using similar conditions to those used for small molecules with C18 columns (100–150 × 2.1 mm i.d.) packed with sub-2 μm particles and pore size in the range of 100–120 Å. UHPLC systems are mainly coupled to tandem MS with either QqQ or hybrid instruments, such as QqTOF/MS or LTQ-Orbitrap MS. Peak capacities higher than 600 have been reported for tryptic digests under high-resolution conditions (e.g., with two or three 100 mm columns in series and $t_{grad} > 90$ min) (127). Because of the large number of peaks detected, data processing by either manual or automatic peptide sequencing with bioinformatics tools has become the primary challenge. Briefly, the identity of the original protein is retrieved by comparing the mass spectra of the peptide to theoretical peptide masses calculated from a proteomic database or produced by *in silico* digestion. Finally, absolute quantitation can be achieved by using affinity tags and stable isotope labels (e.g., ICAT, iTRAQ). Peptide mapping is particularly challenging for mAb analysis because of the presence of numerous other peptides in the sample. In addition, only a portion of the protein sequence is identified, and this limited sequence coverage induces a significant loss of information about PTMs. For this reason, the analysis of intact mAb appears to be a good alternative. An analysis was performed in a detailed study of mAb deamidation under extreme conditions that used UHPLC-QqTOF/MS to achieve chromatographic resolution of the deamidation products while maintaining relatively short analysis times (C18 column 150 × 1 mm i.d., 1.7 μm particle size, $F = 100$ μL/min, $t_{gradient} = 45$ min) (128). Figure 4.10a shows the extracted ion chromatogram obtained from a mAb stressed for one day. Five species were detected, and their respective mass spectra are presented in Figure 4.10b. The peaks represent the non-deamidated protein (NNN, peak 2), three singly deamidated species (peaks 1, 4, 5) with a +1 amu mass increase relative to the parent protein, and one doubly deamidated product (peak 3) with a +2 amu mass increase. Identification of the deamidation variants was achieved by high energy MS1 analysis, while their quantitation was based on the isotopic peak intensities of each form. Kinetic profiling for the deamidation of intact mAb over 2 months was eventually performed, as shown in Figure 4.10c, and the estimated half-life was calculated, based on the overall loss of the intact mAb, to be more than 100 days.

4.7.2.2   HILIC-UHPLC-MS

HILIC is the most recent form of normal-phase chromatography (NP-LC) and is comprised of a hydrophilic stationary phase with a mobile phase consisting of a mixture of water and more than 70% organic solvent (typically acetonitrile). HILIC appears to be a valuable alternative to ion exchange chromatography, as it is directly compatible with MS. This technique has already been used to analyze glycopeptides and glycans from glycoproteins (129, 130), but it has yet to be widely applied for the

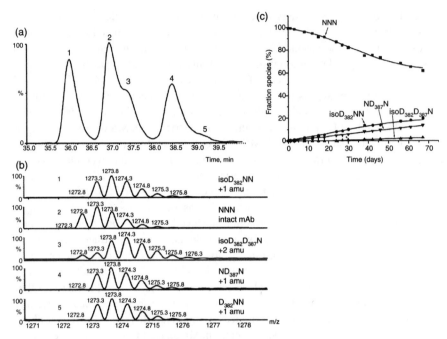

Figure 4.10   Deamidation products of Fc IgG for a sample stressed for 28 h at 37°C and pH 7.4.
(a) Extracted ion chromatogram (EIC), (b) molecular ion isotope envelopes of peaks in the EIC, and
(c) kinetic profile of deamidation. Adapted from Sinha, S., et al. *Protein Sci.* 2009, with permission.

analysis of intact proteins. So far, only one study has reported an HILIC intact protein
analysis, which used UV detection and a conventional particle size (131). Because
HILIC columns packed with sub-2 μm particles have been recently launched, interest
in its use for intact protein analysis should grow quickly in the fields of protein
characterization and stability. However, the possible adsorption of proteins onto the
HILIC material still needs to be considered. Finally, the dissolution solvent is a key
parameter in HILIC for determining the peak shape integrity, and the elution must be
performed with large amounts of organic solvent (i.e., >70% acetonitrile). Organic
solvent solubility may, therefore, be an issue for protein analysis as denaturation
could occur, and alternative mobile phases and dissolution solvents will certainly
be needed.

4.8 CONCLUSIONS

This chapter demonstrates the benefits of UHPLC-MS over conventional HPLC-MS
in terms of throughput and resolution. However, when coupling UHPLC to MS detection, there are a number of technical issues that should be considered. In our opinion,
the latest generation of MS devices is fully compatible with UHPLC technology.

Indeed, the peak widths generated by UHPLC are quite thin and the acquisition rate of the instrument should be sufficient. This is particularly critical for quadrupole-based instruments as the different channels are gathered sequentially, whereas, on a TOF/MS device, a wide range of m/z ratios can be recorded simultaneously.

Despite several significant advances that can reduce the dwell times of quadrupole instruments to only 1 ms, TOF devices (latest generation of TOF/MS have acquisition rates as high as 100–200 Hz) can be more naturally combined with UHPLC. However, because the quantitative performance and sensitivity of TOF/MS is still not equal to that of QqQ instruments operating in the SRM mode, the latter remains the gold standard for targeted bioanalytical assays. In addition, the resolution of TOF/MS devices (between 10,000 and 40,000 FWHM) is not considered sufficient for multiresidue screening, and the official regulatory texts still recommend the use of QqQ devices because they provide a sufficient number of identification points through their capability to fragment ions. However, UHPLC combined with TOF/MS appears to be an optimal choice for untargeted analyses, such as metabolite identification and metabolomic assays. Indeed, this platform is particularly useful as it combines the high chromatographic resolution of UHPLC with the increased sensitivity, mass resolution, and routine mass accuracy (<5 ppm) of TOF/MS.

UHPLC-MS also possesses attractive features that enable the analysis of proteins in their intact form, providing a real advantage over the classical bottom-up approach.

It is noteworthy that, because of the significant reduction in analysis time with UHPLC technology, the main limitation of the entire analytical process has now become the sample preparation and data processing—the latter being particularly important when working with complicated untargeted analyses, such as metabolomics.

REFERENCES

1. MAURER, H.H. *Anal. Bioanal. Chem.* 2007, 388: 1315–1325.
2. CABRERA, K. *J. Sep. Sci.* 2004, 27: 843–852.
3. GUIOCHON, G. *J. Chromatogr. A.* 2007, 1168: 101–168.
4. WEI, R., LI, G., SEYMOUR, A.B. *Anal. Chem.* 2010, 82: 5527–5533.
5. HEINISCH, S., ROCCA, J.L. *J. Chromatogr. A.* 2009, 1216: 642–658.
6. GUILLARME, D., HEINISCH, S. *Sep. Purif. Rev.* 2005, 34: 181–216.
7. GUILLARME, D., RUSSO, R., RUDAZ, S., BICCHI, C., VEUTHEY, J.L. *Curr. Pharm. Anal.* 2007, 3: 221–229.
8. CUNLIFFE, J.M., MALONEY, T.D. *J. Sep. Sci.* 2007, 30: 3104–3109.
9. GRITTI, F., GUIOCHON, G. *J. Chromatogr. A.* 2011, 1228: 2–19.
10. RUTA, J., GUILLARME, D., RUDAZ, S., VEUTHEY, J.L. *J. Sep. Sci.* 2010, 33: 2465–2477.
11. SWARTZ, M.E. *J. Liq. Chromatogr. R. T.* 2005, 28: 1253–1263.
12. MAZZEO, J.R., NEUE, U.D., KELE, M., PLUMB, R.S. *Anal. Chem.* 2005, 77: 460A–467A.
13. GUILLARME, D., GRATA, E., GLAUSER, G., WOLFENDER, J.L., VEUTHEY, J.L., RUDAZ, S. *J. Chromatogr. A.* 2009, 1216: 3232–3243.
14. GUILLARME, D., NGUYEN, D.T.T., RUDAZ, S., VEUTHEY, J.L. *Eur. J. Pharma. Biopharm.* 2008, 68: 430–440.
15. NOVAKOVA, L., MATYSOVA, L., SOLICH, P. *Talanta.* 2006, 68: 908–918.
16. FENN, J.B., MANN, M., MENG, C.K., WONG, S.F., WHITEHOUSE, C.M. *Science.* 1989, 246: 64–71.

17. FENN, J.B., MANN, M., MENG, C.K., WONG, S.F., WHITEHOUSE, C.M. *Mass Spectrom. Rev.* 1990, 9: 37–70.
18. HOPFGARTNER, G., BOURGOGNE, E. *Mass Spectrom. Rev.* 2003, 22: 195–214.
19. SIMPSON, H., BERTHEMY, A., BUHRMAN, D., BURTON, R., NEWTON, J., KEALY, M., WELLS, D., WU, D. *Rapid Commun. Mass SP.* 1998, 12: 75–82.
20. THEVIS, M., GUDDAT, S., SCHAENZER, W. *Steroids.* 2009, 74: 315–321.
21. GUILLARME, D., SCHAPPLER, J., RUDAZ, S., VEUTHEY, J.L. *Trends Anal. Chem.* 2010, 29: 15–27.
22. VOYKSNER, R.D., COLE, R.B. (ed.). *Combining Liquid Chromatography with Electrospray Mass Spectrometry.* New York: Wiley, 1997.
23. COLE, R.B. (ed.). *Electrospray Ionization Mass Spectrometry: Fundamentals, Instrumentation and Applications.* New York: Wiley, 1997.
24. CECH, N.B., ENKE, C.G. *Mass Spectrom. Rev.* 2001, 20: 362–387.
25. HERBERT, C.G., JOHNSTONE, R.A.W. *Mass Spectrometry Basics.* Boca Raton, FL: CRC Press, 2003.
26. NOVAKOVA, L., VEUTHEY, J.L., GUILLARME, D. *J. Chromatogr. A.* 2011, 1218: 7971–7981.
27. DE VILLIERS, A., LAUER, H., SZUCS, R., GOODALL, S., SANDRA, P. *J. Chromatogr. A.* 2006, 1113: 84–91.
28. DE HOFFMANN, E., STROOBANT, V. *Mass Spectrometry Principles and Applications*, 3rd edition. New York: Wiley, 2007.
29. GUILLARME, D., NGUYEN, D.T.T., RUDAZ, S., VEUTHEY, J.L. *Eur. J. Pharm. Biopharm.* 2007, 66: 475–482.
30. LESTREMAU, F., WU, D., SZÜCS, R. *J. Chromatogr. A.* 2010, 1217: 4925–4933.
31. NGUYEN, D.T.T., GUILLARME, D., RUDAZ, S., VEUTHEY, J.L. *J. Chromatogr. A.* 2006, 1128: 105–113.
32. NGUYEN, D.T.T., GUILLARME, D., HEINISCH, S., BARRIOULET, M.P., ROCCA, J.L., RUDAZ, S., VEUTHEY, J.L. *J. Chromatogr. A.* 2007, 1167: 76–84.
33. DELATOUR, C., LECLERCQ, L. *Rapid Commun. Mass SP.* 2005, 19: 1359–1362.
34. SCHAPPLER, J., NICOLI, R., NGUYEN, D.T.T., RUDAZ, S., VEUTHEY, J.L., GUILLARME, D., *Talanta.* 2009, 78: 377–387.
35. MORDEHAI, A., FJELDSTED, J. Agilent jet stream thermal gradient focusing technology, 5990-3494EN. Palo Alto, CA: Agilent Technologies, Inc., 2009.
36. FEKETE, S., FEKETE, J. *J. Chromatogr. A.* 2011, 1218: 5286–5291.
37. DEVENTER, K., VAN EENOO, P., DELBEKE, F.T. *Rapid Commun. Mass SP.* 2005, 19: 90–98.
38. PRASAD, B., GARG, A., TAKWANI, H., SINGH, S. *Trend Anal. Chem.* 2011, 30: 360–387.
39. BALOGH, M. "A High Speed Combination Multi-mode Ionization Source for Mass Spectrometers," PCT Int. Appl., WO 2003102537 A2 20031211, 2003, 38 pp.
40. GRATA, E., GUILLARME, D., GLAUSER, G., BOCCARD, J., CARRUPT, P.A., VEUTHEY, J.L., RUDAZ, S., WOLFENDER, J.L. *J. Chromatogr. A.* 2009, 1216: 5660–5668.
41. NOVAKOVA, L., VLCKOVA, H. *Analytica Chimica Acta.* 2009, 656: 8–35.
42. SHAH, V.P. *J Am. Assoc. Pharma. Scientists.* 2007, 9: E43–E47.
43. STARK, T., WOLLMAN, N., LOSCH, S., HOFMANN, T. *Anal. Chem.* 2011, 83: 3398–3405.
44. ALDER, L., LUEDERITZ, S., LINDTNER, K., STAN, H.J. *J. Chromatogr. A.* 2004, 1058: 67–79.
45. DEL MAR RAMIREZ FERNANDEZ, M., WILLE, S.M.R., DI FAZIO, V., GOSSELIN, M., SAMYN, N. *J. Chromatogr. B.* 2010, 878: 1616–1622.
46. GOSETTI, F., CHIUMINATTO, U., ZAMPIERI, D., MAZZUCCO, E., ROBOTTI, E., CALABRESE, G., GENNARO, M.C., MARENGO, E. *J. Chromatogr. A.* 2010, 1217: 7864–7872.
47. JACOB, C.C., GAMBOA DA COSTA, G. *J. Chromatogr. B.* 2011, 879: 652–656.
48. BADOUD, F., GRATA, E., PERRENOUD, L., SAUGY, M., RUDAZ, S., VEUTHEY, J.L. *J. Chromatogr. A.* 2010, 1217: 4109–4119.
49. BADOUD, F., GRATA, E., BOCCARD, J., GUILLARME, D., VEUTHEY, J.L., RUDAZ, S., SAUGY, M. *Anal. Bioanal. Chem.* 2011, 400: 503–516.
50. BADOUD, F., GRATA, E., PERRENOUD, L., AVOIS, L., SAUGY, M., RUDAZ, S., VEUTHEY, J.L. *J. Chromatogr. A.* 2009, 1216: 4423–4433.

51. DECOSTERD, L.A., ROCHAT, B., PESSE, B., MERCIER, T., TISSOT, F., WIDMER, N., BILLE, J., CALANDRA, T., ZANOLARI, B., MARCHETTI, O. *Antimicrob. Agents Ch.* 2010, 54: 5303–5315.

52. KO, D.H., JUN, S.H., PARK, K.U., SONG, S.H., KIM, J.Q., SONG, J. *J. Inherit. Metab. Dis.* 2011, 34: 409–414.

53. HOOF, G.P., MEESTERS, R.J.W., VAN KAMPEN, J.J.A., VAN HUIZEN, N.A., KOCH, B., AL HADITHY, A.F.Y., VAN GELDER, T., OSTERHAUS, A.D.M.E., GRUTERS, R.A., LUIDER, T.M. *Anal. Bioanal. Chem.* 2011, 400: 3473–3479.

54. DEGLON, J., THOMAS, A., MANGIN, P., STAUB, C. *Anal. Bioanal. Chem.* 2011, 402: 2485–2498.

55. NICOLI, R., MARTEL, S., RUDAZ, S., WOLFENDER, J.L., VEUTHEY, J.L., CARRUPT, P.A., GUILLARME, D. *Expert Opin. Drug Disc.* 2010, 5: 475–489.

56. NICOLI, R., CURCIO, R., RUDAZ, S., VEUTHEY, J.L. *J. Med. Chem.* 2009, 52: 2192–2195.

57. STEWART, N.A., BUCH, S.C., CONRADS, T.P., BRANCH, R.A. *Analyst.* 2011, 136: 605–612.

58. BADMAN, E.R., BEARDSLEY, R.L., LIANG, Z., BANSAL, S. *J. Chromatogr. B.* 2010, 878: 2307–2313.

59. ANDREWS, G.L., SIMONS, B.L., YOUNG, J.B., HAWKRIDGE, A.M., MUDDIMAN, D.C. *Anal. Chem.* 2011, 83: 5442–5446.

60. BROWN, A., BICKFORD, S., HATSIS, P., AMIN, J., BELL, L., HARRIMAN, S. *Rapid Commun. Mass Sp.* 2010, 24: 1207–1210.

61. PLUMB, R.S., RAINVILLE, P.D., POTTS III, W.B., CASTRO-PEREZ, J.M., JOHNSON, K.A., WILSON, I.D. *Rapid Commun. Mass Sp.* 2007, 21: 4079–4085.

62. MACDONALD, C., SMITH, C., MICHOPOULOS, F., WEAVER, R., WILSON, I.D. *Rapid Commun. Mass Sp.* 2011, 25: 1787–1793.

63. CROCKFORD, D.J., MAHER, A.D., AHMADI, K.R., BARRETT, A., PLUMB, R.S., WILSON, I.D., NICHOLSON, J.K. *Anal. Chem.* 2008, 80: 6835–6844.

64. THORNGREN, J.O., OSTERVALL, F., GARLE, M. *J. Mass Spectrom.* 2008, 43: 980–992.

65. KAUFMANN, A., BUTCHER, P., MADEN, K., WIDMER, M. *Analytica Chimica Acta.* 2007, 586: 13–21.

66. KASPRZYK-HORDERN, B., DINSDALE, R.M., GUWY, A.J. *Anal. Bioanal. Chem.* 2008, 391: 1293–1308.

67. GERVAIS, C., BROSILLON, S., LAPLANCHE, A., HELEN, C. *J. Chromatogr. A.* 2008, 1202: 163–172.

68. PASTOR MONTORO, E., ROMERO GONZALEZ, R., GARRIDO FRENICH, A., HERNANDEZ TORRES, M.E., MARTINEZ VIDAL, J.L. *Rapid Commun. Mass Sp.* 2007, 21: 3585–3592.

69. STOLKER, A.A.M., RUTGERS, P., OOSTERINK, E., LASAROMS, J.J.P., PETERS, R.J.B., VAN RHIJN, J.A., NIELEN, N.W.F. *Anal. Bioanal. Chem.* 2008, 391: 2309–2322.

70. CAI, Z., ZHANG, Y., PAN, H., TIE, X., REN, Y. *J. Chromatogr. A.* 2008, 1200: 144–155.

71. ROMERO-GONZALEZ, R., GARRIDO FRENICH, A., MARTINEZ VIDAL, J.L. *Talanta.* 2008, 76: 211–225.

72. TAYLOR, M.J., KEENAN, G.A., REID, K.B., URIA FERNANDEZ, D. *Rapid Commun. Mass Sp.* 2008, 22: 2731–2746.

73. GARRIDO FRENICH, A., MARTINEZ VIDAL, J., PASTOR-MONTORO, E., ROMERO-GONZALEZ, R. *Anal. Bioanal. Chem.* 2008, 390: 947–959.

74. KASPRZYK-HORDERN, B., DINSDALE, R.M., GUWY, A.J. *Talanta.* 2008, 74: 1299–1312.

75. PETROVIC, M., GROS, M., BARCELO, D. *J. Chromatogr. A.* 2006, 1124: 68–81.

76. ANASTASSIADES, M., LEHOTAY, S.J., STAJNBAHER, D., SCHENK, F.J. *J. AOAC Int.* 2003, 86: 412–431.

77. AGUILERA-LUIZ, M.M., PLAZA-BOLANOS, P., ROMERO-GONZALEZ, R., MARTINEZ-VIDAL, J.L., GARRIDO FRENICH, A. *Anal. Bioanal. Chem.* 2011, 399: 2863–2875.

78. ROMERO-GONZALEZ, R., GARRIDO FRENICH, A., MARTINEZ-VIDAL, J.L., PRESTES, O.D., GRIO, S.L. *J. Chromatogr. A.* 2011, 1218: 1477–1485.

79. PETERSSON, P., FRANK, A., HEATON, J., EUERBY, M.R. *J. Sep. Sci.* 2008, 31: 2346–2357.

80. GUILLARME, D., GRATA, E., GLAUSER, G., WOLFENDER, J.L., VEUTHEY, J.L., RUDAZ, S. *J. Chromatogr. A.* 2009, 1216: 3232–3243.

81. GRIMALT, S., SANCHO, J.V., POZO, O.J., HERNANDEZ, F. *J. Mass Spectrom.* 2010, 45: 421–436.

82. BATT, A.L., KOSTICH, M.S., LAZORCHAK, J.M. *Anal. Chem.* 2008, 80: 5021–5030.

83. FARRE, M., GROS, M., HERNANDEZ, B., PETROVIC, M., HANCOCK, P., BARCELO, D. *Rapid Commun. Mass Sp.* 2008, 22: 41–51.

84. KAUFMANN, A., BUTCHER, P., MADEN, K., WALKER, S., WIDMER, M. *Analytica Chimica Acta.* 2010, 673: 60–72.
85. TING-WANG, I., TING-FENG, Y., YANG CHEN, C. *J. Chromatogr. B.* 2010, 878: 3095–3105.
86. MARCHI, I., RUDAZ, S., VEUTHEY, J.L. *Talanta.* 2009, 78: 1–18.
87. MARCHI, I., RUDAZ, S., SELMAN, M., VEUTHEY, J.L. *J. Chromatogr. B.* 2007, 845: 244–252.
88. SPAGOU, K., WILSON, I.D., MASSON, P., THEODORIDIS, G., RAIKOS, N., COEN, M., HOLMES, E., LINDON, J.C., PLUMB, R.S., NICHOLSON, J.K., WANT, E.J. *Anal. Chem.* 2011, 83: 382–390.
89. ISSAQ, H.J., ABBOTT, E., VEENSTRA, T.D. *J. Sep. Sci.* 2008, 31: 1936–1947.
90. LINDON, J.C., NICHOLSON, J.K. *Trend Anal. Chem.* 2008, 27: 194–204.
91. BINO, R.J., HALL, R.D., FIEHN, O., KOPKA, J., SAITO, K., DRAPER, J., NIKOLAU, B.J., MENDES, P., ROESSNER-TUNALI, U., BEALE, M.H., TRETHEWEY, R.N., LANGE, B.M., WURTELE, E.S., SUMNER, L.W. *Trend Plant Science.* 2004, 9: 41–425.
92. WANT, E.J., NORDSTROM, A., MORITA, H., SIUZDAK, G. *J. Proteome Res.* 2007, 6: 459–468.
93. DUNN, W.B., ELLIS, D.I. *Trend Anal. Chem.* 2005, 24: 285–294.
94. BOCCARD, J., VEUTHEY, J.L., RUDAZ, S. *J. Sep. Sci.* 2010, 33: 290–304.
95. MICHOPOULOS, F., THEODORIDIS, G., SMITH, C.J., WILSON, I.D. *J. Proteome Res.* 2010, 9: 3328–3334.
96. PLUMB, R.S., GRANGER, J.H., STUMPF, C.L., JOHNSON, K.A., SMITH, B.W., GAULITZ, S., WILSON, I.D., CASTRO-PEREZ, J. *Analyst.* 2005, 130: 844–849.
97. GIKA, H.G., THEODORIDIS, G.A., WILSON, I.D. *J. Sep. Sci.* 2008, 31: 1598–1608.
98. GIKA, H.G., THEODORIDIS, G.A., EXTANCE, J., EDGE, A.M., WILSON, I.D. *J. Chromatogr. B.* 2008, 871: 279–287.
99. GIKA, H.G., MACPHERSON, E., THEODORIDIS, G.A., WILSON, I.D. *J. Chromatogr. B.* 2008, 871: 299–305.
100. NORDSTROM, A., O'MAILLE, G., QIN, C., SIUZDAK, G. *Anal. Chem.* 2006, 78: 3289–3295.
101. GUY, P.A., TAVAZZI, I., BRUCE, S.J., RAMADAN, Z., KOCHHAR, S. *J. Chromatogr. B.* 2008, 871: 253–260.
102. WONG, M.C.Y., LEE, W.T.K., WONG, J.S.Y., FROST, G., LODGE, J. *J. Chromatogr. B.* 2008, 871: 341–348.
103. DUNN, W.B., BROADHURST, D., BROWN, M., BAKER, P.N., REDMAN, C.W.G., KENNY, L.C., KELL, D.B. *J. Chromatogr. B.* 2008, 871: 288–298.
104. ZHAO, X., FRITSCHE, J., WANG, J., CHEN, J., RITTIG, K., SCHMITT-KOPPLIN, P., FRITSCHE, A., HARING, H.U., SCHLEICHER, E.D., XU, G., LEHMANN, R. *Metabolomics.* 2010, 6: 362–374.
105. PEREIRA, H., MARTIN, J.F., JOLY, C., SEBEDIO, J.L., PUJOS-GUILLOT, E. *Metabolomics.* 2010, 6: 207–218.
106. JIANG, Z., SUN, J., LIAN, Q., CAI, Y., LI, S., HUANG, Y., WANG, Y., LUO, G. *Talanta.* 2011, 84: 298–304.
107. SU, Z.H., LI, S.Q., ZOU, G.A., YU, C.Y., SUN, Y.G., ZHANG, H.W., ZOU, Z.M. *J. Pharmaceut. Biomed.* 2001, 55: 533–539.
108. MASSON, P., COUTO ALVES, A., EBBELS, T.M.D., NICHOLSON, J.K., WANT, E.J. *Anal. Chem.* 2010, 82: 7779–7786.
109. WANT, E.J., WILSON, I.D., GIKA, H., THEODORIDIS, G., PLUMB, R.S., SHOCKCOR, J., HOLMES, E., NICHOLSON, J.K. *Nature Protocols.* 2010, 5: 1005–1018.
110. ZELENA, E., DUNN, W.B., BROADHURST, D., FRANCIS-MCINTYRE, S., CARROLL, K.M., BEGLEY, P., O'HAGAN, S., KNOWLES, J.D., HALSALL, A., WILSON, I.D., KELL, D.B. *Anal. Chem.* 2009, 81: 1357–1364.
111. VANHOENACKER, G., SANDRA, P. *J. Sep. Sci.* 2006, 29: 1822–1835.
112. NGUYEN, D.T.T., GUILLARME, D., HEINISCH, S., BARRIOULET, M.P., ROCCA, J.L., RUDAZ, S., VEUTHEY, J.L. *J. Chromatogr. A.* 2007, 1167: 76–84.
113. GUILLARME, D., HEINISCH, S., ROCCA, J.L. *J. Chromatogr. A.* 2004, 1052: 39–51.
114. ALPERT, A.W. *J. Chromatogr. A.* 1990, 499: 177–196.
115. JANDERA, P. *Analytica Chimica Acta.* 2011, 692: 1–25.
116. HEMSTROM, P., IRGUM, K. *J. Sep. Sci.* 2006, 29: 1784–1821.

117. McCALLEY, D.V. *J. Chromatogr. A.* 2007, 1171: 46–55.
118. RUTA, J., RUDAZ, S., McCALLEY, D.V., VEUTHEY, J.L., GUILLARME, D. *J. Chromatogr. A.* 2010, 1217: 8230–8240.
119. GRATA, E., BOCCARD, J., GUILLARME, D., GLAUSER, G., CARRUPT, P.A., FARMER, E.E., WOLFENDER, J.L., RUDAZ, S. *J. Chromatogr. B.* 2008, 871: 261–270.
120. GLAUSER, G., GUILLARME, D., GRATA, E., BOCCARD, J., THIOCONE, A., CARRUPT, P.A., VEUTHEY, J.L., RUDAZ, S., WOLFENDER, J.L. *J. Chromatogr. A.* 2008, 1180: 90–98.
121. XIE, G.X., NI, Y., SU, M.M., ZHANG, Y.Y., ZHAO, A.H., GAO, X.F., LIU, Z., XIAO, P.G., JIA, W. *Metabolomics.* 2008, 4: 248–260.
122. XIE, G.X., PLUMB, R.S., SU, M.M., XU, Z., ZHAO, A.H., QIU, M., LONG, X., LIU, Z., JIA, W. *J. Sep. Sci.* 2008, 31: 1015–1026.
123. DAN, M., SU, M.M., GAO, X.F., ZHAO, T., ZHAO, A.H., XIE, G.X., QIU, Y., ZHOU, M., LIU, Z., JIA, W. *Phytochemistry.* 2008, 69: 2237–2244.
124. STAUB, A., GUILLARME, D., SCHAPPLER, J., VEUTHEY, J.L., RUDAZ, S. *J. Pharmaceut. Bimed.* 2011, 55: 810–822.
125. EVERLEY, R.A., CROLEY, T.R. *J. Chromatogr. A.* 2008, 1192: 239–247.
126. JI, C., SADAGOPAN, N., ZHANG, Y., LEPSY, C. *Anal. Chem.* 2009, 81: 9321–9328.
127. KÖCHER, T., SWART, R., MECHTLER, K. *Anal. Chem.* 2011, 83: 2699–2704.
128. SINHA, S., ZHANG, L., DUAN, S., WILLIAMS, T.D., VLASAK, IONESCU, R., TOPP, E.M. *Protein Sci.* 2009, 18: 1573–1584.
129. LAM, M.P.Y., SIU, S.O., LAU, E., MAO, X., SUN, H.Z., CHIU, P.C.N., YEUNG, W.S.B., COX, D.M., CHU, I.K. *Anal. Bioanal. Chem.* 2010, 398: 791–804.
130. NEUE, K., MORMANN, M., PETER-KATALINIC, J., POHLENTZ, G. *J. Proteome Res.* 2011, 10: 2248–2260.
131. TETAZ, T., DETZNER, S., FRIEDLEIN, A., MOLITOR, B., MARY, J.L. *J. Chromatogr. A.* 2011, 1218: 5892–5896.

Chapter 5

The Potential of Shell Particles in Fast Liquid Chromatography

Szabolcs Fekete and Jenő Fekete

5.1 INTRODUCTION

Higher separation efficiency and faster speed have always been of great interest in high performance liquid chromatography (HPLC) and have become increasingly important in recent years mainly driven by the challenges of either more complex samples or an increase in the numbers of samples. Pharmaceutical industry is particularly interested in using rapid and efficient procedures for qualitative and quantitative analysis to cope with a large number of samples and to reduce the time required for delivery of results. Reducing analysis time and guarantying the quality of the separation in HPLC requires high kinetic efficiency. A general approach to increase the separation power is to enhance the column efficiency.

In the recent development of particle technology targeted for liquid chromatography, the use of core-shell (or superficially porous, fused core, shell, pellicular) particles has received considerable attention. Shell particles manifest the advantages of porous particles as well as some benefit of nonporous particles.

The concept of shell particles was imagined by Horváth and co-workers in the late 1960s (1, 2). They were initially intended for the analysis of macromolecules such as peptides and proteins. Later Kirkland presented that 30–40 μm diameter superficially porous packing provides much faster separations compared with the large fully porous particles used earlier in liquid chromatography (3). The rational behind this concept was to improve column efficiency by shortening the diffusion path that the analyte molecules must travel and to improve their mass transfer kinetics. Several brands of superficially porous particles were developed, and they became popular in the 1970s. However, the major improvements in the manufacturing of high-quality, fully

Ultra-High Performance Liquid Chromatography and Its Applications, First Edition. Edited by Quanyun Alan Xu.
© 2013 John Wiley & Sons, Inc. Published 2013 by John Wiley & Sons, Inc.

porous particles that took place in the same time, particularly by making them finer and more homogeneous, hampered the success of shell particles, which eventually disappeared. Recently, the pressing needs to improve analytical throughputs forced the particle manufacturers to find a better compromise between the demands for higher column efficiency and the need for columns that can be operated with the conventional instruments for liquid chromatography (with moderate column back pressures) (4). This led to the development of a new generation of columns packed with shell particles (5). Now core-shell packing materials are commercially available in various diameters (5 μm, 2.7 μm, 2.6 μm, and 1.7 μm) with different shell thickness (0.5 μm, 0.35 μm, 0.25 μm, 0.23 μm, and 0.15 μm (Table 5.1) (6–9). The thickness of the porous layer plays a major role in governing the porosity of the particles (9).

This chapter gives an insight about the theory to what is behind the success of shell particles and presents a summary of the latest applications from different fields.

Table 5.1   Particle Structure and Stationary Phase Chemistry of Currently Commercially Available Shell Packings

Vendor	Column/Product Name	Average Particle Diameter (μm)	Shell Thickness (μm)	Stationary Phase Chemistry
Agilent	Poroshell 300	5	0.25	C18, C8, C3
Agilent	Poroshell 120	2.7	0.50	EC-C18, SB-C18
Advanced Material Technology	Halo	2.7	0.50	C18, C8, HILIC, RP-amide, phenylhexyl, pentafluorophenyl
Advanced Material Technology	Halo Peptide-ES 160 Å	2.7	0.50	C18
Sigma-Aldrich	Ascentis Express	2.7	0.50	C18, C8, HILIC, RP-amide, phenylhexyl, pentafluorophenyl
Sigma-Aldrich	Ascentis Express Peptide-ES 160 Å	2.7	0.50	C18
Phenomenex	Kinetex	2.6	0.35	C18, XB-C18, C8,
		1.7	0.23	HILIC, pentafluorophenyl
Thermo Scientific	Accucore	2.6	0.50	C18, aQ, RP-MS, HILIC, phenylhexyl, pentafluorophenyl
Macherey-Nagel	Nucleoshell	2.7	0.5	RP-18, HILIC
Sunniest	SunShell	2.6	0.5	C18

5.2 COLUMN EFFICIENCY

The peak dispersion in chromatography is generally characterized by the theoretical plate height (H) and the number of theoretical plates (N). The treatment of the mass transfer processes and the distribution equilibrium between the mobile and stationary phase in a column lead to equations that link the theoretical plate height as the crucial column performance parameter to the properties of the chromatographic systems, such as the linear velocity of the mobile phase, the viscosity, the diffusion coefficient of analyte, the retention coefficient of analyte, column porosity, etc.

Van Deemter first proposed an equation that described the column performance as a function of the linear velocity (10). Since then several plate height and rate models were derived for liquid chromatography by numerous researchers. The most accepted equations were introduced by Giddings, Snyder, Huber and Hulsman, Kennedy and Knox, Horvath and Lin, and Yang (11–16).

Knox suggested a three-term equation to describe the dependency of the theoretical plate height of a column as a function of linear velocity:

$$H = Au^{1/3} + \frac{B}{u} + Cu \tag{5.1}$$

where A, B, and C are constants, determined by the magnitude of band-broadening due to eddy dispersion, longitudinal diffusion, and mass transfer resistance, respectively. The constant A depends on the quality of the column packing. The B and C terms of the plate height equation depend on analyte retention. The B term increases with analyte retention as more time is available for diffusion to take place in the stationary phase (surface diffusion). The C term expresses the resistance to mass transfer.

The application of reduced parameters is common in chemical engineering to compare the performance of columns in unit operations. It is useful to convert H and u to dimensionless parameters according to the following simple formulas:

$$h = \frac{H}{d_p} \tag{5.2}$$

$$v = \frac{ud_p}{D_M} \tag{5.3}$$

where h is the reduced plate height, d_p is the particle size of the column packing material, v is the reduced linear velocity, and D_M is the analyte diffusion coefficient in the mobile phase. The particular advantage of this approach is the ability to compare the performance of columns packed with particles of different sizes or structures (morphology). With the use of the reduced parameters, similar equations can be written as Eq. (5.1). These equations are commonly used to fit functions on

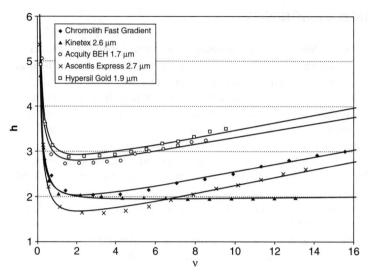

Figure 5.1 h-v plots of 2.6 μm shell type (Kinetex), 2.7 μm shell type (Ascentis Express), sub-2 μm totally porous particles (Acquity BEH and Hypersil Gold), and a monolith column (Chromolith Fast Gradient). Experiments were conducted on 5-cm-long narrow bore columns in 48/52 ACN/H2O at 35°C, DM = 1.15 × 10−5 cm²/s. From Olah, E., et al. *J. Chromatogr. A.* 2010, with permission.

the experimental data. Equation (5.4). shows a Knox type, while (Eq. 5.5). shows a van Deemter type formula:

$$h = av^{1/3} + \frac{b}{v} + cv \qquad (5.4)$$

$$h = a + \frac{b}{v} + cv \qquad (5.5)$$

Generally the *H-u* or *h-v* plots are used to compare the kinetic efficiency of the chromatographic columns. Figure 5.1 demonstrates an example of *h-v* plots of different type modern stationary phases. The characteristic properties of these plots are the minimum plate height value (H_{min} or h_{min}) and the observed optimum in linear velocity (u_{opt} or v_{opt}). The position of the minimum on the *H-u*, or *h-v* curve, and the optimum linear velocity can be determined by the use of differential calculus. The optimum linear velocity occurs when the slope of the *H* (or *h*) versus *u* (or *v*) curve is zero—that is, when d*H*/d*u* = 0 (or d*h*/d*v* = 0). This condition is satisfied in the van Deemter form when:

$$u_{opt} = \sqrt{\frac{B}{C}} \qquad (5.6)$$

The value of *H* at the optimum linear velocity can be obtained by substituting the value of *u* given in Eq. (5.6) into Eq. (5.1), (5.4) or (5.5).

In a practical point of view, h_{min} and u_{opt} are characteristic properties of the columns, but in real life we always work over the optimum (at higher linear velocity).

Van Deemter, Knox, and other h-v plots unfortunately lack permeability considerations. Alternative approaches like kinetic plots are neat tools for visualizing the compromise between separation speed and efficiency (16). It is very straightforward to map the kinetic performance potential of a given chromatographic support type by taking a representative set of H-u data and replotting them as H^2/K_{v0} versus $K_{v0}/(uH)$ instead of as H versus u (K_{v0} is the unretained component-based column permeability). Multiplying both quantities with the same proportionality constant (being the ratio of the available pressure drop, ΔP, and the mobile phase viscosity, η), the obtained values correspond directly to the minimal t_0 time needed in a column taken exactly long enough to yield a given number of N theoretical plates at the available pressure drop. N and t_0 can be calculated according to the following equations, which have been introduced by Desmet et al. (17):

$$N = \frac{\Delta P}{\eta}\left(\frac{K_{V0}}{u \cdot H}\right) \tag{5.7}$$

$$t_0 = \frac{\Delta P}{\eta}\left(\frac{K_{V0}}{u^2}\right) \tag{5.8}$$

For the construction of kinetic plots, certain defining experimental parameters are used, including the maximum operating pressure (ΔP_{max}), column reference length and flow resistance or permeability (K_{v0}), temperature, mobile phase viscosity (η), and the diffusion coefficient of the analyte in the mobile phase (D_M). Figure 5.2

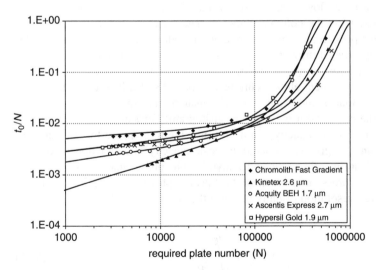

Figure 5.2 t_0/N – N type kinetic plots of 2.6 μm shell type (Kinetex), 2.7 μm shell type (Ascentis Express), sub-2 μm totally porous particles (Acquity BEH and Hypersil Gold), and a monolith column (Chromolith Fast Gradient). These plots were constructed based on the H-u data of Figure 5.1. From Olah, E., et al., *J. Chromatogr. A.* 2010, with permission.

shows the presentation of kinetic data as t_0/N versus the required plate number (N). With the help of these plots, one can easily choose the suitable column that offers the fastest separation for a given plate number. Column permeability can be determined experimentally using the following relation:

$$K_{V,0} = \frac{u \eta L}{\Delta P} \tag{5.9}$$

in which ΔP is the pressure drop over the column with length L, and u is the linear velocity.

In the obtained plot, each data point corresponds to a column with a different length, that is, the length that yields the maximally allowable pressure drop ΔP_{max} for the u value under consideration. This is automatically the fastest way to achieve the corresponding number of plates.

The hydrodynamic properties expressed by the column pressure versus flow dependency provide an insight into the flow behavior. From these data the column permeability can be calculated. The Darcy law (for laminar flow) expresses that the back pressure of a column increases in inverse proportion to the square of the particle diameter:

$$\Delta P = \frac{\phi \eta L u}{d_p^2} \tag{5.10}$$

in which ϕ is the flow resistance of the packed bed. (This value is generally between 500 and 1000 in reversed phase liquid chromatography.)

In gradient elution mode the column efficiency depends on practical conditions, such as gradient time, flow rate, and so on; therefore, new sets of parameters are used for characterizing the column efficiency. There are a number of measures of chromatographic performance in gradient elution mode, such as peak width, resolution, or peak capacity. Of these, resolution and peak capacity are relevant to gradient systems. Peak capacity, a concept first described by Giddings, was soon put to good use by Horvath for gradient elution chromatography (18, 19). It is a measure of the separation power that includes the entire chromatographic space together with the variability of the peak width over the chromatogram. For samples, which contain many components, peak capacity is a useful measure to compare the efficiency of different columns. In gradient separation, peak capacity is a function of column efficiency, gradient time, flow rate, and analyte/mobile phase characteristics.

The general expression of the theoretical peak capacity P_c in liquid chromatography, assuming a resolution of unity between the successively eluted peaks, can be written as (20):

$$P_C = 1 + \int_{t_I}^{t_F} \frac{1}{4\sigma} dt \tag{5.11}$$

where t_I is the retention time of the first eluted peak, t_F is the retention time of the last eluted peak, dt is a dummy time variable, and σ is the time standard deviation of a peak. For the practical comparison of different columns' efficiency in gradient elution mode, two experimental formulas are often used. The conditional peak capacity is directly related to the average peak resolution and is computed from experimental data as (21):

$$n_C = \frac{t_{R,n} - t_{R,1}}{w} \tag{5.12}$$

where $t_{R,n}$ and $t_{R,1}$ are the retention times of the last and the first eluting peaks, and w is the average peak width (at four sigma). Here n_c is called conditional peak capacity, because it depends strongly on all of the experimental conditions of the gradient elution program including temperature, flow rate, initial and final mobile phase compositions, as well as the column parameters and the sample's properties (22). If the peak width pattern over a chromatogram is similar, as it is in most reversed-phase gradient separations, one can use the following popular formula of experimental peak capacity (23–25):

$$P_C = 1 + \frac{t_g}{4\sigma} \tag{5.13}$$

where t_g is the gradient run time. Figure 5.3 shows the change in peak capacity as the flow rate and gradient duration are varied (20).

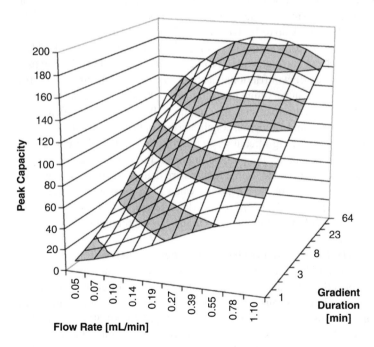

Figure 5.3   Peak capacity as a function of flow rate and gradient duration. From Neue, U.D. *J. Chromatogr. A.* 2005, with permission.

5.3   FAST LIQUID CHROMATOGRAPHY

5.3.1   Trends in Fast Liquid Chromatography

Reducing analysis time and guarantying the quality of a separation in liquid chromatography requires high kinetic efficiency. This means that peak widths must be as narrow as possible. The speed of analysis can be increased by different approaches. In liquid chromatography a revolution has started by using sub-2 μm porous particles, monolith columns, and sub-3 μm shell particles.

On very fine particles (sub-2 and sub-1 μm), due to the narrow peaks, sensitivity and separation are improved at the cost of pressure. The main advantage of this approach is that the analysis time could be reduced to a 1- or 2-min interval without the loss of resolution and sensitivity (26–28). Most commercial HPLC instruments have a maximum operating pressure limitation of 400 bar, leading to the common practice of using short columns packed with small particles to speed up the analysis (29, 30). Knox and Saleem were the first to discuss the compromise between speed (pressure drop) and efficiency (31). To overcome the pressure limitations of modern HPLC, the groups of Jorgenson (32, 33) and Lee (34) constructed dedicated instrumentation and columns to allow analysis at very high pressures. New nomenclatures have come with the terms *ultrahigh pressure liquid chromatography, ultrahigh performance liquid chromatography, or very high pressure liquid chromatography* (UHPLC, UPLC, VHPLC or vHPLC). It was done so to describe the higher back pressure requirement. The first system for ultrahigh pressure separation was released in the year of 2004. The new hardware was able to work up to 1000 bar (15,000 psi), and the system was called ultra performance liquid chromatography (UPLC^{TM}). Since then several UHPLC systems are commercialized and can work up to 1200–1300 bar (18,000–19,500 psi).

A critical aspect is the effect of frictional heating, causing significant temperature gradients within the columns at very high pressures ($p > 400$ bar). The radial temperature gradient, due to the heat dissipation at the column wall, can cause significant loss in plate count (35, 36). Gritti and Guiochon concluded that both longitudinal and radial temperature gradients are more significant when the column length is decreased (37).

On the other hand, the smaller the particle diameter, the greater the difficulty in preparing a well-packed column bed. Some reasons are that the particle aggregation, frit blockage, or particle fracture are all issues when high pressure is required to pack sub-2 μm particles (29). Guo and co-workers found that the efficiency of sub-2 μm particles for small molecules is not as high as it was theoretically predicted and widely cited (38).

Temperature in HPLC also offers a chance to shorten the analysis time. Elevating the temperature greatly reduces the viscosity of mobile phase, increases the mass transfer, and therefore allows the use of high flow rates. Analysis time can be shortened without the loss of resolution through column heating (39–42). If we consider only the pressure drop, systems with a maximum pressure capability of 400 bar can then be used with the sub-2 μm columns without overpressuring the pump. Preheating of the mobile phase is essential to avoid band-broadening. This technique is often called high temperature liquid chromatography (HTLC). However, this strategy suffers from limitations such as the small number of stable packing materials at temperatures higher than 80–90°C as well as the potential degradation of thermolabile analytes. Therefore, until now, the pharmaceutical industry has not considered this approach in everyday routine.

The third possibility to enhance the separation speed is the reduction of the intrinsic flow resistance of the column. Increasing the external porosity and the flow-through pore size of the packing could lead to fast separations. The monolith approach, originally initiated by the work of Hjertén et al. (43), Svec and Fréchet (44), Horváth and co-workers (45), and Tanaka and co-workers (46), already leads to a number of well performing, commercially available polymeric and silica monolith columns (47, 48). Commercial silica monolith columns are the Chromolith (Merck KgGa, Darmstadt, Germany) and Onyx (Phenomenex, Torrance, California, based upon technology licensed from Merck). Monoliths may have larger external porosity than packed beds. Hence, they have a higher permeability than columns packed with particles having the same size as the skeleton elements of these monolithic columns. The mass-transfer kinetics of analytes is faster through monolithic columns than through packed columns of comparable geometry and similar domain size (49). The kinetic efficiency of commercially available monolith columns is comparable to columns packed with 3–4 μm totally porous particles. On monolithic columns, the analysis time can be shortened by enhancing the flow rate of the mobile phase. Operations applying high flow rates cannot be used for mass spectrometric (MS) detection. This latter argument was the starting point of decreasing the monolith column dimensions. First the length was cut to 25 mm, after the inner diameter was decreased from 4.6 mm to 3 mm and 2 mm. Applying these narrow bore monolith columns, high speed separation and MS detection can also be done with moderate flow. The first generation of monolith columns is very promising but has some limitations: we are currently waiting for the new generation of commercially available silica-based monolith columns.

The concept of superficial or shell type stationary phases was introduced by Horváth and co-workers in the late 1960s (1, 2). Horváth applied 50 μm glass bead particles covered with styrene-divinylbenzene-based ion exchange resin and became known as pellicular packing material. Later Kirkland presented that 30–40 μm diameter superficially porous packings (1 μm phase thickness, 100 Å pores) provided much faster separations compared with the large porous particles used earlier in liquid chromatography (3). Later the core diameter was reduced, and the thickness of active layer was cut to 0.5 μm and was used for fast separation of peptides and proteins (5). Now fused-core packing materials are commercially available in different diameters (5 μm, 2.7 μm, 2.6 μm, and 1.7 μm). The 5 μm Poroshell™ particles consist of a 4.5 μm nonporous core and a 0.25 μm porous silica layer, and the 2.7 μm Halo™ or Ascentis Express™ and Poroshell-120 particles consist of a 1.7 μm nonporous core and a 0.5 μm porous silica layer. Sub-3 μm and sub-2 μm shell particles with a very thin porous layer were released in 2009 (2.6 μm and 1.7 μm Kinetex™ particles). This Core-Shell™ technology provides particles, which consist of a 1.9 μm or 1.24 μm nonporous core and a 0.35 μm or 0.23 μm porous silica layer, respectively. Other vendors launched similar sub-3 μm shell packings in 2011 (Accucore, Nucleoshell, SunShell). Studies have proven (50) that in the case of 2.7 μm fused-core packing (Halo, Ascentis Express), the peak broadening at high flow rates is larger than expected. It might be explained by the rough surface of particles in which the mass transfer rate is reduced through the outer stagnant liquid film (film diffusion) (6).

5.3.2 Adjusting Conventional Methods to Fast Separations (Geometrical Method Transfer)

To increase sample throughput, a conventional HPLC method can be easily scaled down to a UHPLC method. When transferring a separation from conventional HPLC to UHPLC, comparable method parameters must be used to maintain equivalent separations. To keep selectivity while scaling an analysis, column properties and operating conditions should stay consistent, while other parameters are optimized. It must be emphasized that the same or very similar stationary phase chemistry must be used to keep the selectivity. Differences in selectivity can be seen when the stationary phase chemistry of the two columns is not matched (not equivalent phases). However, when decreasing particle size and column dimensions, it is equally important that certain operating conditions be adjusted properly. By applying scaling factors, the modifications of the column length and/or particle dimension allow for fast analytical methods without losing resolution or sensitivity (51). This approach is sometimes called geometrical transfer (52, 53).

To avoid a detrimental extra-column band-broadening and maintain an equivalent level of sensitivity and to achieve the same separation factor in the scaled down separation, it is necessary to adapt the injection volume to the column dimension. The injected volume can be determined according to the following equation:

$$V_{i2} = V_{i1} \frac{d_{C2}^2 \cdot L_2}{d_{C1}^2 \cdot L_1}$$ (5.14)

V_i is the injection volume, d_C is the column diameter, and L is the column length. For a successful method transfer, the reduced linear velocity of the mobile phase (v) must be kept constant, because this value is independent of the column geometry. Hence, for a geometrical transfer, the new flow rate (F) can be calculated with the following equation:

$$F_2 = F_1 \left(\frac{d_{C2}}{d_{C1}} \right)^2 \frac{d_{p1}}{d_{p2}}$$ (5.15)

In linear or multilinear gradient elution, the gradient profile can be decomposed as the combination of two parts: isocratic and gradient segments. For both parts, the gradient volume should be scaled in proportion to the column volume to yield identical elution patterns. The rules for efficient gradient transfer, introduced by Snyder and Dolan (53) and updated recently by Carr and Schellinger (54), should be strictly followed. For the isocratic step (and also for the equilibrating time), the ratio between isocratic step time (t_{iso}) and column dead time (t_0) should be adapted with the help of the following equation:

$$t_{iso2} = t_{iso1} \frac{F_1}{F_2} \frac{L_2}{L_1} \left(\frac{d_{C2}}{d_{C1}} \right)^2$$ (5.16)

For slope segments, the initial and final gradient composition (%B) must be constant, and the new gradient time (t_{g2}) expressed as:

$$t_{g2} = \frac{\%B_{final1} - \%B_{final2}}{slope2} \quad (5.17)$$

Considering the equations, the new (transferred, scaled down) gradient times can be expressed as:

$$t_{g2} = t_{g1}\frac{F_1}{F_2}\frac{L_2}{L_1}\left(\frac{d_{C2}}{d_{C1}}\right)^2 \quad (5.18)$$

An example of successful method transfer can be seen in Figure 5.4. A pharmaceutical mix was separated within 30 min by using a conventional HPLC separation. A more efficient small column with the same stationary phase chemistry was applied for the

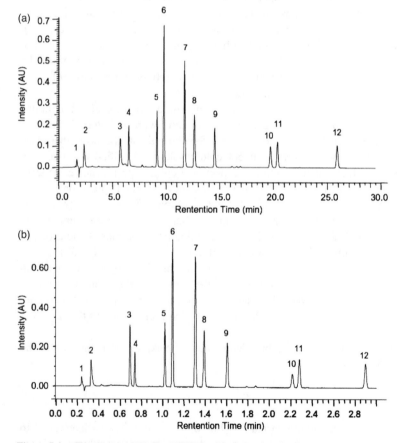

Figure 5.4   Conventional HPLC to UHPLC method transfer. Gradient separation of a pharmaceutical formulation containing the main product (peak 6) and eleven impurities in gradient mode with HPLC (a) and UHPLC (b) system. From Guillarme, D., et al. *Eur. J. Pharm. Biopharm.* 2008, with permission.

fast UHPLC separation, keeping the selectivity. After a correct transformation of chromatographic parameters, a 3-min-long fast UHPLC separation was achieved for the same sample (52).

5.4 THE IMPACT OF EXTRA-COLUMN BAND-BROADENING IN FAST LIQUID CHROMATOGRAPHY

The advent of new high-performance columns, packed with sub-2 μm particles or with the new generation of sub-3 μm shell particles, raises the problem that periodically plagued chromatography: how to minimize the contributions to band-broadening that take place in the extra-column volumes of the instrument compared to the band-broadening occurring in the column. The requirements imposed by the new generation of high-performance columns are more drastic than they ever were because (1) modern high-performance columns are more efficient than former ones; (2) accordingly, shorter columns are needed, because part of the gain in column efficiency is traded for shorter analysis time and 5-cm-long columns are becoming popular; and (3) modern columns tend to have smaller inner diameters, in part to reduce the disreputable consequences of the heat effect caused by friction of the mobile phase stream against the bed and in part to satisfy the requests of mass spectrometry for low mobile phase flow rate.

Extra-column band spreading affects the measured performance of columns packed with smaller particles, especially for columns with an internal diameter smaller than the classical standard of 4.6 mm. Several published reports that characterize the performance of very efficient columns have identified extra-column effects as a major factor that negatively impacts the column performance (55–59). These reports used either UPLC/UHPLC systems with low band spread or conventional HPLC systems with corrections for extra-column effects. In other reports, sub-2 μm particle columns with a small internal diameter have been compared to columns packed with 2–3 μm particles in larger diameter columns without taking into account the detrimental effects of extra-column band spreading (60–62).

There are two types of extra-column contributions. The first is volumetric in nature and derives from the injection volume, the detector volume, and the volume of the connection tubing between the injector and detector. The second extra-column contribution to band spreading stems from time-related events, such as the sampling rate and the detector time constant.

Generally, extra-column band-broadening could be expressed as the sum of the main dispersion sources:

$$\sigma_{V,ext}^2 = \sigma_i^2 + \sigma_d^2 + \sigma_t^2 \qquad (5.19)$$

where $\sigma_{V,ext}^2$ is the extra-column variance, while σ_i^2, σ_d^2, and σ_t^2 are variances due to the injector, the detector, and the tubing, respectively. Thus, extra-column variance depends on the injected volume V_i, the tubing radius r and length l, the flow-cell

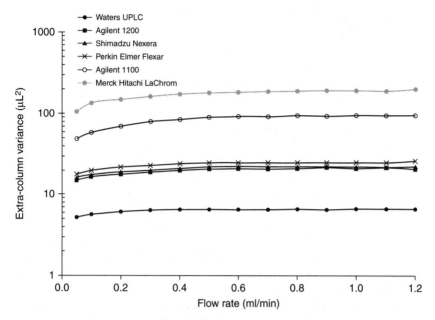

Figure 5.5 Plots of extra-column variance versus mobile phase flow rate. *Instruments:* Waters Acquity UPLC, Agilent 1200, Agilent 1100, Shimadzu Nexera, Perkin Elmer Flexar, Merck Hitachi LaChrom. The possible maximum acquisition rate was set on each instrument (10 Hz on Merck Hitachi LaChrom system, 80 Hz on Agilent 1100 and 1200 systems, and 100 Hz on Shimadzu Nexera, Perkin-Elmer Flexar, and Waters Acquity systems). From Fekete, S., Fekete, J. *J. Chromatogr. A.* 2011, with permission.

volume V_c, the detector time constant τ, and the flow rate F and can be estimated according to the following generally used equation given below:

$$\sigma^2_{V,ext} = K_i \frac{V_i^2}{12} + K_c \frac{V_c^2}{12} + \tau^2 F^2 + \frac{r^4 \cdot l \cdot F}{7.6 \cdot D_M} \tag{5.20}$$

in which K_i and K_c are constant and linked to the injection mode and the detector cell geometry, respectively.

According to a recent study, the commercially available LC systems can be classified in three groups: (1) optimized systems for fast separation with very low dispersion ($\sigma^2_{V,ext} < 10\ \mu L^2$); (2) hybrid LC systems recommended for both fast and conventional separations ($\sigma^2_{V,ext} = 10\text{--}30\ \mu L^2$); and (3) conventional LC systems with an extra-column variance over $50\ \mu L^2$ (Figure 5.5). These major differences in extra-column peak variance have a significant impact on measured column performance and achievable analysis time (63). Further improvements in instrument design (smaller dispersion) are necessary to take the full advantage of the most recent very efficient small columns. Today it is not always possible to utilize the potential of these small columns. The loss in efficiency can reach 30%–55% with commercially available optimized UHPLC systems (63).

5.5 SHELL PARTICLES, THE INFLUENCE OF SHELL THICKNESS

Shell particles are made of a solid, nonporous core surrounded by a shell of a porous material that has properties similar to those of the fully porous materials conventionally used in HPLC. Figure 5.6 shows the general structure of shell particles.

The volume of the porous shell surrounding a particle is $4\pi(R_e^3-R_i^3)/3$, where R_e and R_i are the radius of the particle and of its solid core, respectively. The volume fraction of the porous material of the shell in the column is $1-(R_i/R_e)^3$. When the layer of porous material around the solid core is thin, this fraction becomes close to $3(1-(R_i/R_e))$. When the thickness of this layer becomes significant, the volume fraction of porous material in the particle becomes large and eventually tends toward unity. For example, the shell thickness of the Halo or Ascentis Express particles ($R_e = 1.35$ μm, $R_i = 0.85$ μm) is 0.5 μm while the shell thickness of the Kinetex particles ($R_e = 1.3$ μm, $R_i = 0.95$ μm) is 0.35 μm. The volume fractions of the porous shells in these particles are 75% and 63%, respectively. Figure 5.7 demonstrates the relationship between the fractional volume and the relative shell thickness ($e = (R_e-R_i)/R_e$) (4).

The most important differences between the characteristics of columns packed with core shell and with porous particles is in their hold-up times, retention factors, and their loading capacities.

Generally, the acceptable loading capacity for a packed analytical column can be estimated as (64):

$$M_{\text{max}}^{10\%} = \frac{2}{N}\left[\frac{k_\infty + 1}{k_\infty}\right]^2 \beta V_S \tag{5.21}$$

where $M_{\text{max}}^{10\%}$ is the maximum loading capacity for a solute for a 10% plate height increase, N is the theoretical plate number for the column, k_∞ is the retention factor at infinite dilution, β is a constant dependent of the phase ratio and type of solute, and V_s is the volume of the stationary phase for the packings. This equation shows that the sample loading capacity is proportional to the stationary phase volume. The stationary

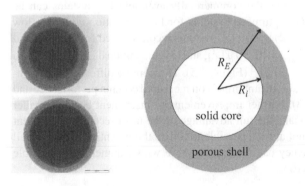

R_E

R_i

solid core

porous shell

Figure 5.6   Scanning electron microscopic (SEM) images (left-hand side) and schematic structure (right-hand side) of core-shell particles. From Omamogho, J.O., et al. *J. Chromatogr. A.* 2011, with permission.

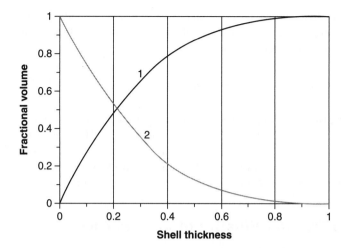

Figure 5.7   Fractional volumes in the core-shell particles as a function of the relative shell thickness (solid-core to core-shell ratio (ρ)), fractional volume of the shell (1) and fractional volume of the core (2). From Guiochon, G. *J. Chromatogr. A.* 2011, with permission.

phase volume in reversed-phase liquid chromatography is in turn proportional to the carbon-loading percentage and the density of the packing.

The separation power of shell particles increases with decreasing shell thickness if the strength of the mobile phase is decreased to compensate for the retention change caused by the decreased surface area of the stationary phase (65). On the other hand, if the mobile phase remains the same, the retention factor will decrease, and therefore, the resolution of low molecular size compounds may also decrease. The smaller the diffusivity of the solutes, the larger the increase of the separation power is, compared to that of fully porous particles. On one hand, the theory suggests that the thickness of the porous layer should be decreased drastically (down to $\rho = 0.90$–0.95) to increase the separation efficiency of the columns for large molecular size compounds. On the other hand, there is a strict limitation in decreasing the thickness of the porous layer, because decreasing the shell thickness decreases markedly the loadability of the column, making column overload easily, broadening the bands, and decreasing the separation efficiency (65). Therefore, the optimum shell thickness in reality is likely to be a compromise between efficiency, sample loading capacity, and analyte retention and is strongly sample dependent. Overload problems are likely to be more severe for both sub-2 μm porous as well as shell particles, due to the very high efficiencies produced by both types of column (66).

5.6 THE EFFICIENCY OF COLUMNS PACKED WITH SHELL PARTICLES

Different plate height models are written as the sum of different contributions to peak broadening such as (1) longitudinal diffusion, (2) eddy dispersion, (3) the external film mass transfer, (4) the transparticle mass transfer resistance, and currently (5) the frictional heat term. Therefore the overall reduced theoretical plate height (reduced HETP, h) generally can be written as:

$$h = h_{eddy} + h_{longitudinal} + h_{film} + h_{transparticle} + h_{heat} \qquad (5.22)$$

The initial idea of preparing shell particles (pellicular) was to increase the column efficiency by reducing the mass transfer resistance across the particles. However, now it seems that transparticle mass transfer resistance is far from being the dominant contribution to band-broadening in HPLC (67). The columns packed with the new generation of shell particles are also enormously successful, but for other reasons (4).

5.6.1 The Mass Transfer Kinetics of Shell Particles

The mass transfer resistance term is the sum of two coefficients such as the external or film mass transfer coefficient, h_{film}, and the transparticle mass transfer coefficient, $h_{transparticle}$. The former accounts for the difficulties encountered by analyte molecules to penetrate into the network of mesopores inside the particles, and the latter, for the time that it takes for them to diffuse across this network, once they have entered into it. The transparticle mass transfer resistance for shell particles was derived by Kaczmarski and Guiochon (68). According to this theory the intraparticle diffusivity depends on the ratio ($\rho = R_i/R_e$) of the diameter of the solid core to that of the particle in a core–shell particle (Figure 5.6). As this ratio increases, the mass transfer kinetics becomes faster through the shell particles.

The transparticle mass transfer resistance is given by the following equation (68):

$$h_{transparticle} = \frac{\varepsilon_e}{1 - \varepsilon_e} \frac{k_1^2}{(1 + k_1)^2} \frac{1}{30\Omega} \frac{1 + 2\rho + 3\rho^2 - \rho^3 - 5\rho^4}{(1 + \rho + \rho^2)^2} v \qquad (5.23)$$

where Ω is the ratio of the intraparticle diffusivity of the sample through the porous shell to the bulk diffusion coefficient. This equation is consistent with the one applied for totally porous particles when $\rho = 0$. As ρ increases, the apparent intraparticle diffusivity of the probe studied increases.

The Wilson-Geankoplis correlation provides a convenient estimation of the film mass transfer coefficient, k_f :

$$Sh = \frac{k_f d_p}{D_m} = \frac{1.09}{\varepsilon_e^{2/3}} v^{1/3} \qquad (5.24)$$

where Sh is the Sherwood number, d_p is the average particle size, D_m is the molecular diffusivity, ε_e is the external porosity, and $v = u d_p/D_m$ is the reduced interstitial linear velocity of the mobile phase. The external film mass transfer term was derived from

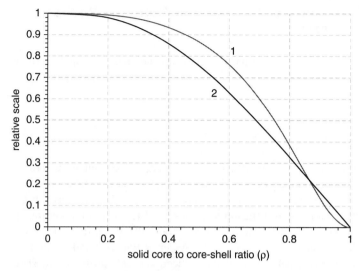

Figure 5.8   Variation of the external film mass transfer (1) and the transparticle mass transfer resistance (2) as a function of solid-core to core-shell ratio (ρ).

the Laplace transform of the general rate model equations (69). It can be expressed as:

$$h_{film} = \frac{\varepsilon_e}{1 - \varepsilon_e} \frac{k_1^2}{(1 + k_1)^2} \frac{1}{3Sh} v \tag{5.25}$$

where k_1 is the retention factor of the analyte, and for shell particles it can be written as (69):

$$k_1 = \frac{\varepsilon_e}{1 - \varepsilon_e} \left(\varepsilon_P + \frac{1 - \varepsilon_P(1 - \rho^3)}{1 - \rho^3} K_a \right) (1 - \rho^3) \tag{5.26}$$

where ε_p is the porosity of the porous shell of the particle, and K_a is the Henry constant of adsorption on the walls of the porous shells. Remarkably, this contribution does seem to be dependent on the structure of the particles.

Figure 5.8 demonstrates the effect of the solid-core to core-shell ratio (ρ) on the external film mass transfer and on the transparticle mass transfer resistance.

According to some recent experimental measurements, the mass transfer kinetic is mostly accounted for the external film mass transfer resistance across the thin layer of the mobile phase surrounding the external surface area of the particles (4). This suggests that the initial idea of preparing shell or superficially porous particles with the purpose to increase the column efficiency by reducing the mass transfer resistance across the particles might provide only modest practical gains for the separation of low or medium molecular weight compounds (4).

5.6.2 The Longitudinal Diffusion in Columns Packed with Shell Particles

The reduced longitudinal diffusion term ($h_{longitudinal}$) for shell particles can be written by the following equation (70):

$$h_{longitudinal} = 2\frac{\gamma_e + (1 - \varepsilon_e)\dfrac{1 - \rho^3}{\varepsilon_e}\Omega}{\nu} \tag{5.27}$$

where γ_e is the obstruction factor for diffusion in the interparticle volume. According to this equation the longitudinal diffusion is more favorable as the thickness of the porous shell is reduced. The presence of a solid core inside the particles has a direct consequence on the longitudinal diffusion term observed for a column, because it decreases this contribution to the plate height by about 20% when the ratio of the core to the particle diameter is $\rho = 0.63$ (Halo, Ascentis Express) and about 30% when $\rho = 0.73$ (Kinetex) (4, 70). However, the reduced internal porosity of the shell particles brings a limited improvement in their efficiency because the longitudinal diffusion coefficient of columns packed with shell particles is smaller than that of columns packed with fully porous particles. This causes at best a gain of 0.2 reduced plate height (h) unit, that is, a 10% increase in the total column efficiency compared to that of columns packed with fully porous particles (4, 70).

Recently Desmet and Deridder transformed the effective medium theory (EMT), which applied thermal and electrical conductivity, to determine longitudinal diffusion in chromatography (71). EMT equations can be applied for fully porous, porous shell, spherical, and cylindrical particles. The theory considers the column as a binary medium, which consists of an interstitial void with a volumetric fraction ε_e and particles with a volumetrical fraction of $1 - \varepsilon_e$.

Transposing the conductivity with permeability in the basic Maxwell expression, the following equation can be derived, which describes the B term of band-broadening:

$$B = 2\gamma_{eff}(1 + k) = \frac{2}{\varepsilon_T}\frac{1 + 2\beta_1(1 - \varepsilon_e)}{1 - \beta_1(1 - \varepsilon_e)} \tag{5.28}$$

where γ_{eff} is the proportion of effective diffusion coefficient and diffusion coefficient in the mobile phase.

Polarizability constant can be defined as:

$$\beta_1 = \frac{\alpha_{part} - 1}{\alpha_{part} + 2} \tag{5.29}$$

where α_{part} is the relative particle permeability.

In case of shell particles, the column is not a binary but a ternary medium with a mobile phase, solid, impermeable core, and the shell part. The fraction of the non-retained species and the mobile phase is ε_{pz}, while the fraction of the whole particle is ε_{part}. The fraction of the whole particle for the spherical particle can be expressed by the following equation:

$$\varepsilon_{part} = (1 - \rho^3)\varepsilon_{pz} \tag{5.30}$$

Desmet and Deridder (71) recently defined the intra-particle diffusion as:

$$D_{part} = \frac{2}{2 + \rho^3}D_{pz} \tag{5.31}$$

It was implied that the solid core reduced the B term not more than 34% in comparison with the fully porous particle (72, 73).

As a conclusion, it can be stated that recent core-shell particles manifest a gain of approximately 20%–30% in the longitudinal diffusion.

5.6.3 The Eddy Dispersion in Columns Packed with Shell Particles

The eddy dispersion term (h_{eddy}) includes sources of four different origins, differing in the length scale considered, for example, the transchannel ($i = 1$), the short-range inter-channel ($i = 2$), the long-range inter-channel ($i = 3$), and the transcolumn flow heterogeneities ($i = 4$). For the eddy dispersion, a general expression is given by (74, 75):

$$h_{eddy} = \sum_{i=1}^{i=4} \frac{1}{\dfrac{1}{2\lambda_i} + \dfrac{1}{\omega_i \nu}} \cong 2\sum_{i=1}^{i=4} \lambda_i \tag{5.32}$$

The values of $\lambda_1 - \lambda_3$ were estimated by Giddings (74). The value of λ_4 can be derived from the flow distribution across the column diameter. For quadratic flow profile distributions, the following expression was derived as (74):

$$\lambda_4 = \frac{2}{45}\frac{L}{d_p}\omega_{\beta,c}^2 \tag{5.33}$$

where $\omega_{\beta,c}$ is the relative flow velocity difference between the center and the wall of the column, and L is the column length.

According to several experimental results, the eddy diffusion terms of the column packed with shell particles is significantly smaller ($\sim$30%–40%) than that of the column packed with fully porous particles (6, 8, 70, 76–78).

It is still unclear whether this is due to the particle size distribution (PSD) of shell particles, which is significantly narrower than that of fully porous particles (Figure 5.9). Some recent studies, which focused on particles with a different design such as the superficially porous particles, have suggested that particles displaying a very narrow PSD can lead to unprecedented low minimal plate heights (60, 67). It is,

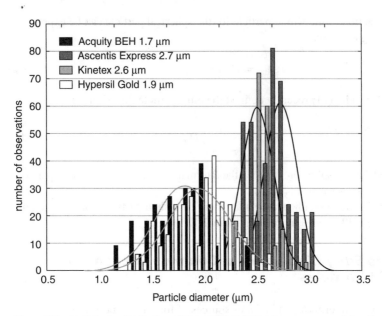

Figure 5.9   Particle size distribution of Waters UPLC BEH 1.7 μm porous particles, Hypersil Gold 1.9 μm porous particles, Ascentis Express 2.7 μm shell particles, and Kinetex 2.6 μm shell particles. From Olah, E., et al. *J. Chromatogr. A.* 2010, with permission.

however, uncertain whether this finding can be purely related, because there are also other factors that might influence the packing quality. Superficially porous particles have a higher density, and some of them are rougher than fully porous particles (60, 79). This might also have had an influence on the achieved packing quality, apart from the PSD.

A strong (nearly linear) correlation has been observed between the width of the particle size distribution of several commercially available HPLC particle types (both fully porous and superficially porous) and some commonly used parameters that reflect the quality of a packing, namely the minimum reduced plate height (77). These observations have been made despite the fact that the studied particles have a number of other differences besides PSD, such as particle porosity, pore size, pore structure, and bonding conditions. Covering a wide group of fully porous as well as porous-shell particles, these observations confirm the most recent views in the field, stating that there is a strong relation between the particle size distribution of the particles and the quality of the packing.

5.6.4 The Frictional Heating in Columns Packed with Shell Particles

Frictional heating generates heat everywhere across the column, resulting in a heterogeneous temperature distribution along and across it, which can severely decrease its efficiency. The additional reduced plate height due to this frictional heating was investigated from theoretical and experimental points of view (80–87). This additional h term was measured for narrow-bore columns (2.1 mm × 150 mm) packed with 1.7 μm Kinetex-C18 shell particles and with BEH-C18 fully porous particles. The large h_{heat} values observed with the fully porous particles are directly related to the smaller heat conductivity of a bed packed with fully porous particles than that of a bed packed with shell particles (85, 87). As a result, the radial temperature gradients are markedly larger across the columns packed with porous particles than across the columns packed with shell particles. The difference between the thermal conductivities of the two different types of packing is due to the thermal conductivity of the solid silica core being much larger than that of porous silica impregnated with an organic eluent like acetonitrile (4).

5.6.5 Limited Efficiency When Shell Particles Packed in Narrow-Bore Columns

The brands of shell packing materials made of fine particles are available in both conventional (4.6 mm and 3 mm i.d.) and narrow-bore (2.1 mm i.d.) columns. It is a general observation that the efficiency of the former tends to be markedly higher than that of the latter. It was shown that the landmark performance of columns packed with the Kinetex 2.6 μm particles ($h_{min} = 1.2$–1.3) is only limited to the standard bore column (4.6 mm i.d.); however, when packed in a narrow-bore column (2.1 mm i.d.), the reduced plate height of 1.9 was the minimum achieved (8). This suggests that the packing of narrow-bore columns does not provide comparable packed bed homogeneity to that of the standard-bore columns. Gritti and Guiochon studied the mass transfer kinetics of the Kinetex 1.7 μm C18 material packed in a 2.1 mm i.d. column, and the minimum reduced plate height above 2.0 was obtained (88). This provides further suggestion that the problematic situation of packing narrow-bore columns is compounded when the packing materials are very fine such as the sub-2 μm particles. The difference in efficiency is accounted for a contribution to the column HETP of the long-range eddy diffusion term that is larger in the 2.1 mm than in the 4.6 mm i.d. columns (89). While the associated relative velocity biases are of comparable magnitude in both types of columns, the characteristic radial diffusion lengths are in the order of 100 and 40 μm in the wall regions of narrow-bore and conventional columns, respectively (89).

Another observation is that the 4.6 mm i.d. beds packed with 2.6–2.7 μm shell particles are more homogeneous than those of the 2.1 mm i.d. narrow-bore beds packed with 1.7 μm fully porous particles (89). The external roughness of the shell particles might explain the origin of this advantageous property because the shear stress that takes place during the slurry packing process is stronger between rugged particles than between smooth ones. Therefore, particles move less by respect to each other, and the amount of strain occurring through the bed is smaller. Thus, the distribution of the external porosity throughout the bed of rugged particles is more homogeneous from the center (low packing stress) to the wall of the column (high packing stress) than through beds of smooth particles (90).

5.6.6 The Success of the New Generation Shell Particles

Most of the success of this new generation of sub-3 μm particles lies in the unexpectedly low minimum reduced plate height (h_{min}) in the range of $h_{min} = 1.2$–1.7 instead of 2.0–2.5 for the same columns packed with fully porous particles (4). Unusually, the exceptional performance of 4.6 mm i.d. columns packed with the last generation of sub-3 μm shell particles is not caused by the reduction of the sample diffusion path across these particles, which was the initial motivation for commercializing shell particles since the early 1970s (4). The actual advantages of columns packed with these new shell particles lie in the decrease of both the longitudinal diffusion coefficient (-20%–30%) and the eddy dispersion term (-30%–40%). We can conclude that recent shell packings can be applied for very efficient separations of both small- and macro-molecules.

5.7 FAST SEPARATIONS BY APPLYING THE NEW GENERATION OF SHELL PARTICLES

The key concept of porous-shell particles is to increase both the efficiency and the separation speed by reducing the mass transfer across the particles while keeping their diameter large enough to avoid pressure limitations at high linear velocities (91). Generally, kinetic plots are used to represent the compromise between separation speed and efficiency (see Section 5.2). The kinetic plots show at a glance in what range of efficiencies a given support can yield faster separations than another support type.

Several studies compared the efficiency of columns packed with sub-3 μm core-shell, sub-2 μm porous particles, and monolithic columns in isocratic elution mode (6, 8, 9, 76). According to these results, currently the columns packed with sub-3 μm core-shell particles provide the most favorable plate time (t_0/N) values (or impedance time) and therefore offer the shortest analysis time when small test compounds, pharmaceutical test analytes, peptides, or proteins are separated (8, 76).

In everyday practice, chromatographers usually work above the optimum linear velocity. Therefore, it is necessary to emphasize that in the range of high linear velocities, the shell type Kinetex column performs higher plate numbers than other core-shell packings (Ascentis Express or Halo), or columns packed with totally porous sub-2 μm particles, and offers more efficient separation when fast analysis is required. The nominal particle size of the Kinetex material is 2.6 μm; therefore, the pressure drop is much lower compared to sub-2 μm columns, thus higher linear velocity can be applied during the analysis.

When comparing the 2.7 μm Ascentis Express or Halo particles (0.5 μm porous silica layer) to the 2.6 μm Kinetex particles (0.35 μm porous silica layer), the

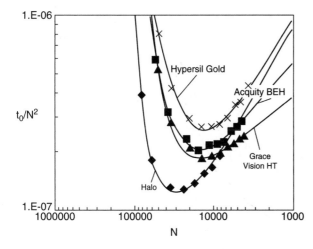

Figure 5.10 Impedance time plots of a steroid separation obtained with columns packed with fully porous sub-2 μm particles (Grace Vision HT C18, Waters Acquity BEH C18, and Hypersil Gold C18) and with a column packed with 2.7 μm shell particles (Halo C18). From Fekete, S., et al. *J. Pharm. Biomed. Anal.* 2009, with permission.

latter perform lower C terms for both small and large molecules. The intra-particle diffusivity depends on the ratio (ρ) of the diameter of the solid core to that of the particle in a core-shell particle. As this ratio increases, the mass transfer kinetic becomes faster through the shell particles than it is through totally porous particles.

Please note that the achievable plate time values depend on the maximum allowable pressure drop. A column can offer faster separation if it has a stationary phase with stronger mechanical stability (higher ΔP_{max}) than the column that has lower mechanical stability.

Figure 5.10 shows an example of the achievable separation speed of pharmaceuticals. Impedance time plots of a steroid separation obtained with columns packed with fully porous sub-2 μm particles (Grace Vision HT C18, Waters Acquity BEH C18, and Hypersil Gold C18) and with a column packed with 2.7 μm shell particles (Halo C18) are compared (6). This figure obviously shows that the larger particle diameter shell particles offer faster analysis than the smaller fully porous particles. In practice, the columns packed with 2.6–2.7 μm shell particles provide the same or sometimes slightly higher plate counts than the totally porous 1.5–1.9 μm particles. Sub-3 μm core-shell columns (2.6 μm, 2.7 μm) generate only approximately half the back pressure of the sub-2 μm fully porous particles under their own optimal flow rates. Therefore, it is easy to see that the shell packing with equivalent or higher kinetic efficiency and more favorable permeability can offer significantly faster separation than the totally porous packings.

Another example is shown in Figure 5.11. This figure demonstrates the separation speed in the case of a peptide separation (76). This example also confirms the excellent kinetic efficiency of recent core-shell technology. Please note that in this case the achievable separation time of a sub-2 μm shell packing was compared to a sub-2 μm fully porous and a sub-3 μm other shell packing. The columns packed with shell particles outperformed the fully porous material in this example.

As a conclusion it can be stated that the columns packed with the new generation of shell particles can provide faster separations than is possible using columns packed

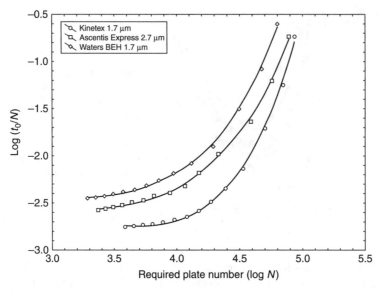

Figure 5.11 t_0/N–N type kinetic plots of a polypeptide. Experiments were performed on 5-cm-long narrow-bore columns: 1.7 μm Waters Acquity BEH C18, 2.7 μm Ascentis Express C18, and 1.7 μm Kinetex C18. From Fekete, S., et al. *J. Pharm. Biomed. Anal.* 2011, with permission.

with fully porous particles. Shell particles propose very fast separations for small pharmaceuticals, peptides, and even for moderate sized proteins.

5.8 APPLICATIONS OF COLUMNS PACKED WITH THE LATEST GENERATION OF SHELL PARTICLES

The first new generation shell packing was debuted in 2006. Since then, several applications were published, and the new core-shell technology is frequently referred in papers and presentations. This fact also proves the success of recent core-shell technology. This section reviews the latest applications of shell columns in the field of pharmaceutical, bio, environmental, and food analysis.

5.8.1 Pharmaceutical and Bioanalytical Applications

There are challenges to maximizing the exclusive patented lifetime of new medicines. It is, therefore, essential that compounds unlikely to reach the market are eliminated as early as possible (92). This demands the ability to make informed decisions regarding compounds in accordance with tight deadlines. A contributory part of the required information is a knowledge of the metabolism and pharmacokinetic properties of potential drug candidates with the implication that fast generic methods for pharmaceutical bioanalysis are essential (92). In a drug discovery bioanalysis laboratory, a huge number of samples derived from studies with a high number of widely differing compounds require analysis within short timelines, although often there will be not a lot of samples per compound. The lack of time available for custom method development means that generic methodologies have to be established, and as these methods have to retain and elute a wide diversity of analyte chemistries, which exhibit a range of lipophilicities, gradient LC procedures are required (92). The use of shell particles has dramatically improved chromatographic peak efficiencies over fully porous particles in reversed-phase chromatography (6, 79, 93) as well as in hydrophilic interaction (HILIC) separation mode (5, 94), both in gradient and in isocratic elution mode. The use of shell particles is a relatively recent trend in chromatographic separation; several pharmaceutical and bioanalytical applications can be found in the literature.

Mycotoxin zearalenon and its related metabolites were determined in urine using an Ascentis Express C18 shell type analytical column (95). A new method of high-throughput LC-MS/MS using a HALO C18 analytical column was developed to quantify posaconazole in human plasma (96). Imipramine and desipramine were assayed in rat plasma using a conventional as well as a HALO C18 analytical column (97). Shell type stationary phases were confirmed to be valuable tools also for bioanalytical purposes, as thousands of protein-precipitated plasma extracts could be measured with acceptable precision and accuracy (92). An LC-MS/MS method for remifentanil quantification from human plasma was developed and validated using a C18 shell type stationary phase. This method was applied fruitfully for remifentanil determination in clinical samples (98). A screening method for the determination of efavirenz in human saliva has been developed and validated based on high performance liquid chromatography tandem mass spectrometry. The analytes were separated on a Kinetex C18 column. The total run time of the analysis was 8.4 min (99). A sensitive bioanalytical assay for the quantitative determination of tamoxifen and five of its metabolites in serum was described through a core-shell column. The method was introduced to support clinical studies in which patient-specific dose optimization was performed based on serum concentrations of tamoxifen and its metabolites (100).

Applications to the HPLC impurity profiling of drug substance candidates were performed using a column packed with the latest generation of shell particles (Figure 5.12). The chromatographic performance of Ascentis Express C18 column was compared to porous sub-2 μm particle columns. It was found that the shell particles, bonded with C18 alkyl chains, had a very similar selectivity to the sub-2 μm Zorbax C18 phase, but provided a better shape selectivity. Solute capacity and the overall retention were slightly compromised compared to the porous sub-2 μm particles. The key advantages of the shell particle columns for pharmaceutically relevant analyses are their substantially lower back pressures, which allow them to be used at much higher flow rates than porous sub-2 μm particle phases for fast LC applications, or longer columns can be used to improve the efficiency without exceeding the capabilities of conventional HPLC equipment (101). The serum pharmacokinetics of kynurenic acid amide was studied. Chromatographic separations with the HPLC-MS system were performed on a Kinetex C18 column applying isocratic elution (102). Very fast and efficient impurity profiling of ethinylestradiol-containing tablets were

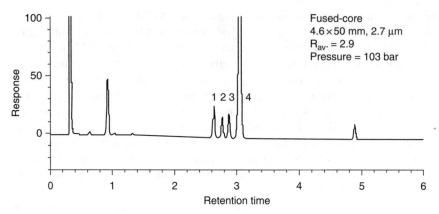

Figure 5.12   Chromatograms of a pharmaceutical intermediate and its impurities. From Abrahim, A., et al. *J. Pharm. Biomed. Anal.* 2010, with permission.

developed by using a short, narrow bore Kinetex column (8). A fast isocratic separation, which can be applied for simultaneous determination of nine steroid API residues in support of cleaning control analysis in the pharmaceutical formulation area (Pilot Plant), was also reported by the same group (Figure 5.13) (8). A systematic study concluded that 5-cm-long narrow-bore columns packed with sub-3 μm shell

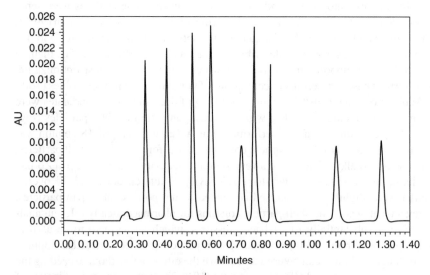

Figure 5.13   Column: Kinetex C18 2.6 μm (150 × 3.0 mm); mobile phase: acetonitrile-water 47–53 (v/v%); flow: 1.8 ml/min (p = 516 bar); column temperature: 60°C; injection volume: 1μl; detection: 215 nm; analytes: steroids (neutral polar API) swabbed from stainless steel model surface. From Olah, E., et al. *J. Chromatogr. A.* 2010, with permission.

particles offer the chance of very fast and efficient gradient separations in routine pharmaceutical analysis (103).

Very efficient quantitative methods for pharmacokinetic studies with clopidogrel were reported. The chromatographic separations were performed on Ascentis Express C8 and Ascentis Express RP Amide phases (104). A simple and enantioselective method for the separation and determination of carnitine enantiomers in dietary supplements and pharmaceutical formulation samples is described. The method is based on achiral liquid chromatographic separation of carnitine enantiomers from interferences and direct circular dichroism (CD) detection (105). A new RP-HPLC method using Ascentis Express C8 column for the determination of process impurities and degradation products of atazanavir sulfate drug substances was also reported (106). High-quality, ultra-fast bioanalytical LC-MS/MS methods were developed using short columns packed with fused-core particles and high (1.0– 3.0 mL/min) flow rates. Generic HPLC methods developed with these columns are very useful for a wide range of analytes in early discovery studies when method development is undesirable or often unnecessary. These generic methods have also been adopted for all LC-MS/MS screening assays used in drug metabolism studies (107). A fast chromatographic separation (applying partially porous packings) coupled with condensation nucleation light scattering detection (CNLSD) for the quantification of polysorbate 20 and unbound PEG from protein solutions was developed and reported. Adequate separation of these compounds was achieved within 7 min. The time-reducing and solvent-saving characteristics of the fast separation is exceptionally beneficial, compared to the most widely used conventional HPLC technique. Generally the separation of polysorbate and PEG compounds from protein origin peaks can be achieved within 10 min depending on the characteristic of protein and other excipients (matrix components) (108).

5.8.2 Food Analytical Applications

A multiresidue method for the analysis in egg matrices of residues of nine quinolones used in veterinary medicine has been developed (109). Two chromatographic columns were compared in that study: a conventional fully porous (Inertsil C8) column and a Kinetex C18 core-shell column. Separation with the latter resulted in a significant reduction in solvent consumption and in analysis time, combined with good resolution for all analytes and better sensitivity (109).

Another group presented that supplanting a conventional fully porous particle column with a solid core particle column reduced the analysis time from 15 to 5 min and significantly improved the resolution when paralytic shellfish toxins were analyzed (110).

An improved high-resolution mass spectrometry-based multiresidue method for veterinary drugs in various food matrices was reported by Kaufmann et al. (111). Their new method covers more than 100 different veterinary drugs. Validated matrices included muscle, kidney, liver, fish, and honey. A 15-cm-long Kinetex C18 column was applied in this study (111).

A Kinetex PFP (penta-fluoro-phenyl) column was used with appropriate selectivity and efficiency to analyze a total set of 237 samples of malting barley, malt, hop, wort, and beer (112).

For the analysis of fungicides in various samples such as grapes, musts, pomaces, lees, distilled spirits, and wines, a specific and sensitive method was described. The core-shell type analytical column (Kinetex C18) performed adequate separation (113). The Kinetex C18 column showed success also in the quantification of rutin, epicatechin, catechin, and epicatechin gallate in buckwheat groats (114).

Corticosteroids (such as prednisolone, methylprednisone, flumetasone, dexamethasone, and methylprednisolone) in raw bovine milk and pig fat samples were recently identified and quantified with new methods by applying Ascentis Express C18 columns (115,116).

The antioxidant profiles of various espresso coffees were established using HPLC with UV-absorbance detection and two rapid, simultaneous, online chemical assays that enabled the relative reactivity of sample components to be screened (117). The assays were based on (1) the color change associated with reduction of the 2,2-diphenyl-1-picrylhydrazyl radical and (2) the emission of light (chemiluminescence) upon reaction with acidic potassium permanganate. Results from the two approaches were similar and reflected the complex array of antioxidant species present in the samples. However, some differences in selectivity were observed. Chromatograms generated with the chemiluminescence assay contained more peaks, which was ascribed to the greater sensitivity of the reagent toward minor, readily oxidizable sample components (Figure 5.14).

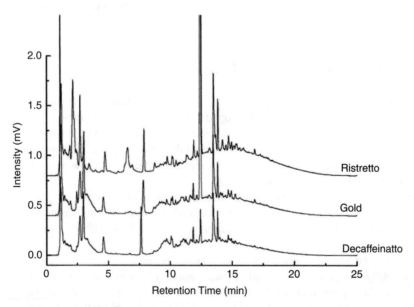

Figure 5.14   Chromatograms for separation on Kinetex column and UV-absorbance detection, of Ristretto, Gold, and Decaffeinatto café espresso samples. From Mnatsakanyana, M., et al. *Talanta*. 2010, with permission.

A sensitive, selective, and fast LC-MS/MS method and a simple sample treatment were proposed for the analysis of amprolium in food samples. Hydrophilic interaction liquid chromatography allowed the retention of this cationic coccidiostat. The method provided an efficient chromatographic analysis in less than 3 min using isocratic elution mode (118).

A new LC/APPI-MS/MS method was developed to identify and quantify simultaneously the major mycotoxins included in EU 1881/2006 Regulation (119). The authors used a 10-cm-long narrow bore Kinetex C18 chromatographic column.

5.8.3 Environmental Analytical Applications

A new method was developed for the separation of three known and ten new anthraquinone pigments. In addition, five new pigments were determined by fourier transformation mass spectrometric detection (FTMS) as co-eluting impurities (120). The analyses were performed on a Kinetex C18 column.

A fast and sensitive HPLC-MS method was worked out and validated for the analysis of fifteen prescription pharmaceuticals and four of their metabolites in influent wastewater (121). The selected pharmaceuticals belonged to various classes, such as angiotensin-converting enzyme inhibitors, angiotensin receptor antagonists, calcium antagonists, β-blockers, antidepressants, analgetics, anticonvulsants, platelet anti-aggregants, and cholesterol-lowering agents. The chromatographic separation was optimized to achieve suitable retention times, good resolution for analytes susceptible of mass spectrometric cross talk, and high sensitivity in one single run. All compounds eluted within 9 min.

The results obtained on Passiflora tincture samples also confirm that columns packed with the latest generation of shell particles achieve significant improvement in total analysis time and chromatographic efficiency (122). These properties are particularly relevant for LC-MS systems leading to a reduction in the number of co-eluting species by providing a higher spectral purity for a wider knowledge of the analyzed samples.

Two recent studies systematically compared the efficiency of several reversed-phase columns (fully porous, partially porous, and monolithic) in the separation of microcystins nodularins and cyclic peptidic hepatotoxins (123, 124). Many of the tested short, narrow-bore, reversed-phase columns produced excellent results in those separations even on a traditional low-pressure gradient HPLC system. The solvent-saving benefits can be received without compromising the quality of the separation. The typical run time ranged between 3 and 4 min.

5.8.4 Multidimensional Separations

Comprehensive separation of complex mixtures is a difficult challenge due to the presence of several components that vary from polar to nonpolar and from very low to high concentrations and that show diverse physicochemical properties (acid-base properties, stability, solubility, detectability). The potential of conventional separation techniques such as liquid chromatography and detection approaches like UV or mass spectrometric detection are limited. In the last decade, comprehensive multidimensional separation techniques such as LC×LC have been developed and reported. These multidimensional techniques offer an enormous separation power and are therefore ideally suited for the analysis of very complex mixtures. In contrast to multidimensional separation techniques (LC×LC), where a particular fraction of the first-dimension separation is transferred and re-separated on the second-dimension column, in LC×LC the entire first dimension is analyzed in the second-dimension separation. Comprehensive LC×LC offers various advantages over both multidimensional off-line and conventional separation techniques, especially with respect to enhanced peak capacity, automation potential, reproducibility, and shorter analysis time.

There is growing interest related to rapid screening and full characterization of the constituents of plants with medicinal properties. The high content in polyphenols accounts for in vitro and in vivo antioxidant activity of the extracts obtained from plants; on the other hand, the high complexity of the samples extracted, depending on the method employed, may preclude complete resolution by conventional HPLC techniques. For this purpose, a comprehensive two-dimensional liquid chromatography (LC×LC) system, comprised of an RP-Amide first dimension and a partially porous octadecylsilica column in the second dimension, has been compared with a one-dimensional system (125). The chromatographic methods optimized in this research allowed the complete resolution and full characterization of polyphenols and xanthines in mate extracts.

A recent study focused on the application of a two-dimensional LC to pharmaceutical analysis and addressed the specific problem of separating co-eluting impurities/degradation products that maybe "hidden" within the peak of the active pharmaceutical ingredient and thus may escape detection by conventional methods (126). A comprehensive two-dimensional liquid chromatograph (LC×LC) has been constructed from commercially available HPLC equipment (Figure 5.15). This system utilizes two independently configurable second-dimension binary pumping systems to deliver independent flow rates, gradient profiles, and mobile phase compositions to dual shell type secondary columns. Very fast gradient separations (30 sec of total cycle time) were achieved without excessive back pressure and without compromising optimal first-dimension sampling rates. The sensitivity of the interface has been demonstrated for the analysis of a 1 mg/ml standard mixture containing 0.05% of a minor component. The ultimate sensitivity is limited by the inherent second-dimension baseline noise, which has contributions from both the gradient program and the pulses from the switching valve. The peak capacity of the present second-dimension separation is still limited, but could be further increased by employing state-of-the-art HPLC equipment, such as pumps designed to deliver very fast gradients with minimal dead volume and UV detectors with high sensitivity micro-flow cells. This instrument has been applied to the RP-LC×RP-LC analysis of real-world levels of pharmaceutical degradant products by exploiting differences in pH to change the selectivity of the second-dimension separation.

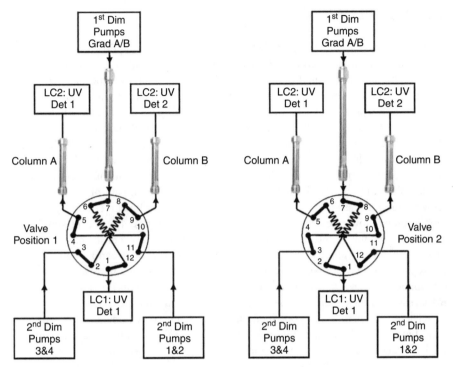

Figure 5.15   Schematic diagram of constructed LC×LC system employing two parallel second-dimension columns A and B with independently configurable gradient profiles and mobile phases. The interface consists of a two-position/twelve-port interface valve fitted with two active storage loops. The flow path for even cycles of the switching valve (position 1) is shown on the left, while the right-hand schematic (position 2) shows the flow path for odd cycles. Loop 1 is connected to ports 6 and 12, and Loop 2 is connected to ports 2 and 8. From Alexander, A.J., Ma, L. *J. Chromatogr. A.* 2009, with permission.

A polar PEG micro-column used in the first dimension and a porous-shell fused core 2.7 μm particle Ascentis Express C18 column used in the second dimension provided a highly orthogonal two-dimensional system for the separation of natural antioxidants (127). Using parallel segmented gradients in both the first and the second dimension, various classes of natural antioxidants showed different structurally related retention behavior. Phenolic acids were relatively weakly retained on both the PEG and the C18 columns, whereas flavone glycosides were weakly retained on the PEG column, but relatively strongly retained on the C18 column; flavone aglycones were relatively strongly retained on both columns. Using a porous-shell C18 column in the second dimension at high flow rate (4.8 mL/min) and elevated temperature (60°C) in combination with optimized segmented parallel gradients in the two dimensions allowed the decrease of the time of LC×LC analysis to 30 min (a time comparable to the single dimension separation). In this setup, the total peak capacity was approximately 280 in 30 min.

5.9 CONCLUSION

In the recent development of particle technology targeted for liquid chromatography, the use of shell particles (core-shell, fused core, or superficially porous) has received considerable attention. Shell particles manifest the advantages of porous and some benefit of nonporous particles. Various types of core-shell packing materials are now commercially available (5 μm, 2.7 μm, 2.6 μm, and 1.7 μm). The kinetic efficiency of columns packed with these shell particles increases as the shell thickness is decreased. However, the optimum shell thickness in reality is likely to be a compromise between efficiency, sample loading capacity, and analyte retention. Now it seems that the structure of the third generation shell particles is very close to its optimum in terms of the column efficiency and loadability. The short diffusion path of shell particles allows superior mass transfer kinetics and leads to higher kinetic efficiency at high mobile phase flow rates compared to fully porous particles. The recent sub-3 μm core-shell columns (2.6 μm, 2.7 μm) generate approximately half the back pressure of sub-2 μm fully porous particles. Therefore, high flow rates can be applied, and the separation time can be considerably shortened. The unique narrow particle size distribution of core-shell particles benefits in column packing. Probably this is the main reason of their great success in separation of small analytes.

The new generation of core-shell particles became very popular in pharmaceutical, biomedical, food, and environmental analysis in the last few years and allows for faster and more efficient separations than was possible before.

REFERENCES

1. HORVATH, C., PREISS, B.A., LIPSKY, S.R. *Anal. Chem.* 1967, 39: 1422–1428.
2. HORVATH, C., LIPSKY, S.R. *J. Chromatogr. Sci.* 1969, 7: 109–116.
3. KIRKLAND, J.J. *Anal. Chem.* 1969, 41: 218–220.
4. GUIOCHON, G. *J. Chromatogr. A.* 2011, 1218: 1915–1938.
5. KIRKLAND, J.J., TRUSZKOWSKI, F.A., DILKS Jr, C.H., ENGEL, G.S. *J. Chromatogr. A.* 2000, 890: 3–13.
6. FEKETE, S., FEKETE, J., GANZLER, K. *J. Pharm. Biomed. Anal.* 2009, 49: 64–71.
7. GRITTI, F., GUIOCHON, G. *J. Chromatogr. A.* 2011, 1218: 907–921.
8. OLAH, E., FEKETE, S., FEKETE, J., GANZLER, K. *J. Chromatogr. A.* 2010, 1217: 3642–3653.
9. OMAMOGHO, J.O., HANRAHAN, J.P., TOBIN, J., GLENNON, J.D. *J. Chromatogr. A.* 2011, 1218: 1942–1953.
10. VAN DEEMTER, J.J., ZUIDERWEG, F.J., KLINKENBERG, A. *Chem. Eng. Sci.* 1956, 5: 271–289.
11. GIDDINGS, J.C. *J. Chromatogr.* 1961, 5: 61–70.
12. SNYDER, L.R., KIRKLAND, J.J. *Introduction to Modern Liquid Chromatography*, 2nd ed. New York: John Wiley & Sons, 1979.
13. HUBER, J.F.K., HULSMAN, J.A.R.J. *Anal. Chim. Acta.* 1967, 38: 305–313.
14. KENNEDY, G.J., KNOX, J.H. *J. Chromatogr. Sci.* 1972, 10: 549–556.
15. HORVATH, C., LIN, H.-J. *J. Chromatogr. A.* 1978, 149: 43–70.
16. YANG, Y., VELAYUDHAN, A., LADISH, C.M., LADISH, M.R. *J. Chromatogr. A.* 1992, 598: 169–180.
17. DESMET, G., CLICQ, D., GZIL, P. *Anal. Chem.* 2005, 77: 4058–4070.
18. HORVATH, C., LIPSKY, S.R. *Anal. Chem.* 1967, 39: 1893.
19. GIDDINGS, J.C. *Anal. Chem.* 1967, 39: 1027–1028.

20. NEUE, U.D. *J. Chromatogr. A*. 2005, 1079: 153–161.
21. WANG, X., STOLL, D.R., SCHELLINGER, A.P., CARR, P.W. *Anal. Chem*. 2006, 78: 3406–3416.
22. DOLAN, J.W., SNYDER, L.R., DJORDJEVIC, N.M., HILL, D.W., WAEGHE, T.J. *J. Chromatogr. A*. 1999, 857: 1–20.
23. NEUE, U.D., MAZZEO, J.R. *J. Sep. Sci*. 2001, 24: 921–929.
24. NEUE, U.D., CARMODY, J.L. CHENG, Y.-F., LU, Z., PHOEBE, C.H., WHEAT, T.E., *Adv. Chromatogr*. 2001, 41: 93–136.
25. NEUE, U.D., CHENG, Y.-F., LU, Z. in S. Kromidas (ed.), *HPLC Made to Measure: A Practical Handbook for Optimization*, Weinheim: Wiley-VCH, 2006, p. 59–70.
26. NEUE, U.D. *HPLC Columns*, New York: Wiley-VCH, 1997, p. 42–49.
27. SWARTZ, M.E. MURPHY, B., *American Laboratory*. 2005, 37: 22–30.
28. SWARTZ, M.E. *J. Liquid Chrom*. 2005, 28: 1253–1263.
29. PHILLIPS, D.J., CAPPARELLA, M., NEUE, U.D., EL FALLAH, Z. *J. Pharm. Biomed. Anal*. 1997, 15: 1389–1395.
30. GERBER, F., KRUMMEN, M., POTGETER, H., ROTH, A., SIFFRIN, C., SPOENDLIN, C. *J. Chromatogr. A*. 2004, 1036: 127–133.
31. KNOX, J.H., SALEEM, M. *J. Chromatogr. Sci*. 1969, 7: 614–622.
32. MACNAIR, J.E., LEWIS, K.C., JORGENSON, J.W. *Anal. Chem*. 1997, 69: 983–989.
33. MACNAIR, J.E., PATEL, K.D., JORGENSON, J.W. *Anal. Chem*. 1999, 71: 700–708.
34. WU, N., LIPPERT, J.A., LEE, M.L. *J. Chromatogr. A*. 2001, 911: 1–12.
35. DE VILLIERS, A., LAUER, H., SZUCS, R., GOODALL, S., SANDRA, P. *J. Chromatogr. A*. 2006, 1113: 84–91.
36. GRITTI, F., GUIOCHON, G. *J. Chromatogr. A*. 2007, 1138: 141–157.
37. GRITTI, F., GUIOCHON, G. *Anal. Chem*. 2008, 80: 5009–5020.
38. GUO, Y., SRINIVASAN, S., GAIKI, S., LIU, Y. *Chromatographia*. 2008, 68: 19–25.
39. BLACKWELL, J.A., CARR, P.W. *J. Liquid Chrom*. 1991, 14: 2875–2889.
40. ZHU, C., GOODALL, D.M., WREN, S.A.C. *LCGC North America*. 2005, 23: 54–72.
41. CLAESSENS, H.A., VAN STRATEN, M.A. "Reduction of Analysis Times in HPLC at Elevated Column Temperatures," Eindhoven University of Technology, 2004.
42. JACKSON, P.T., CARR, P.W. *Chemtech*. 1988, 28: 29–37.
43. HJERTÉN, S., LIAO, J.L., ZHANG, R. *J. Chromatogr*. 1989, 473: 273–275.
44. SVEC, F., FRECHET, J.M. *J. Anal. Chem*. 1992, 64: 820–822.
45. GUSEV, I., HUANG, X., HORVATH, C. *J. Chromatogr. A*. 1999, 885: 273–290.
46. MINAKUCHI, H., NAGAYAMA, H., SOGA, N., ISHIZUKA, N., TANAKA, N. *J. Chromatogr. A*. 1998, 797: 121–131.
47. OBERACHER, H., PREMSTALLER, A., HUBER, C.G. *J. Chromtogr. A*. 2004, 1030: 201–208.
48. IKEGAMI, T., DICKS, E., KOBAYASHI, H., MORISAKA, H., TOKUDA, D., CABRERA, K., TANAKA, N. *J. Sep. Sci*. 2004, 27: 1292–1302.
49. GUIOCHON, G. *J. Chromatogr. A*. 2007, 1168: 101–168.
50. GRITTI, F., GUIOCHON, G. *J. Chromatogr. A*. 2007, 1166: 30–46.
51. GUILLARME, D., NGUYEN, D.T., RUDAZ, S., VEUTHEY, J.L. *Eur. J. Pharm. Biopharm*. 2007, 66: 475–482.
52. GUILLARME, D., NGUYEN, D.T., RUDAZ, S., VEUTHEY, J.L. *Eur. J. Pharm. Biopharm*. 2008, 68: 430–440.
53. DOLAN, J.W., SNYDER, R.L. *J. Chromatogr. A*. 1998, 799: 21–34.
54. SCHELLINGER, A.P., CARR, P.W. *J. Chromatogr. A*. 2005, 1077: 110–119.
55. HEINISCH, S., DESMET, G., CLICQ, D., ROCCA, J.L. *J. Chromatogr. A*. 2008, 1203: 124–136.
56. USHER, K.M., SIMMONS, C.R., DORSEY, J.G. *J. Chromatogr. A*. 2008, 1200: 122–128.
57. CABOOTER, D., BILLEN, J., TERRYN, H., LYNEN, F., SANDRA, P., DESMET, G. *J. Chromatogr. A*. 2008, 1204: 1–10.
58. CABOOTER, D., DE VILLIERS, A., CLICQ, D., SZUCS, R., SANDRA, P., DESMET, G. *J. Chromatogr. A*. 2007, 1147: 183–191.
59. GRITTI, F., FELINGER, A., GUIOCHON, G. *J. Chromatogr. A*. 2006, 1136: 57–72.

60. DeStefano, J.J., Langlois, T.J., Kirkland, J.J. *J. Chromatogr. Sci.* 2008, 46: 254–260.
61. Butchart, K., Potter, T., Wright, A., Brien, A. *Intl. Labmate.* 2007, 7: 1–3.
62. Anspach, J.A., Rahn, P.C. "High Speed Separations." 30th International Symposium and Exhibit on High Performance Liquid Phase Separations and Related Techniques (HPLC), San Francisco, CA, 2006.
63. Fekete, S., Fekete, J. *J. Chromatogr. A.* 2011, 1218: 5286–5291.
64. Poppe, H., Kraak, J.C. *J. Chromatogr.* 1983, 255: 395–414.
65. Horváth, K., Gritti, F., Fairchild, J.N., Guiochon, G. *J. Chromatogr. A.* 2010, 1217: 6373–6381.
66. McCalley, D.V. *J. Chromatogr. A.* 2011, 1218: 2887–2897.
67. Gritti, F, Leonardis, I., Abia, J., Guiochon, G. *J. Chromatogr. A.* 2010, 1217: 3819–3843.
68. Kaczmarski, K., Guiochon, G. *Anal. Chem.* 2007, 79: 4648–4656.
69. Guiochon, G., Felinger, A., Katti, A., Shirazi, D. *Fundamentals of Preparative and Nonlinear Chromatography*, 2nd ed. Boston, MA: Academic Press, 2006.
70. Gritti, F., Leonardis, I., Shock, D., Stevenson, P., Shalliker, A., Guiochon, G. *J. Chromatogr. A.* 2010, 1217: 1589–1603.
71. Desmet, G., Deridder, S. *J. Chromatogr. A.* 2011, 1218: 32–45.
72. Deridder, S., Desmet, G. *J. Chromatogr. A.* 2011, 1218: 46–56.
73. Desmet, G., Broeckhoven, K., De Smet, J., Deridder, S., Baron, G.V., Gzil, P. *J. Chromatogr. A.* 2008, 1188: 171–188.
74. Giddings, J. *Dynamics of Chromatography*. New York: Marcel Dekker, 1965.
75. Gritti, F., Guiochon, G. *AIChE J.* 2011, 57: 333–345.
76. Fekete, S., Ganzler, K., Fekete, J. *J. Pharm. Biomed. Anal.* 2011, 54: 482–490.
77. Cabooter, D., Fanigliulo, A., Bellazzi, G., Allieri, B., Rottigni, A., Desmet, G. *J. Chromatogr. A.* 2010, 1217: 7074–7081.
78. Fekete, S., Fekete, J., Ganzler, K. *J. Pharm. Biomed. Anal.* 2009, 50: 703–709.
79. Gritti, F., Cavazzini, A., Marchetti, N., Guiochon, G. *J. Chromatogr. A.* 2007, 1157: 289–303.
80. Kotska, J., Gritti, F., Guiochon, G., Kaczmarski, K. *J. Chromatogr. A.* 2010, 1217: 4704–4712.
81. Gritti, F., Guiochon, G. *J. Chromatogr. A.* 2009, 1216: 1353–1362.
82. Gritti, F., Guiochon, G. *J. Chromatogr. A.* 2007, 1166: 47–60.
83. Gritti, F., Guiochon, G. *Anal. Chem.* 2009, 81: 3365–3384.
84. Gritti, F., Guiochon, G. *Anal. Chem.* 2008, 80: 6488–6499.
85. Gritti, F., Guiochon, G. *J. Chromatogr. A.* 2008, 1206: 113–122.
86. Gritti, F., Guiochon, G. *J. Chromatogr. A.* 2010, 1217: 1485–1495.
87. Gritti, F., Guiochon, G. *Chem. Eng. Sci.* 2010, 65: 6310–6319.
88. Gritti, F., Guiochon, G. *J. Chromatogr. A.* 2010, 1217: 5069–5083.
89. Gritti, F., Guiochon, G. *J. Chromatogr. A.* 2011, 1218: 1592–1602.
90. Yew, B.G., Ureta, J., Shalliker, R.A., Drumm, E.C., Guiochon, G. *AIChE J.* 2003, 49: 642–664.
91. Fanigliulo, A., Cabooter, D., Bellazzi, G., Tramarin, D., Allieri, B., Rottigni, A., Desmet, G. *J. Sep. Sci.* 2010, 33: 3655–3665.
92. Mallett, D.N., Ramírez-Molinab, C. *J. Pharm. Biomed. Anal.* 2009, 49: 100–107.
93. Marchetti, N., Cavazzini, A., Gritti, F., Guichon, G. *J. Chromatogr. A.* 2007, 1163: 203–211.
94. McCalley, D.V. *J. Chromatogr. A.* 2008, 1193: 85–91.
95. de Andres, F., Zougagh, M., Castaneda, G., Ros, A. *J. Chromatogr. A.* 2008, 1212: 54–60.
96. Cunliffle, J.M., Noren, C.F., Gates, R.N., Clement, R.P., Shen, J.X. *J. Pharm. Biomed. Anal.* 2009, 50: 46–52.
97. Song, W., Pabbisetty, D., Groeber, E.A., Sttenwyk, R.C., Fast, D.M. *J. Pharm. Biomed. Anal.* 2009, 50: 491–500.
98. Saida, R., Pohanka, A., Andersson, M., Becka, O., Rehim, M.A. *J. Chromatogr. B.* 2011, 879: 815–818.
99. Therona, A., Cromartyb, D., Rheedersa, M., Viljoena, M. *J. Chromatogr. B.* 2010, 878: 2886–2890.
100. Teunissen, S.F., Jager, N.G.L., Rosing, H., Schinkel, A.H., Schellens, J.H.M., Beijnen, J.H *J. Chromatogr. B.* 2011, 879: 1677–1685.

101. ABRAHIM, A., SAYAH, M.A., SKRDLA, P., BEREZNITSKI, Y., CHEN, Y., WU, N. *J. Pharm. Biomed. Anal.* 2010, 51: 131–137.
102. ZÁDORI, D., ILISZ, I., KLIVÉNYI, P., SZATMÁRI, I., FÜLÖP, F., TOLDI, J., VÉCSEI, L., PÉTER, A. *J. Pharm. Biomed. Anal.* 2011, 55: 540–543.
103. FEKETE, S., FEKETE, J. *Talanta.* 2011, 84: 416–423.
104. SILVESTRO, L., GHEORGHE, M.C., TARCOMNICU, I., SAVU, S., SAVU, S.R., IORDACHESCU, A., DULE, C. *J. Chromatogr. B.* 2010, 878: 3134–3142.
105. ANDRÉS, F.D., CASTANEDA, G., RÍOS, Á. *J. Pharm. Biomed. Anal.* 2010, 51: 478–483.
106. CHITTURI, S.R., SOMANNAVAR, Y.S., PERURI, B.G., NALLAPATI, S., SHARMA, H.K., BUDIDET, S.R., HANDA, V.K., VURIMINDI, H.B. *J. Pharm. Biomed. Anal.* 2011, 55: 31–47.
107. BADMAN, E.R., BEARDSLEY, R.L., LIANG, Z., BANSAL, S. *J. Chromatogr. B.* 2010, 878: 2307–2313.
108. FEKETE, S., GANZLER, K., FEKETE, J. *J. Chromatogr. A.* 2010, 1217: 6258–6266.
109. JIMENEZ, V., COMPANYO, R., GUITERAS, J. *Talanta.* 2011, 85: 596–606.
110. DEGRASSE, S.L., DEGRASSE, J.A., REUTER, K. *Toxicon.* 2011, 57: 179–182.
111. KAUFMANN, A., BUTCHER, P., MADEN, K., WALKER, S., WIDMER, M. *Analytica Chimica Acta.* 2010, 673: 60–72.
112. BELAKOVA, S., BENEŠOVA, K., MIKULIKOVA, R., SVOBODA, Z. *Food Chem.* 2011, 126: 321–325.
113. RODRIGUEZ, R.M.G., GRANDE, B.C., GANDARA, J.S. *Food Chem.* 2011, 125: 549–554.
114. KALINOVA, J., VRCHOTOVA, N. *Food Chem.* 2011, 127: 602–608.
115. TOLGYESI, A., TOLGYESI, L., SHARMA, V.K., SOHN, M., FEKETE, J. *J. Pharm. Biomed. Anal.* 2010, 53: 919–928.
116. TOLGYESI, A., SHARMA, V.K., FEKETE, J. *J. Chromatogr. B.* 2011, 879: 403–410.
117. MNATSAKANYANA, M., GOODIE, T.A., CONLAN, X.A., FRANCIS, P.S., McDERMOTT, G.P., BARNETT, N.W., SHOCK, D., GRITTI, F., GUIOCHON, G., SHALLIKER, R.A. *Talanta.* 2010, 81: 837–842.
118. VILLALBA, A.M., MOYANO, E., GALCERAN, M.T., *J. Chromatogr. A.* 2010, 1217: 5802–5807.
119. CAPRIOTTI, A.L., FOGLIA, P., GUBBIOTTI, R., ROCCIA, C., SAMPERI, R., LAGANÀ, A. *J. Chromatogr. A.* 2010, 1217: 6044–6051.
120. LKOVA, E.S., MAN, P., KOLARIK, M., FLIEGER, M. *J. Chromatogr. A.* 2010, 1217: 6296–6302.
121. TARCOMNICU, I., NUIJS, A.L.N., SIMONS, W., BERVOETS, L., BLUST, R., JORENS, P.G., NEELS, H., COVACI, A. *Talanta.* 2011, 83: 795–803.
122. PIETROGRANDE, M.C., DONDI, F., CIOGLI, A., GASPARRINI, F., PICCIN, A., SERAFINI, M. *J. Chromatogr. A.* 2010, 1217: 4355–4364.
123. SPOOF, L., NEFFLING, M.R., MERILUOTO, J. *Toxicon.* 2010, 55: 954–964.
124. NEFFLING, M.R., SPOOF, L., MERILUOTO, J. *Analytica Chimica Acta.* 2009, 653: 234–241.
125. DUGO, P., CACCIOLA, F., DONATO, P., JACQUES, R.A., CARAMAO, E.B., MONDELLO, L. *J. Chromatogr. A.* 2009, 1216: 7213–7221.
126. ALEXANDER, A.J., MA, L. *J. Chromatogr. A.* 2009, 1216: 1338–1345.
127. CESLA, P., HAJEK, T., JANDERA, P. *J. Chromatogr. A.* 2009, 1216: 3443–3457.

Chapter 6

UHPLC Determination of Drugs of Abuse in Human Biological Matrices

Fabio Gosetti, Eleonora Mazzucco, and Maria Carla Gennaro

6.1 INTRODUCTION

All over the world, drug consumption and abuse are dramatically and continuously increasing. According to the 2008 World Drug Report, in 2007 almost 5% of the world population had, at least once, misused drugs, and 0.6% of the world adults have been severely drug addicted (1). In 2009, according to other statistics, 3.9% of European adults have used cocaine (2). Recent analyses of river and sewage water in different towns and countries of the world have evidenced widespread drug use as well as *prescription drug abuse*: this term refers to the consumption of medications not directly prescribed or taken for reasons or in dosages different from those prescribed. It is often impossible to differentiate between the two uses: for example, benzodiazepines, therapeutically prescribed as depressants to produce sedation, are frequently misused, and their large and frequent use leads, as illicit drugs do, to physical diseases and psychological dependences. Commonly abused prescription medications include opioids (used for pain), central nervous system depressants such as barbiturates and benzodiazepines (used for anxiety and sleep disorders), and stimulants such as amphetamines used for ADHD (attention-deficit hyperactivity disorder) and narcolepsies. Consumption of drugs is widely diffused also among young people; the substances taken are often hallucinogenic, change perception and behavior, and give apparently positive effects such as euphoria, relaxation, and enhanced empathy. But the side effects are moodiness, intolerance, and dangerous consequences for the health; both psychological and physical dependence often lead to death.

Furthermore, drug misuse causes serious social problems such as violence, criminal activity, motor vehicle accidents, homicides, and suicides.

Ultra-High Performance Liquid Chromatography and Its Applications, First Edition. Edited by Quanyun Alan Xu.
© 2013 John Wiley & Sons, Inc. Published 2013 by John Wiley & Sons, Inc.

Also a great number of other compounds, used in sport activities to enhance performance, are prohibited because they are dangerous to health. More than 200 species are included in the list of forbidden substances prepared by the World Anti-Doping Agency (WADA) and updated each year. The European Union (EU) has elaborated the EU Drugs Strategy 2005–2012 and the EU action plan (2009–2012). A chapter auspicates the international collaboration and underlines the need of balanced approaches of the countries to face the problem together. However, at present, each European state applies its own legislation sanctions as it concerns trade, treatment, prevention, as well as prohibition and punishment of illicit activities (3, 4).

In this review the most diffused drugs and illicit substances are summarized in their principal classes, based on both chemical structure and action (5–16). General information of the class, their main components, and the most relevant metabolites are reported. Some substances are listed both in the chemical class and in the class concerning their effects on the organism.

Because very often drugs and illicit substances are taken under their complex mixtures, the simultaneous determination in biological matrices of many drugs, their metabolites, and illicit substances can offer a wide overview of the patient situation and often helps to understand which precursor drugs the subject has consumed. For more information, it is helpful to take into account the simultaneous consumption of prescribed pharmaceuticals. Many of the analytical methods considered here are devoted to the simultaneous determination of drugs, metabolites, illicit substances, and pharmaceuticals.

6.2 CLASSES OF DRUGS AND ILLICIT SUBSTANCES

6.2.1 Amphetamines

Amphetamines and derivatives act to stimulate the central nervous system; they generally increase alertness and decrease fatigue and appetite. Street names for amphetamines are bennies, black beauties, crosses, hearts, and turnaround, and for methamphetamine, chalk, crank, crystal, fire, glass, go fast, ice, meth, and speed. The most known drug of this class is ecstasy or 3,4-methylenedioxymethamphetamine (MDMA) also known as Adam, clarity, Eve, lover's speed, and peace. About 3.1% of the European adults would use ecstasy and related compounds. During the last decades the abuse of MDMA, a prototype of designer drugs, has considerably increased. The consequences are rapid and irregular heartbeat, appetite reduction, weight loss, and heart failure. Also amphetamine (A); 2-ethylidene-1,5-dimethyl-3,3-diphenylpyrrolidine   (EDDP);   N,N-dimethylamphetamine (DMA);   N,N-dimethylamphetamine N-oxide (DMANO);   ephedrine;   p-hydroxyamphetamine; mefenorex; methamphetamine (MA); 2-methylamino-1-(3,4-methylene-dioxyphenyl)butane (MBDB); 3,4-methylenedioxyamphetamine (MDA); 3,4-methylenedioxyethylamphetamine (MDEA); 4-methoxy-amphetamine (PMA); norephedrine; phentermine; and pseudoephedrine are very diffused. In particular DMA is consumed by quite a large percentage of MA users at the street level; although DMA is listed in different countries among the most dangerous (17), the penalty for DMA possession and trafficking is often less severe than for MA. However, because DMA is not available as a prescription drug and has no documented medical use, its presence in urine indicates with certainness its illicit usage.

6.2.2 Benzodiazepines

Benzodiazepines reduce pain and anxiety, give feeling of well-being, and lower inhibitions. The consequences of their use are confusion, fatigue, impaired coordination, respiratory arrest, and addiction. Some street names for benzodiazepines (with the exception of flunitrazepam) are candy, downers, and sleeping pills. The most used benzodiazepines are alprazolam, 7-aminonitrazepam, 7-aminoclonazepam, 7-aminoflunitrazepam, bromazepam, brotizolam, carbamazepine, chlordiazepoxide, citalopram, clobazam, clonazepam, clotiazepam, delorazepam, diazepam, estazolam, flunitrazepam, flurazepam, α-hydroxy-alprazolam, lorazepam, lormetazepam, medazepam, midazolam, nitrazepam, nordiazepam, oxazepam, pinazepam, prazepam, temazepam, and triazolam. In particular flunitrazepam (some street names: forget-me pill, Mexican valium, and rope) is a hypnotic used for short-term treatment of chronic insomnia but also used as a date rape drug or for robbery.

6.2.3 Cannabinoids

Cannabis is often detected in the blood of drivers and is considered the major cause of road crashes after alcohol. It is often implicated in forensic cases and workplace accidents. Δ^9–tetrahydrocannabinol (THC) is the major psychoactive constituent of *Cannabis sativa*, whose main metabolites formed in blood are 11-nor-9-carboxy-Δ^9–tetrahydrocannabinol (THC-COOH) and 11-hydroxy-Δ^9–tetrahydrocannabinol (THC-OH). Marijuana (also ganja, grass, herb, Mary Jane, pot, green, trees, smoke, sinsemilla, and weed) and hashish (or boom, gangster, hash, hemp) are made up of dried parts of the *Cannabis sativa* hemp plant; they give euphoria and relaxation but also slowed reaction times, impaired coordination, anxiety, panic attack, and psychosis and cause respiratory infections, mental health decline, and addiction.

6.2.4 Cocaine Alkaloids

The most known is cocaine, which belongs to the tropane alkaloid family and is obtained from the leaves of the plant *Erythroxylon coca*, used for centuries by Peruvian Indians to increase endurance and improve well-being. Cocaine is a stimulant of the central nervous system, acts as an appetite suppressant, and produces a euphoric state similar to that induced by amphetamines. Since the mid-twentieth century the recreational use of cocaine has been very diffused worldwide. Some street names are blow, C, candy, Charlie, coke, rock, snow, and toot. Cocaine is available on the street both as the free base (crack), predominantly used for smoking, and as its hydrochloride salt, mainly employed for intravenous injection and nasal insufflations. The rates of infants prenatally exposed to cocaine have been estimated between 2.6% and 11% of all the live births (18). Prenatal cocaine consumption has been associated to premature labor, placental abruption, low birth parameters (weight, length), microcephaly, malformations, increased risk of sudden death syndrome, and encephalopathy. The main cocaine metabolites are benzoylecgonine (BE), cocaethylene, ecgonine methylester (EME), and norcocaine.

6.2.5 Designer Drugs

Designer drugs are psychoactive substances especially designed to circumvent drug laws and are prepared by modifying the molecular structure of existing drugs. They generally present the structural features of phenylethylamine and piperazine derivatives and produce hallucinogenic visual effects similar to those of lysergic acid and mescaline, together with emotional empathic responses similar to those given by ecstasy. Their effects have not been completely studied, but the reports of acute intoxications associated with consumption of m-chlorophenyl-piperazine and 2,5-dimethoxy-4-(n)-propylthiophenethylamine support the hypothesis that future consumption of designer drugs is going to represent a serious threat to human health (19).

The most known designer drugs are carphedone; 4-chloro-2,5-dimethoxyphenethylamine (2C-C); *RS*-6-(5-chloropyridin-2-yl)-7-oxo-6,7-dihydro-5H-pyrrolo[3,4-b]pyrazin-5-yl-4-methylpiperazine-1-carboxylate (zopiclone or Z-drug); 4-(N,N-dimethylamino sulfonyl)-7-fluoro-2,1,3-benzoxadiazole (DBD-F); dimethyltryptamine (DMTA); diphenydramine; 4-ethyl-2,5-dimethoxyphenethylamine (2C-E); 4-ethylthio-2,5-dimethoxyphenethylamine (2C-T-2); 4-fluoroamphetamine (4-FMP); indan-2-amine; 4-iodo-2,5-dimethoxyphenethylamine (2C-I); 1-(4-iodo-2,5-dimethoxyphenyl)propan-2-amine (DOI); 4-isopropylthio-2,5-dimethyoxyphenethylamine (2C-T-4); meperidine; 2-methylamino-1-(3,4-methylenedioxyphenyl)butan-1-one (Bk-MBDB); 1-(3,4-methylenedioxybenzyl)piperazine (MDBP); 1-(3,4-methylenedioxyphenyl)butan-2-amine (BDB); N-methyl-3,4-methylenedioxymethamphetamine (MMDA-2); N-methyl-1-(3,4-methylenedioxyphenyl)butan-2-amine (HMDMA); N-methyl-1-(3,4-methylenedioxyphenyl)butan-3-amine (HMDMA); methylphenidate; 4-methoxymethamphetamine (PMMA); 1-(4-methoxyphenyl)piperazine (4-MPP); 2-(6-methyl-2-p-tolylimidazol[1,2-9]pyridine-3-yl)acetamide (*zolpidem*); normeperidine; 4-phenylbutylamine (4-PBA); 3-phenyl-1-peopyl-amine (3-PPA); and 2,4,6-trimethoxyamphetamine (TMA-6).

6.2.6 Ketamine

Particular   interest   is   devoted   to   ketamine,   2-(2-chlorophenyl)-2-methylaminocyclohexan-1-one, legitimately and largely used for its anesthetic properties in clinical and veterinary practices. However, the misuse of ketamine—often associated to ecstasy—as recreational drug has over the last 10 years increased its popularity worldwide. The street names are K, special K, and vitamin K. The users experience hallucinations and a particular cataleptic state called the K-hole, associated to stupor, sedation, amnesia, difficulty in fighting, feeling of being separate from the body and environment. Ketamine is the most diffused among the so-called club drugs, used by teenagers and young adults in bars, nightclubs, concerts, and parties. Ketamine also has been recently implicated in drug-facilitated sexual assault (DFSA), but unfortunately DFSA incidents are generally denounced about 24 h after, when only negligible traces of the drug are present in tissues and fluids of the victim. Club drugs including ecstasy, MA, LSD (lysergic acid diethylamide), and GHB (gamma hydroxybutyrate or *G*, Georgia home boy, soap) also have been associated with sexual assaults.

6.2.7 Lysergic Acid Diethylamide

LSD gives altered states of perception and feeling, hallucinations, nausea, and panic. Street names are acid, blotter, cubes, microdot, yellow sunshine, and blue heaven. It is sold as tablets, capsules, or liquid or adsorbed on paper. LSD produces unpredictable psychological effects with "trips" lasting about 12 h. The analysis of LSD in biological fluids is a challenge because it is generally taken in small doses, unstable, and rapidly metabolized. Fortunately, iso-LSD and the metabolites N-dimethyl LSD (nor-LSD) and 2-oxo-3-hydroxy-LSD (O-H-LSD), which do not have psychoactive properties, have longer permanence times in biological fluids, and their detection allows to evidence LSD use.

6.2.8 Opiates and Opioids

Opiates are drugs naturally found in the opium *Poppy Papaver Sonniferum*, and opioids are semi-synthetic drugs prepared from the natural opiates. Both act on the central nervous system producing cough suppression, analgesia, euphoria, sedation, respiratory depression and arrest, nausea, confusion, unconsciousness, coma, and addiction. In 2007, the number of opioid users in Europe was estimated between 1.2 and 1.5 million people (2).

The most known opioids, also known as narcotics, are morphine (also Miss Emma, monkey, and white stuff) and heroin, which is processed from morphine and usually appears as a white or brown powder or as a black sticky substance that is injected, sniffed, or smoked. Heroin accounts for the greatest share of drug use and mortality in the European Union.

Methadone is an example of synthetic opioids, and it is used in the management of heroin dependence because it reduces the addiction of heroin when given daily.

Other opioids are (the more common street names are in brackets) acetyl codeine; buprenorphine; codeine (Captain Cody, Cody, and schoolboy); codeine-glucuronide; dextromethorphan; dextromoramide; dihydrocodeine (DHC); EDDP; ethylmorphine;

ethylmorphine-glucuronide; fentanyl (apache, China girl, dance fever, murder 8, and Tango and Cash); hydrocodone; hydromorphone; methadone; 6-monoacetylmorphine (6-MAM); morphine-3-glucoronide (M3G); morphine-6-glucuronide (M6G); nalbuphine; narceine; norbuprenorphine; norcodeine; norfentanyl; noscapine; opium (big 0, black stuff, gum, and hop); oxycodone; noroxycodone; oxymorphone; papaverine; pentazocine; pethidine; pholcodine; propoxyphene; thebaine; tramadol; and trimeperidine.

6.2.9 Diuretics, β-Blockers, and Stimulants

These substances are generally determined in biological matrices together with the drugs of abuse because the mixtures consumed by the users often contain some of them. Diuretics, β-blockers, and stimulants are prohibited by the WADA due to their documented potential misuse as performance-enhancing substances.

Diuretics and β-blockers are clinically used to control high blood pressure, while stimulants are used for a number of applications including the management of narcolepsy and ADHD and as nasal decongestants. Diuretics act by stimulating the overproduction of urine in the kidneys to excrete excess fluids and electrolytes from the body and are often illicitly taken by athletes to mask the misuses of performance-enhancing substances, such as anabolic steroids, or to "make weight" in weight class sports such as boxing. Examples of diuretics are acetazolamide, altizide, amiloride, bendroflumethiazide, benzthiazide, bumetanide, canrenoic acid, canrenone, chlorothiazide, chlorthalidone, clopamide, cyclopenthiazide, cyclothiazide, dichlorphenamide, ethacrynic acid, furosemide, hydrochlorothiazide, hydroflumethiazide, indapamide, methylclothiazide, meticrane, metolazone, norpethidine, piretanide, polythiazide, probenecid, salamid, spironolactone, torasemide, triamterene, trichloromethiazide, and xipamide.

β-blockers act by relaxing muscles. This effect makes their misuse attractive for athletes competing in sports that require balance and dexterity such as gymnastics, archery, and biathlon. The most common β-blockers are acebutolol, acetazolamide, alprenolol, atenolol, betaxolol, bisoprolol, butametanide, carteolol, carvedilol, cateolol, celiprolol, esmolol, labetalol, levobunolol, metipranolol, metolazone, nadolol, oxprenolol, pindolol, propanolol, sotalol, and timolol.

Stimulants enhance the central nervous system activity by mimicking the action of brain neurotransmitters and, due to the ability to increase awareness and counteract fatigue, are misused by athletes participating in sports that require high levels of physical and mental endurances. Examples of stimulants are A, amfetaminil, amiphenazole, andrafinil, anfepramone, BE, benzylpiperazine, bupropion, carphedon, caffeine, chlorphentermine, clobenzorex, cocaine, cropropamide, crotetamide, didesmethylsibutramine, DMA, ephedrine, etamivan, etaphedrine, ethylamphetamine, ethylefrine, famprofazone, fenbutrazate, fencamfamine, fenetylline, fenfluramine, fenproporex, furfenorex, heptaminol, hydroxybromantan, isometheptene, MDA, MDMA, mefonerex, mephentermine, mesocarb, mesocarb-OH, mesocarb-di-OH, MA, methoxyphenamine, para-methylamphetamine, methylecgonine, methylephedrine, methylphenidate, modafinilic acid (modafinil), monodesmethylsibutramine, nikethamide, norfenfluramine, oxilofrine, pemoline, pentetrazol, phendimetrazine, phenpromethamine, phentermine, phenylpropanolamine (cathine), pholedrine, pipradol, prolintane, propylhexedrine, pseudoephedrine, ritalinic acid, sibutramine, and strychnine.

6.3 DRUG METABOLIZATION IN THE HUMAN BODY

To obtain complete and reliable analytical results concerning the presence and the amounts of drugs and illicit substances in biological matrices, it is strictly necessary to know the metabolism pathway undergone in the organism by the substances consumed. It is important, therefore, to get information not only on the chemical structures of the metabolites formed, but also on the permanence time of the drugs and of their metabolites in the biological matrices undergone to analysis. The analytical methods must be addressed to determine not only the drug (or its traces) but also its metabolites. The combined information about drug and metabolite presence can furnish useful information about the substances taken, while the relative amounts of drug/metabolites help distinguish if the residues found are due to illicit consumption or to the medical treatments.

Taking into account that very often complex mixtures of drugs and illicit substances are consumed, the analysis is often a challenge. To obtain significant results, the permanence time as well as the conditions of pH and temperature at which the substances searched for are more stable is important information to collect. For example, the sample for heroin determination must be analyzed as rapidly as possible, because heroin is unstable at room temperature. The opioid is also unstable at basic pH values. Nevertheless, a recent UHPLC–MS method obtained a good chromatographic separation also by using a mobile phase at basic pH values (7).

Heroin is rapidly metabolized to the active and specific metabolite 6-MAM and further to morphine and conjugated morphine. In turn, morphine is quickly converted to its principal metabolite M3G and, more slowly and in smaller amounts, to M6G. In general, the metabolism pathway of opiates is complex, because the taking place of several reactions of interconversion makes it difficult to certainly evaluate what a patient has consumed. For example, it is hard to understand if the patient, who has been prescribed diamorphine or morphine sulphate, has also consumed illicit heroin. The diagnosis cannot be done on the basis of morphine and 6-MAM contents, because their presence can be due to prescribed medications. Acetyl codeine would be a very useful marker for illicit heroin consumption, but unfortunately has a short half-life of only 237 min. Also through the determination of codeine and norcodeine, it is possible to evaluate whether the patient has taken illicit heroin as a source of morphine, but the interpretation is still difficult if codeine has been used in addition to diamorphine or illicit heroin.

After DMA consumption, DMANO is the main and specific metabolite found in urine together with MA and A (17). In addition DMANO has a longer detection-time window in urine than DMA and is the most efficacious indicator to confirm the use of DMA. The simultaneous identification of DMA, DMANO, MA, and A in urine can give a good picture of a DMA and MA abuse situation. The possibility that the presence of DMA or DMANO in human fluids may originate from DMA impurities contained in MA intake can be excluded because DMA impurity concentrations in MA are always lower than 1%. Also from the concentration ratios of DMA/DMANO and MA/A found in urine, it is difficult to ascertain whether a person has taken pure DMA, DMA/MA mixtures, or DMA and MA separately. On the other hand, the simultaneous presence of DMA, DMANO, MA, and A could also be due to the intake of prescribed drugs such as *l-deprenyl* (antiparkinsonian and antidepressant) or famprofazone (analgesics and antipyretic) that both produce MA or A as metabolites.

In turn, cocaine is rapidly metabolized in blood and urine to inactive BE (the main metabolite), EME, ecgonine, and norcocaine, but low amounts of precursor cocaine can be still found in urine. When cocaine is consumed with alcohol, cocaethylene is preferentially formed in the body.

To increase the window of detection for LSD use, because it is rapidly metabolized, recent methods target the metabolites that have longer half-lives. Half-life of nor-LSD is 10 h in urine, compared with that of LSD at about 5 h. The diastereomer iso-LSD is formed during the production of illicit LSD from lysergic acid but also if LSD is exposed to basic aqueous solution and high temperatures.

Also the metabolite dehydronorketamine has a longer duration time in urine than its precursor ketamine and the other metabolite norketamine. Dehydronorketamine can be detected after 7–10 days from the consumption of a modest dose of ketamine and is therefore a very useful diagnostic metabolite.

6.4   HUMAN MATRICES ANALYZED

The matrices analyzed to identify and quantify drugs of abuse can be obtained through invasive techniques, such as serum and plasma, and through less invasive ones, such as urine, hair, oral fluid, fingerprints, and sweat.

When dealing with drug analysis in biological matrices, it must be remembered that most of the drugs are characterized by a short permanence time, which also depends on dose and form of consumption, kind of matrix, and individual metabolism. In the human body, drugs are generally metabolized to water-soluble compounds and eliminated through urine, but their presence can be later detected as derivative species in plasma, urine, sweat, and hair—often with different permanence times. Cannabis THC can be detected only up to 5 h from consumption in plasma and up to 10 h in urine, while its metabolite THC-COOH reaches persistence times in plasma up to 20–57 h in occasional users and 3–13 days in regular users (20).

6.4.1   Whole Blood, Plasma, and Serum

Sample collection is invasive and needs medical personnel, and the volume collected is lower than for urine. However, it presents the advantage that the sample is not subjected to modification or substitution.

6.4.2 Fingerprints, Sebum, and Sweat

Anabolic steroids, organic pollutants, and cocaine have been easily detected in sebum and sweat (21). The potentiality of fingerprint for the determination of drugs and metabolites is conditioned by a correct preparation, because obviously the analysis must only concern the drugs taken and their metabolites, not drugs transferred to the finger from touching a contaminated surface or handling the drug; a direct deposit left by a fingerprint on a surface would be a combination of endogenous substances secreted by the skin and of exogenous substances transferred to the hands from the objects previously touched. As an example of correct sampling, a fingerprint is collected from the index finger, previously cleaned with an ethanol wipe and then rubbed over forehead and face.

6.4.3 Hair

Hair is often used as an alternative or additional biological matrix. With respect to blood and urine, hair is a more stable matrix and, for many substance, shows longer detection-time windows (from weeks to years). Hair analysis can give a historical picture of drug intake, and especially the segmented hair analysis can provide useful information on state and evolution of drug abuse. Hair analysis is largely employed in clinical and forensic applications, for example, to evaluate a pattern of chronic drug use and in the investigations of poisoning cases.

6.4.4 Meconium

Examples of meconium used as a biological matrix are also found, in particular, for the determination of cocaine and BE (18).

6.4.5 Oral Fluid

Oral fluid (OF), or saliva, is used for drugs of abuse especially for tests in workplace and on individuals suspected of driving under drug influence. Because the illicit substances can be detected for short periods of time (12–24 h) after consumption, OF is suitable for detecting recent drug use, e.g., for roadside testing. A major advantage of using OF instead of blood is the noninvasive nature of the sampling procedure, which does not require the presence of medical personnel. Furthermore OF can be collected under direct observation, which makes it difficult to substitute or adulterate the sample. In addition, being composed of 99.4% water, 0.3% proteins, and 0.3% mucin, OF is a simpler matrix than urine and plasma, and the matrix effect is generally lower. Drugs are usually found in OF at concentrations highly correlated with those present in plasma and are less susceptible to dilution by fluid intake. Some drugs are better detected in OF than in blood: for example, heroin, which has a short half-life in blood, can be well detected in OF through its metabolite 6-MAM. Codeine is easily found both in plasma and in OF, while A and MA, if taken orally, are found at higher concentrations in OF than in plasma. MDMA and its metabolite MDA are easily found in OF of MDMA users, as well as the metabolite 7-aminoflunitrazepam is easily detected in OF of flunitrazepam users. Also cocaine and its major metabolite BE are well detectable in OF of cocaine users.

6.4.6 Urine

The determination of several abuse drugs and illicit substances is frequently performed in urine because large volumes of samples can be available and collection is easy and not invasive. However, urine samples might be subjected to adulteration, through substitution, dilution, or addition of "masking agents." A method clinically accepted and used to evidence possible adulterations or dilutions in urine samples is based on the evaluation of species typical of the urine composition, as, for example, the creatinine content. Also a urine sample of a tobacco smoker that does not contain cotinine can be suspected of substitution.

In urine relatively long detection-time windows are generally observed for drug metabolites, while the precursors are often present at trace amounts.

6.5 PRETREATMENT AND ANALYSIS

Detection and determination of unknown drugs of abuse in biological matrices represent a challenge for the analytical laboratory due to the high amount of substances possibly present and the different potential matrix effects. Sample pretreatment of biological matrices can reduce the effects of the interfering species, preconcentrate the analytes of interest, and reduce, when LC-MS techniques are used, possible effects of signal suppression or enhancement.

A review published by the authors has been devoted to discuss the signal suppression or enhancement in high performance liquid chromatography (HPLC) tandem mass spectrometry (22). The different mechanisms that can give rise to ion suppression or enhancement in electrospray ionization (ESI) and atmospheric pressure chemical ionization (APCI) sources have been discussed, and the possible actions that can be taken to overcome these effects have been considered and discussed. They can be summarized in pretreatment and extraction processes, modifications of chromatographic and/or mass spectrometry conditions, and selection of the calibration strategy in the quantification step. The conclusion, based on the results obtained by different authors (as many as 82 references) dealing with the problem, is that the occurrence of signal suppression or enhancement in MS detection cannot be attributed to only one cause but to a synergic effect of all the conditions involved: analyte chemical properties, matrix components, clean-up procedure, chromatographic conditions of both stationary and mobile phases, kind and features of mass spectrometer (as the ionization technique and positive ion or negative ion mode), the equipment, and the different source design (23–27). All the hypotheses proposed by the different authors to explain ion suppression/enhancement phenomena are reasonable and worth consideration, but, very likely, each mechanism takes place or can take place as a function of the experimental conditions involved. Because of the different reactions and mechanisms postulated, very likely the cause is never only one, but more effects simultaneously occur. In particular when comparing the matrix effect in the most used ionization sources ESI and APCI, the examples reported indicate that it is not possible to generalize and expect, for example, to observe signal suppression in ESI and signal enhancement in APCI. It is peculiar in fact that a different behavior for analytes of similar structure, under the same conditions of ionization, instrument, chromatography, cleanup, and kind of matrix has been observed (22). On the other hand, another example showed a different matrix effect behavior (suppression or enhancement) when the same analytes in the same matrix were determined under different extraction, chromatographic and mass spectrometry conditions (28–32). It can perhaps also be that among all the potential causes that lead to matrix effects, one prevails and becomes the only responsible one, but this situation can hardly be predicted. The conclusion is that more effects can simultaneously occur, and it is not possible to generally predict the final effect.

In each analysis, to minimize the matrix effect, the best strategy must be chosen and optimized, and under these conditions the recovery for each analyte must be evaluated.

These conclusions also hold for the works considered here and devoted to the UHPLC determination of abuse drugs and other illicit substances in biological matrices.

So, for example, a study concerning the UHPLC-MS/MS determination of seventeen illicit drugs in oral fluid compares the performances of ESI, APCI, and atmospheric pressure photo ionization (APPI) (1). The ion suppression of most analytes on ESI (28%–78%) was lower than on both APCI and APPI. According to the authors, a possible explanation of the phenomenon is that APCI and APPI probes evaporate inlet solutions and ionize analytes via gas-phase chemistry, and consequently they are less affected than ESI by macromolecules present in samples and that are not evaporated. Oral fluids likely contain many salts and small molecules partitioned from plasma or nasal pharynx rather than macromolecules, which can lead to increased ion suppression on APCI and APPI. Nevertheless, the authors suggest that these results cannot be generalized for all the oral fluid UHPLC-MS analyses. Figure 6.1 shows a typical UHPLC-ESI MS chromatogram of a standard solution mixture of the seventeen illicit drugs considered, which includes amphetamines, benzodiazepines, cocaine alkaloids, and opioids.

In the pretreatment step of the matrix, liquid-liquid extraction (LLE), off-line solid phase extraction (SPE), and online SPE are the most used techniques. The online SPE strategy seems to be the best choice to improve method sensitivity, to shorten pretreatment and analysis times, and to increase the number of the samples analyzed in the same time. An online SPE UHPLC-MS/MS method was developed for the simultaneous identification and determination in human urine of forty-two therapeutic drugs and drugs of abuse, belonging to different chemical classes (8). The online extraction procedure is based on the use of a cationic extraction column coupled with a LC-MS hybrid mass spectrometer through a two-way switching valve, which

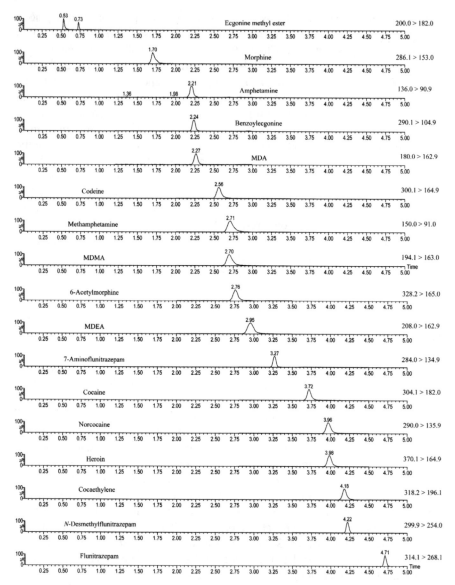

Figure 6.1   The chromatogram of 17 seventeen illicit drugs on a HSS-T3 column (0.1 ng/mL, 4 μL injection on ESI+). Wang, I.-T., Feng, Y.-T., Chen, C.-Y. *J. Chromatogr. B.* 2010, Figure 2, with permission.

allows the loading and the injection steps. Figure 6.2 shows a typical chromatogram, recorded for a urine sample of a drug-dependent patient, found positive to the I level screening test. In the sample the following substances have been identified and quantified: cocaine and its metabolites BE and EME, diazepam, nordazepam, and THC-COOH.

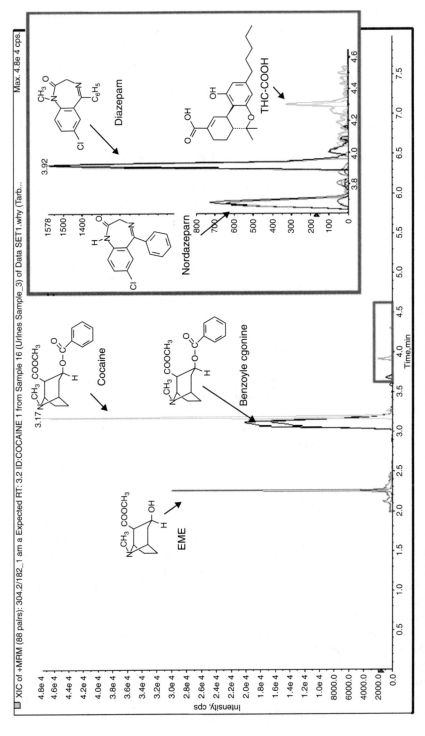

Figure 6.2 Online SPE UHPLC-MS/MS chromatogram of urine sample diluted 1/10 (v/v) in SPE loading solution. From Chiuminatto, U., et al. *Anal. Chem.* 2010, Figure 5, with permission.

The analysis of drugs in biological matrices is a very important task in clinical-medical and forensic fields. Generally, drugs of abuse are screened in a level I test performed through immunochemical techniques (enzyme-linked immunosorbent assay (ELISA), radioimmunoassay (RIA), and fluorimetric polarization immunoassay (FPIA)). Positive samples are further confirmed with a level II test characterized by higher sensitivities to minimize the number of false positive responses. It is worth underlining that the screening cutoff values in urine samples are not the same for all the countries; for example, concerning opiates, the threshold concentration is 300 ng/mL^{-1} in Europe and Australia, and 2000 ng/mL^{-1} in the United States (33). The cutoff concentrations used in the screening test of drugs in biological fluids in Europe (34, 35) and in United States (36) are available.

According to the European Monitoring Centre for Drugs and Drug Addiction published in 2009 (37), polydrug patterns and the combined use of different substances are today the norm in Europe (38–40). Therefore, rapid, accurate, precise, and possibly cheap multicomponent methods are required to overcome the long analysis times required by screening and confirming tests. Rapid, widely applicable and automated analytical methods are especially useful during important sport events, when the results are required within 24–48 h.

The techniques more largely employed in the determination of drugs of abuse are liquid chromatography with electrochemical (HPLC-ED) and diode array detectors (HPLC-DAD) and capillary electrochromatography-time-of-flight mass spectrometry (CEC-TOF). More recent methods are based on gas or liquid chromatography hyphenated with mass spectrometry detection, HPLC-MS/MS. HPLC allows to separate complicated mixtures of low and high molecular weight compounds and of different polarities and acid-base properties.

Much more recently the need to reduce analysis times and together to increase sensitivity and resolution has led to the development of ultra-fast separations, that is, UHPLC or ultrahigh pressure liquid chromatography. The improvements are largely due to advancements in the particle size and bridging structure of the column packing, complemented by additional instrumental modifications. The UHPLC technique enables the use of columns packed with small particles (< 2 μm) coupled to chromatographic systems specially designed to run all the optimum linear velocities (high pressure and minimal system volumes). The MS analyzer must be able to quickly acquire data to collect enough data points to define the narrow peaks obtained with the UHPLC systems when running at high linear velocities. In addition, without compromising resolution and separation efficiency, UHPLC enables the reduction of analytical time. As an example, the chromatogram reported in Figure 6.3 concerns the UHPLC-MS determination in whole blood extract of six common amphetamine-type substances (41). The good resolution and sensitivity permitted to work in single ion monitoring, that is, the procedure that uses only one m/z ratio for each analyte.

Nevertheless in MS/MS detection, when possible, two ion transitions are used to increase the confidence of identification, and this allows the use of ion ratios. So, for example, the potential misidentification between the isomers morphine and norcodeine, which structurally differ for the position of a methyl group, was greatly reduced choosing fragment ions that were unique to each molecule and using different

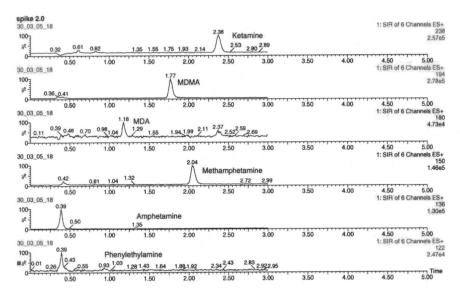

Figure 6.3   Selected ion monitoring chromatogram of four amphetamine-type substances, phenylethylamine, and ketamine from extracted whole blood. From Apollonio, L.G., et al. *J. Chromatogr. B.* 2006, Figure 1b, with permission.

ion ratios. Also hydrocodone/codeine and hydromorphone/morphine/codeine were identified through the use of two transitions (42). However, it is not always possible to use two transitions, when, for example, it is difficult to achieve adequate sensitivity; only one product ion can be used when the possibility of misidentification is relatively low, as, for example, in the identification of buprenorphine and norbuprenorphine (molecular weights 467 and 413 amu, respectively) (42).

Also TOF provides significant advantages in terms of sensitivity, selectivity, and speed. Even if the selectivity provided by full scan of accurate mass by TOF instrument is generally lower than that provided by monitoring MS/MS transitions, UHPLC generates narrower peaks, which reduces the likelihood of unwanted interferences. An UHPLC-TOF MS developed for hair sample analysis, characterized by high selectivity and sensitivity, allows the reliable quantification of fifty-two drugs present at low levels in hair. As an example, Figure 6.4 shows the UHPLC-TOF MS total ion chromatogram and the extracted ion chromatogram of a hair sample positive to six substances belonging to different classes of drugs of abuse and pharmaceuticals (12).

The results of the published UHPLC methods for the determination in human biological matrices of drugs of abuse and other substances, whose determination can offer useful additive information, are presented in Table 6.1. For each manuscript the drugs and the other substances considered, the kind of matrix analyzed, the pretreatment process, the chromatographic and detection conditions used, and the limit of detection (LOD) and limit of quantification (LOQ) values, when available, are reported. As it can be observed most of the methodologies concern the simultaneous determination of a great number of species.

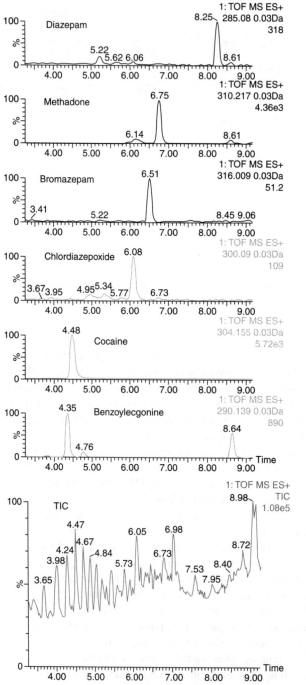

Figure 6.4   Total ion chromatogram and extracted ion chromatograms of a hair sample. From Nielsen, M.K.K., et al. *Forensic Sci. Int.* 2010, Figure 1, with permission.

Table 6.1 List of Analytical Methodologies for the Determination of Drugs of Abuse in Biological Samples

Analytes	Analytical Method	Experimental Conditions	LOD, LOQ	Matrix	Sample Pretreatment	Ref.
17 analytes: 5 amphetamines, 3 benzodiazepines, 5 cocaine alkaloids, and 4 opioids	UHPLC-MS/MS	Stationary phase: Acquity HSS T3. Mobile phase: mixture of 10 mM ammonium acetate aqueous solution and acetonitrile. Gradient elution. Flow rate 0.5 mL/min.	LOD < 4.72 ng/mL with ESI; < 7.25 ng/mL with APCI and < 6.10 ng/mL with APPI. LOQ $\leq$ 1.9 ng/mL for ESI, $\leq$ 2.2 ng/mL for APCI and $\leq$ 2.1 ng/mL for APPI.	OF	Dilution with twice amounts of Milli-Q water.	1
27 analytes: 5 amphetamines, buprenorphine and metabolite, cocaine and metabolite, 2 drugs for substitution treatment, 2 narcotic antagonists, and 13 opioids	UHPLC-MS/MS	Stationary phase: Acquity HSS T3. Mobile phase: mixture of 5 mM ammonium formate aqueous solution and methanol adjusted to operational pH = 3 with formic acid. Gradient elution. Flow rate 0.5 mL/min.	LOQ always < 7.3 µg/L.	Blood serum	Off-line SPE with Oasis MCX (mixed-mode cation exchange) cartridges.	2
103 analytes: 9 β-blockers, 19 diuretics, 8 narcotics (including methadone and buprenorphine), 8 pharmaceuticals, and 59 stimulants (including 11 amphetamines and cocaine)	UHPLC-TOF MS	Stationary phase: Acquity UPLC BEH C18 with Acquity BEH C18 pre-column. Mobile phase: mixture of 0.1% formic acid aqueous solution and methanol 0.1% in formic acid. Gradient elution. Flow rate 0.4 mL/min.	LODs always < 500 ng/mL.	Urine	Centrifugation, 1/1 dilution with ultrapure water.	6

(continued)

Table 6.1 List of Analytical Methodologies for the Determination of Drugs of Abuse in Biological Samples (*Continued*)

Analytes	Analytical Method	Experimental Conditions	LOD, LOQ	Matrix	Sample Pretreatment	Ref.
Opiates: BE, cocaine, codeine, ethylmorphine, morphine, 6-MAM, oxycodone, and pholcodine	UHPLC-MS/MS	Stationary phase: Acquity UPLC BEH C18 with pre-column. Mobile phase: mixture of ammonium bicarbonate aqueous buffer pH 10.2 and methanol. Gradient elution. Flow rate 0.4 mL/min.	LOD range from 0.0010–0.0026 µg/mL and LOQ from 0.010–0.020 µg/mL.	Urine	Off-line SPE with Oasis MCX SPE cartridges.	7
42 analytes: 3 amphetamines, BE, 21 benzodiazepines, 3 cannabinoids, 3 cocaine alkaloids, DMTA, LSD, ketamine, and 8 opiates	Online SPE-UHPLC-MS/MS	Stationary phase: Zorbax Eclipse XDB-C18 at T = 55°C. Mobile phase: mixture of 0.5% formic acid aqueous solution and 0.5% formic acid in acetonitrile. Gradient elution. Flow rate 900 µL/min.	LODs range from 0.20–1.52 µg/L and LOQs from 0.73–5.00 µg/L.	Urine	Enzymatic hydrolysis in acetate buffer at 65°C, online SPE with Strata X-CW (weak cation exchange) cartridge.	8
103 analytes: 9 β-blockers, 19 diuretics, 8 narcotics (including methadone and buprenorphine), 8 pharmaceuticals, and 59 stimulants (including 11 amphetamines and cocaine)	UHPLC-TOF MS	Stationary phase: Acquity UPLC BEH C18 with BEH C18 pre-column. Mobile phase: mixture of 0.1% formic acid aqueous solution and acetonitrile 0.1% in formic acid. Gradient elution. Flow rate 0.4 mL/min.	—	Urine	Different off-line SPE methods: MCX and MAX (mixed-mode anion exchange) cartridges.	9

Analytes	Technique	Conditions	LODs/LOQs	Matrix	Extraction	
A, ephedrine, MA, , MDA, MDEA, MDMA, and PMA	UHPLC-MS/MS	Stationary phase: Acquity UPLC HSS C18. Mobile phase: mixture of 5 mM ammonium formate, 0.05% in formic acid and methanol. Gradient elution. Flow rate 0.3 mL/min.	LODs range from 0.13 ng/mL for MDEA to 0.50 ng/mL for MDMA. LOQs always ≤ 2.5 ng/mL.	Urine	LLE with ethyl acetate.	10
28 analytes: 9 amphetamines, 5 cocaine alkaloids, ketamine, and 13 opioids	UHPLC-MS/MS	Stationary phase: Acquity BEH C18 with pre-column. Mobile phase: mixture of aqueous ammonia buffer pH 9 and acetonitrile. Gradient elution. Flow rate 600 µL/min.	–	Urine	Supported liquid–liquid extraction (SLE), pH 6.9–9.0, ethyl acetate as the solvent.	11
52 analytes: 6 amphetamines, 14 benzodiazepines, 2 cocaine alkaloids, 22 pharmaceuticals, ketamine, and 7 opioids	UHPLC-TOF MS	Stationary phase: Acquity HSS T3. Mobile phase: mixture of 0.05% formic acid aqueous solution and methanol. Gradient elution. Flow rate 0.30 mL/min.	LODs range from 0.01–0.05 ng/mL; LOQs range from 0.05–2.0 ng/mL.	Hair	Extraction with a 25:50:25 (v/v/v) mixture of methanol:acetonitrile: 2 mM ammonium formate containing 8% acetonitrile.	12
50 analytes: 19 β-blockers, 21 diuretics, 2 steroids, and 8 stimulants	UHPLC-MS/MS	Stationary phase: Acquity BEH Shield RP18. Mobile phase: mixture of 0.1% formic acid aqueous solution and methanol. Gradient elution. Flow rate 0.5 mL/min.	LODs range from 3 ng/mL (bisoprolol) to 200 ng/mL (hydroflumethylazide).	Urine	Off-line SPE with Nexus™ cartridge.	13

(continued)

Table 6.1 List of Analytical Methodologies for the Determination of Drugs of Abuse in Biological Samples (*Continued*)

Analytes	Analytical Method	Experimental Conditions	LOD, LOQ	Matrix	Sample Pretreatment	Ref.
130 analytes: 16 amphetamines, cocaine, cotinine, 2 designer drugs, 36 diuretics, 18 pharmaceuticals, nicotine, 20 opioids, and 36 stimulants	UHPLC-MS/MS	Stationary phase: Acquity UPLC BEH shield RP18. Mobile phase: mixture of 10 mM ammonium acetate and methanol. Gradient elution. Flow rate was 0.4 mL/min.	LODs range from 1 ng/mL (BE) to 50 ng/mL (methylclothiazide).	Urine	Dilution with the internal standard.	14
37 analytes: 28 diuretics, 4 pharmaceuticals, and 5 stimulants	UHPLC-MS/MS	Stationary phase: Acquity BEH C18. Mobile phase: mixture of 0.01% formic acid aqueous solution and 0.01% formic acid in acetonitrile. Gradient elution. Flow rate 0.6 mL/min.	LOD < 200 ng/mL (50 ng/mL for BE).	Urine	LLE with ethyl acetate.	15
28 analytes: 5 amphetamines, 4 benzodiazepines, 2 cocaine alkaloids, cotinine, 2 designer drugs, ketamine, LSD, 6 pharmaceuticals, and 6 opioids	UHPLC-TOF MS	Stationary phase: Acquity UPLC HSS T3. Mobile phase: mixture of 0.05% formic acid aqueous solution and methanol. Gradient elution. Flow rate 0.3 mL/min.	LODs always <125 ng/mL.	Urine	Acidic and basic LLE extractions, the solvent being a mixture of dichloromethane: hexane: ether: isoamyl alcohol; 30: 50: 20: 0.5 (v/v/v/v).	16
A, DMA, DMANO, and MA	UHPLC-MS/MS	Stationary phase: Platinum EPS C18. Mobile phase: mixture of 0.01 M ammonium formate aqueous solution and acetonitrile in isocratic elution 77/23 (v/v). Flow rate 0.2 mL/min.	LOD 0.01 μg/mL for A, MA, and DMA; 0.04 μg/mL for DMANO.	Urine	Off-line SPE with C18 cartridge.	17

Analyte	Technique	Stationary/Mobile phase	LODs/LOQs	Matrix	Sample preparation	Ref.
Cocaine and BE	UHPLC-MS/MS	Stationary phase: UPLC HSS T3. Mobile phase: mixture of 0.1% formic acid aqueous solution and acetonitrile 0.1% in formic acid. Gradient elution. Flow rate from 0.5–0.6 mL/min.	For both the analytes LODs = 3 ng/g and LOQs = 30 ng/g.	Meconium	Off-line SPE with Clean-Screen cartridge.	18
14 designer drugs	UHPLC-FD (excitation 450 nm, emission 550 nm)–TOF MS	Stationary phase: Acquity UPLC™ BEH C18. Mobile phase: mixture of 0.1% formic acid aqueous solution and 0.1% formic acid in acetonitrile/methanol 20/80 (v/v). Isocratic elution 40/60 v/v. Flow rate 0.4 mL/min.	For both plasma and urine: LODs between 0.30 and 150 pmol/0.1 mL and LOQs between 1.00 and 500 pmol/0.1 mL.	Plasma and urine	4-(N,N-dimethylamino-sulfonyl)-7-fluoro-2,1,3-benzoxadiazole derivatization at pH 9.3 for borax.	19
THC, 11-OH-THC, and THC-COOH	UHPLC-MS/MS	Stationary phase: Acquity UPLC BEH C18. Mobile phase: mixture of 0.1% formic acid aqueous solution and acetonitrile. Gradient elution. Flow rate 0.7 mL/min.	LODs range from 0.02 (THC) to 0.10 ng/mL (THC-COOH). LOQs range from 0.05–0.20 ng/mL.	Whole blood	Protein precipitation. Off-line SPE with Bond Elut Certify cartridge.	20

(continued)

Table 6.1 List of Analytical Methodologies for the Determination of Drugs of Abuse in Biological Samples (*Continued*)

Analytes	Analytical Method	Experimental Conditions	LOD, LOQ	Matrix	Sample Pretreatment	Ref.
Methadone, EDDP	UHPLC-MS/MS	Stationary phase: Acquity UPLC BEH C18. Mobile phase: mixture of 0.1% formic acid aqueous solution and 0.01% formic acid in methanol. Gradient elution. Flow rate 0.2 mL/min.	LOQ: 0.1 ng for both the analytes.	Fingerprint	Finger pre-cleaned with ethanol and rubbed over forehead and face was pressed against a glass cover slip that was transferred into a glass vial. The fingerprint deposit was dissolved in methanol/dichloromethane 2/5 (v/v) containing 20 mM hydrochloric acid and the internal standards.	21
Ketamine	UHPLC-MS/MS	Stationary phase: Platinum EPS C18 Rocket™. Mobile phase: mixture of 10 mM ammonium formate and acetonitrile in isocratic elution 77/23 (v/v). Flow rate 0.2 mL/min.	LOD = 5 ng/mL.	Urine	Off-line SPE system with C18 cartridge.	38
29 analytes: 9 amphetamines, 9 benzodiazepines, 2 cocaine alkaloids, 6 opioids, THC, zolpidem, and zopiciclone	UHPLC-MS/MS	Stationary phase: Acquity UPLC HSS T3 C18. Mobile phase: mixture of 2 mmol/L ammonium acetate aqueous solution and methanol. Gradient elution. Flow rate 0.4 mL/min.	LOQ = 1.0 μg/kg for cocaine, bromazepam, nitrazepam and 0.5 μg/kg for the others.	OF	Automatic off-line SPE equipped with Bond Elut Certify SPE cartridges.	39

Analytes	Technique	Chromatographic conditions	LOD/LOQ	Matrix	Sample preparation	Ref.
LSD, iso-LSD, nor-LSD, and OH-LSD	UHPLC-MS/MS	Stationary phase: Acquity UPLC BEH C18. Mobile phase: mixture of aqueous ammonium acetate buffer pH 4.0 and acetonitrile. Gradient elution. Flow rate 0.2 mL/min.	In urine: LODs always = 10 pg/mL and LOQs always ≤ 50 pg/mL. In blood: LODs ≤ 10 pg/mL and LOQs ≤ 50 pg/mL.	Blood and urine	LLE. Solvent: dichloromethane/isopropyl alcohol (85/15, v/v) mixture.	40
A, ephedrine, ketamine, MA, MDA, MDEA, MDMA, and pseudoephedrine	UHPLC-UV (λ = 254 nm) - MS	Stationary phase: UPLC BEH C18. Mobile phase: 52% aqueous solution of pyrrolidine (0.5 mL of glacial acetic acid in 500 mL) and 48% (v/v) methanol. Isocratic elution. Flow rate 0.4 mL/min.	–	Whole blood	Off-line SPE, Oasis MCX cartridge.	41
A, acetyl codeine, BE, buprenorphine, codeine, cotinine, dihydrocodeine, EDDP, 6-MAM, methadone, morphine, norcodeine, and norbuprenorphine	UHPLC-MS/MS	Stationary phase: Acquity UPLC HSS T3. Mobile phase: aqueous 5 mmol/L ammonium formate 0.05% in formic acid solution and methanol. Gradient elution. Flow rate 0.35 mL/min.	LOQs between 5 µg/L for acetyl codeine and 35 µg/L for dihydrocodeine. 250 µg/L for norcodeine.	Urine	Enzymatic treatment. Off-line SPE with Strata Screen C cartridge.	42

(continued)

Table 6.1 List of Analytical Methodologies for the Determination of Drugs of Abuse in Biological Samples (*Continued*)

Analytes	Analytical Method	Experimental Conditions	LOD, LOQ	Matrix	Sample Pretreatment	Ref.
20 analytes: heroin and its basic, and neutral impurities	(1) UHPLC-MS/MS and (2) UHPLC-UV (280 nm)	(1) Stationary phase: Acquity UPLC BEH C18. Mobile phase: mixture of aqueous 1% formic acid and acetonitrile. Gradient elution Flow rate 0.300 mL/min. (2) Stationary phase: Acquity UPLC BEH C18. Mobile phase: mixture aqueous 10 mM ammonium bicarbonate solution and acetonitrile. Gradient elution. Flow rate 0.300 mL/min.	LOD < 12.3 pg (MS detection). For five impurities LODs < 18 ng/mL.	–	–	43
40 analytes: 7 amphetamines, 14 benzodiazepines, 2 cocaine alkaloids, 7 designer drugs, ketamine, LSD, 7 opioids, and THC-COOH	UHPLC-MS/MS	Stationary phase: Zorbax™ Eclipse XDB-C18 with a frit guard assembly UPLC™. Mobile phase: mixture of 0.1% formic acid aqueous solution and acetonitrile. Gradient elution. Flow rate 0.6 mL/min.	LOQs range between 7.0 and 33 ng/mL.	Urine	Enzymatic hydrolysis at 60°C for 60 min. For LSD, basic solvent extraction with chlorobutane.	44

Analytes	Technique	Chromatographic conditions	Sample	LOD/LOQ	Notes	Ref
Ketamine, nor-ketamine, and dehydronorketamine	UHPLC-MS/MS	Stationary phase: Acquity UPLC BEH C18. Mobile phase: mixture of 0.1% formic acid aqueous solution and 0.1% formic acid in acetonitrile. Isocratic elution. Flow rate 350 μL/min.	Urine	LOD: 0.03 ng/mL for ketamine and 0.05 ng/ml for nor-ketamine.	Off-line SPE with Bond Elut Certify mixed-mode (cation exchange and C18 cartridges).	45
THC-COOH	UHPLC-MS/MS	Stationary phase: Acquity UPLC BEH C18. Mobile phase: mixture of aqueous 0.1% formic acid solution and acetonitrile. Gradient elution. Flow rate 200 μL/min.	Urine	0.2 ng/mL.	–	46
14 analytes: 3 amphetamines, 3 cannabinoids, clonazepam, cocaine, 3 opioids, and 3 stimulants	UHPLC-MS/MS	Different stationary and mobile phases were tested.	–	LODs range from 0.29–90.0 ng/mL. LOQs range 0.96–300 ng/mL.	–	47

UHPLC-QTOF MS has been employed to detect 103 doping agents belonging to different classes and present in the WADA prohibited list (6, 9). To obtain an overview of the patient clinical situation, it is very useful to also consider the possible interactions of drugs with the most commonly prescribed pharmaceutical products. For this purpose, many methods consider the simultaneous separation of complex mixtures of drugs, illicit substances, and common pharmaceutical compounds. A mixture of up to 130 compounds has been separated (14).

In Table 6.1, when the number of analytes considered is lower than thirteen, the names of all of them are listed, whereas for greater numbers, the classes and the number of the substances belonging to each class are reported.

It is interesting to observe that the panel of the results concerning the kinds and the amounts of the species found in the biological matrices, besides giving direct information, can help to extract other useful information. So, for example, in heroin determination, the drug profile can also involve the determination of trace impurities, important for "strategic" and "tactical" intelligence purposes (43). The so called strategic intelligence gives information on the drug processing and/or on its geographical origin, whereas tactical intelligence tells whether two or more samples come from the same source. As mentioned, heroin derives from acetylation of morphine obtained from opium and can therefore contain impurities that can arise both from opium and from the acetylation process of opium alkaloids. The impurities that can be found in heroin are numerous and generally of basic properties as morphine, codeine, noscapine, papaverine, acetylcodeine, and acetylmorphine; the concentration ratios of the impurities present can be used for intelligence purposes.

As mentioned, the simultaneous analysis for drugs, their metabolites, diuretics, stimulants, β-blockers, and pharmaceuticals offers a more complete overview about the clinical global situation of the patient. An interesting example is represented by the panel of substances found in the urine of a patient (11). The presence of morphine, codeine, noscapine, and 6-MAM indicated a probable consumption of heroin, but the same urine sample also contained an important quantity of methadone, widely used in heroin substitution treatment programs, and its metabolite EDDP. The sample also included fentanyl, which can serve as a direct pharmacological substitute for heroin in opiate-dependent individuals, and a certain amount of tramadol, generally used as an analgesic.

It is useful to remember that the optimization of a methodology comprehensive of both the pretreatment and the analysis must take into account the specific conditions that must be used. For example, the number of the samples to be analyzed must be considered but also the expected proportion of positive samples. Methodologies and laboratory organization are different for (1) a laboratory that provides a regional screening service for abuse for the local mental health trust and receives about 26,000 samples per year (42), (2) laboratories that handle samples for pre-employment screening where the expected proportion of positive samples is much lower, or (3) cases of forensic interest in which it is often very useful to collect the maximum of information about all the species present.

REFERENCES

1. WANG, I.-T., FENG, Y.-T., CHEN, C.-Y. *J. Chromatogr. B.* 2010, 878: 3095–3105.
2. DUBOIS, N., DEBRUS, B., HUBERT, P.H., CHARLIER, C. *Acta Clin. Belg.* 2010, 65: 75–84.
3. EU Drugs Strategy (2005–2012). Brussels: Council of the European Union, 2004.
4. EU Drugs Action Plan (2009–2012). Brussels: Council of the European Union, 2008.
5. National Institute on Drugs Abuse. Avalable at http://www.drugabuse.gov/DrugPages/DrugsofAbuse. html (access on July 2011).
6. BADOUD, F., GRATA, E., PERRENOUD, L., AVOIS, L., SAUGY, M., RUDAZ, S., VEUTHEY, J.-L. *J. Chromatogr. A.* 2009, 1216: 4423–4433.
7. BERG, T., LUNDANES, E., CHRISTOPHERSEN, A.S., STRANDA, D.H. *J. Chromatogr. B.* 2009, 877: 421–432.
8. CHIUMINATTO, U., GOSETTI, F., DOSSETTO, P., MAZZUCCO, E., ZAMPIERI, D., ROBOTTI, E., GENNARO, M.C., MARENGO, E. *Anal. Chem.* 2010, 82: 5636–5645.
9. BADOUD, F., GRATA, E., PERRENOUD, L., SAUGY, M., RUDAZ, S., VEUTHEY, J.-L. *J. Chromatogr. A.* 2010, 1217: 4109–4119.
10. DEL MAR RAMÍREZ FERNÁNDEZ, M., WILLE, S.M.R., DI FAZIO, V., GOSSELIN, M., SAMYN, N. *J. Chromatogr. B.* 2010, 878: 1616–1622.
11. MAQUILLE, A., GUILLARME, D., RUDAZ, S., VEUTHEY, J.-L. *Chromatographia* 2009, 70: 1373–1380.
12. NIELSEN, M.K.K., JOHANSEN, S.S., WEIHE DALSGAARD, P., LINNET, K. *Forensic Sci. Int.* 2010, 196: 85–92.
13. MURRAY, G.J., DANACEAU, J.P. *J. Chromatogr, B,* 2009, 877: 3857–3864.
14. THÖRNGREN, J.-O., ÖSTERVALL, F., GARLE, M. *J. Mass Spectrom.* 2008, 43: 980–992.
15. VENTURA, R., ROIG, M., MONFORT, N., SÁEZ, P., BERGÉS, R., SEGURA, J. *Eur. J. Mass Spectrom.* 2008, 14: 191–200.
16. LEE, H.K., HOB, C.S., IU, Y.P.H., LAI, P.S.J., SHEK, C.C., LOD, Y.-C., BENDSTRUP KLINKE, H., WOOD, M. *Anal. Chim. Acta.* 2009, 649: 80–90.
17. CHENG, W.-C., MOK, V.K.-K., CHAN, K.-K., LI, A.F.-M. *Forensic Sci. Int.* 2007, 166: 1–7.
18. GUNN, J., KRIGER, S., TERRELL, A.R. in GARG, U., HAMMETT-STABLER, C.A. (eds.), *Clinical Applications of Mass Spectrometry*, 1st ed. New York: Humana Press, 2010, p. 165–174.
19. MIN, J.Z., HATANAKA, S., TOYO'OKA, T., INAGAKI, S., KIKURA-HANAJIRI, R., GODA, Y. *Anal. Bioanal. Chem.* 2009, 395: 1411–1422.
20. JAMEY, C., SZWARC, E., TRACQUI, A., LUDES, B. *J. Anal. Toxicol.* 2008, 32: 349–354.
21. JACOB, S., JICKELLS, S., WOLFF, K., SMITH, N. *Drug Metab. Lett.* 2008, 2: 245–247.
22. GOSETTI, F., MAZZUCCO, E., ZAMPIERI, D., GENNARO, M.C. *J. Chromatogr. A.* 2010, 1217: 3929–3937.
23. MEI, H., HSIEH, Y.S., NARDO, C., XU, X., WANG, S., NG, K., KORFMACHER, W.A. *Rapid Commun. Mass Spectrom.* 2003, 17: 97–103.
24. HOLCAPEK, M., VOLNA, K., JANDERA, P., KOLAROVA, L., LEMR, K., EXNER, M., CIRKVA, A. *J. Mass Spectrom.* 2004, 39: 43–50.
25. HSIEH, Y.S., CHINTALA, M., MEI, H., AGANS, J., BRISSON, J., NG, K., KORFMACHER, W.A. *Rapid Commun. Mass Spectrom.* 2001, 15: 2481–2487.
26. MATUSZEWSKI, B.K., CONSTANZER, M.L., CHAVEZ-ENG, C.M. *Anal. Chem.* 1998, 70: 882–889.
27. GANGL, E.T., ANNAN, M., SPOONER, N., VOUROS, P. *Anal. Chem.* 2001, 73: 5635–5644.
28. MATUSZEWSKI, B.K., CONSTANZER, M.L., CHAVEZ-ENG, C.M. *Anal. Chem.* 2003, 75: 3019–3030.
29. MATUSZEWSKI, B.K. *J. Chromatogr. B.* 2006, 830: 293–300.
30. ALDER, L., LUDERITZ, S., LINDTNER, K., STAN, H.J. *J. Chromatogr. A.* 2004, 1058: 67–79.
31. LIANG, H.R., FOLTZ, R.L., MENG, M., BENNET, P. *Rapid Commun. Mass Spectrom.* 2003, 17: 2815–2821.
32. LINDSEY, M.E., MEYER, M., THURMAN, E.M. *Anal. Chem.* 2001, 73: 4640–4646.
33. Scientific Section Division for Operations and Analysis. Rapid On-Site Screening of Drugs of Abuse of United Nations International Drug Control Programme. Vienna. 2001.

34. VERSTRAETE, A.G. *Ther. Drug Monit.* 2004, 26: 200–205.

35. Quality Commission of Italian Toxicology Forensic Group. Guidelines of the Laboratories of Abuse Drug Analysis for Forensic Purposes. 2010. Available at http://www.drugabuse.gov/DrugPages/ DrugsofAbuse.html (access on July 2011).

36. RODNAY, N. Available at http://ezinearticles.com/?Common-Drugs-of-Abuse—Cut-Off-Concentrations-and-Their-Detection-Periods&id=5885533 (accessed July 2011).

37. European Monitoring Centre for Drugs and Drugs Addiction. Polydrug Use: Patterns and Responses. 2009.

38. CHENG, J.Y.K., MOK, V.K.-K. *Forensic Sci. Int.* 2004, 142: 9–15.

39. BADAWI, N., SIMONSEN, K.W., STEENTOFT, A., BERNHOFT, I.M., LINNET, K. *Clin. Chem.* 2009, 55: 2004–2018.

40. CHUNG, A., HUDSON, J., MCKAY, G. *J. Anal. Toxicol.* 2009, 33: 253–259.

41. APOLLONIO, L.G., PIANCA, D.J., WHITTALL, I.R., MAHERA, W.A., KYD, J.M. *J. Chromatogr. B.* 2006, 836: 111–115.

42. DUXBURY, K., ROMAGNOLI, C., ANDERSON, M., WATTS, R., WAITE, G. *Ann. Clin. Biochem.* 2010, 47: 415–422.

43. LURIE, I.S., TOSKE, S.G. *J. Chromatogr. A.* 2008, 1188: 322–326.

44. EICHHORST, J.C., ETTER, M.L., ROUSSEAUX, N., LEHOTAY, D.C. *Clin. Biochem.* 2009, 42: 1531–1542.

45. PARKIN, M.C., TURFUS, S.C., SMITH, N.W., HALKET, J.M., BRAITHWAITE, R.A., ELLIOTT, S.P., OSSELTON, S.P., COWAN, D.A., KICMAN, A.T. *J. Chromatogr. B.* 2008, 876: 137–142.

46. STEPHANSON, N., JOSEFSSON, M., KRONSTRAND, R., BECK, O. *J. Chromatogr. B.* 2008, 871: 101–108.

47. JIANG, G. *Am. Lab.* 2010, 42: 40–42.

Chapter 7

UHPLC in the Analyses of Isoflavones and Flavonoids

Sylwia Magiera and Irena Baranowska

7.1 INTRODUCTION

Flavonoids were described for the first time in 1936 by Szent-Györga. They are natural, bioactive chemical substances extremely widespread in the vegetable kingdom. Virtually, flavonoids accumulate in all parts of the plant, namely in the leaves, flowers, fruit, seeds, and roots, giving them a characteristic taste, odor, and color for a given species. They are indispensable elements in plant physiology because they act as plant hormones and growth regulators, the conveyors of energy in photosynthesis, enzyme inhibitors and precursors, as well as dyes. Moreover, they are a natural shield that protects plant cells against the harmful effects of the environment, for example, against microbes or ultraviolet radiation (1).

As the compounds, which are commonly present in plants, they are a significant element in the human diet, and what is worth mentioning, their main sources are vegetables, fruit, tea, wine, and cocoa. Furthermore, certain seeds of various plants, especially legumes such as certain cereal crops, herbs, and spices, are the rich sources of flavonoids (2). On top of that, flavonoids are included in the composition of herbal preparations that are used in treatment of numerous diseases. Their distinctiveness comprises the pharmacological and biological properties, which affect various physiological processes. Because of their chemical structure they tend to show a high biological activity, and they impact on different ways of cellular metabolism. Flavonoids have an anti-inflammatory, anti-allergic, antithrombotic, antiviral, and anticarcinogenic effect. They have a positive influence on the cardiovascular system; they protect low-density lipoprotein (LDL) against oxidative modification and reduce the lipid content in plasma (3).

Flavonoids are substances that act as polyphenols, containing in its structure a 1500-atomic carbon skeleton consisting of two benzene rings (A and B), connected by

Ultra-High Performance Liquid Chromatography and Its Applications, First Edition. Edited by Quanyun Alan Xu.
© 2013 John Wiley & Sons, Inc. Published 2013 by John Wiley & Sons, Inc.

a chain of propane (this class is denoted as follows: C6–C3–C6), which together with a oxygen atom forms a heterocyclic ring of C pyran (flavanols and anthocyanins) or pyron (flavones, flavanones, flavonols, and isoflavones). Particular flavonoids differ in substituents in the rings, which arise as a result of hydroxylation, methylation, acylation, and glycosylation by means of mono- or oligosaccharides such as glucose, galactose, rhamnose, xylose, and arabinose. In the world of plants, flavonoids may occur in two forms: aglycone and β-glycoside (aglycone connection with a part of sugar—from one to five particles of simple sugars—such as β-*D*-glucose, β-*L*-rhamnose, and β-*D*-galactose) (4).

Because of the differences in the construction of structural composition, flavonoid compounds can be divided into

- flavanones (e.g., naringenin, naringin, hesperetin, hesperidin);
- flavonols (e.g., quercetin, kaempferol, myricetin, fisetin, morin);
- flavones (e.g., apigenin, diosmetin, luteolin);
- anthocyanins (e.g., cyanidin, pelargonidin, malvidin);
- flavanols (e.g., epicatechin, epigallocatechin, catechin);
- isoflavones (e.g., genistein, daidzein).

7.2 UHPLC IN POLYPHENOLIC COMPOUNDS DETERMINATION

In recent years there has been a growing interest in polyphenolic compounds, by and large, due to their positive health benefits. The determination of polyphenolic compounds has become a subject of interest to many research centers, not only because of the preventive properties of these compounds, but also due to the fact that they have an ability to interact with other compounds. Still increasing development of analytical techniques enables the study of polyphenols in vitro and in vivo and the study of mechanisms of action and bioavailability. The aforementioned facts allow an understanding of their role in the human body and how they affect the quality of food.

For these reasons, the identification and determination of these compounds in real samples (e.g., in body fluids, plant extracts of food samples) took on special importance. This requires the development of a wide range of methods for the analysis of polyphenols, which constitutes a major challenge, due to the chemical diversity of polyphenolic compounds. On top of that, a complex composition of real samples and low concentrations of polyphenols require an extraordinary ability of separation, selectivity, and sensitivity of the developed methods.

Over the past few years there has been an impressive development of high performance liquid chromatography in combination with various detectors, both in the development of technology-related equipment, as well as their area of application. The development of techniques of the isolation and determination of polyphenolic compounds in the samples characterized by a complex composition of the matrix is invariably moving toward the automation of the process, limiting labor and time-consuming analysis, lowering the limit of quantification, and finally minimizing the amount of solvents that are used during the analytical process. Latest technologies involve the use of short chromatographic columns with small diameters and filling of small particles (about 2 μm). The application of short columns affects smaller amounts of solvent being consumed and smaller sample volumes, and it allows operation at high pressures. Technology based on the use of such columns, which is called ultrahigh pressure liquid chromatography (UHPLC), allows for the better separation of analytes and the better sensitivity of determinations.

Because of the wide application and biomedical significance of polyphenolic compounds (especially isoflavones) in recent years, there have been numerous studies describing the research over determination of polyphenolic compounds. These studies described procedures for determination of polyphenolic compounds by high performance liquid chromatography (HPLC) (5–8). A few reviews summarized the latest advances in sample preparation and analysis methods comprising topics such as isoflavones in soy and soy products (9) and in plant extracts (10), and polyphenolic compounds in vegetable and fruit samples (11). Because the UHPLC technique is relatively new, few works (12) focused on UHPLC methods for the determination of the polyphenolic compounds in biological fluids, food products, and plant extracts.

A great number of researchers focus in their works on a comparison of UHPLC and HPLC methods for the determination of polyphenolic compounds in samples. The results indicated that the use of UHPLC technology achieved an average tenfold increase in sensitivity of determinations, about fivefold shorter run time, as well as significant improvements in resolution and peak shape. Churchwell et al. compared the HPLC-MS and UHPLC-MS for the determination of soy isoflavones in dietary supplements (13), while Spácil et al., for the analysis of catechins, coumarins, and flavonoids in red wine samples (14). As expected, the results point to the numerous benefits from the application of UHPLC techniques—namely a much greater sensitivity, speed of analysis, and better resolution were obtained. The use of columns with small particles reduced the width of the peaks; therefore, it was possible to separate the diastereomers, which was not possible while using a conventional HPLC method. Ortega et al. used both the HPLC method and UHPLC for the analysis of procyanidins in the cocoa extracts (15). A column filled with silica and a mobile phase consisting of dichloromethane, methanol, water, and acetic acid was used in the HPLC method, while the UHPLC method used HSS T3 column and mobile phase consisted of a mixture of acetonitrile, water, and acetic acid. The application of UHPLC method reduced the analysis time from 80 min to 12.5 min; moreover, it greatly improved separation efficiency and sensitivity as compared to the previously used HPLC method. The similar results were obtained by other researchers, who presented how the combination of small particle size columns, high temperature, and high pressure can shorten the analysis time of procyanidins in cocoa samples and flavonols in green tea samples (16).

Cooper et al. described the UHPLC-MS method, which allowed the separation of six procyanidins within 3 min, while the same mixture of compounds was separated using HPLC within 20 min (17). In yet another work, the comparison of HPLC and UHPLC methods applied to the determination of polyphenols in thirty-three different samples of brandy can be reported (18). Furthermore, good separation of fourteen polyphenols was obtained in just 6.5 min, which significantly reduced analysis time in comparison with HPLC (60 min), and it also considerably reduced solvent consumption: 4.55 mL after applying UHPLC, 60 mL after using HPLC. The advantages of using the UHPLC method were also indicated by Klejdus et al., who described a method for the determination of ten isoflavones in plant extracts. In this case the analysis time was 4 min, whereas the analysis of mixtures of the same compounds using HPLC lasted 40 min (19).

Based on the results of the tests described in the works concerning the use of UHPLC methods for the determination of polyphenols in foods, it can be unequivocally stated that the replacement of HPLC method with new UHPLC methods is also advantageous from the point of view of shortened analysis time, reduced solvent consumption, and improved sensitivity. The application of UHPLC techniques coupled with spectral methods and appropriate methods of sample preparation seems to be a method of choice for a rapid, comprehensive, inexpensive, and precise determination of polyphenolic compounds in various samples.

7.2.1 Determination of Polyphenolic Compounds in Biological Fluids

Determining polyphenols in biological fluids is a difficult task. These difficulties not only arise from a large variety of analytes belonging to a group of polyphenolic compounds and their low concentrations in biological fluids, but also are associated with the interaction of the other components in the sample. The matrices of physiological fluids (urine, blood, plasma) contain a significant amount of accompanying compounds (such as neurotransmitters, amino acids, peptides, urea, creatinine, dyes, uric acid, and vitamins). Their presence can affect the final result of the analysis by "impregnating" the chromatography column (changing the conditions of separation) and their co-elution with analytes (a decrease of selectivity separation). Therefore, before the introduction of sample into the chromatographic column, it is necessary to apply the initial isolation of interfering matrix components and/or pre-concentration of polyphenolic compounds that are being analyzed.

Polyphenolic compounds are determined mainly in the plasma, rather than in whole blood and in urine. The research, in the majority of cases, relate to samples from patients who take dietary supplements containing polyphenols or are on a diet rich in such compounds. Biological fluid samples are collected at different time intervals after the administration of the herbal preparation (e.g., *Hippophae rhamnoides* (20), *Hawthorn leaves flavonoids* (21), *GegenQinlian decoction* (22), *Flos Lonicerae japonicae* (23)) containing different amounts of polyphenols. The isolation and pre-concentration of analytes of these samples usually precedes the enzymatic hydrolysis (24–28) or acid hydrolysis (20), which is based on the conversion of glucuronide and sulfate derivatives of polyphenolic compounds formed in phase II metabolism to the basic forms of polyphenols. To carry out the enzymatic hydrolysis there are two enzymes used—β-glucuronidase and sulfatase with different activity—where hydrolysis is usually carried for about 18 h and at 37°C (24–28).

The isolation and pre-concentration of polyphenolic compounds from biological fluids is usually carried out on solid phase extraction (SPE) columns of different packing materials (24, 25, 33–36). These compounds are usually extracted from urine and plasma samples by means of the Oasis HLB columns (26–30). The sorbents of this column are a copolymer of *m*-divinylbenzene and *N*-vinylpyrrolidone, which determines its broad uses. The application of hydrophile-lipophile balance (HLB) sorbents obtains high recoveries for a number of polyphenolic compounds with different physicochemical properties. The preparation of plasma samples for the analysis was also carried out by liquid-liquid extraction (LLE) where ethyl acetate was used as an extraction solvent (22, 31).

The analysis of the obtained extracts/eluents involves the identification and quantification of individual components. The determination of polyphenolic compounds in body fluids involves, most commonly, the use of high performance liquid chromatography (HPLC); however, lately UHPLC is used more often, which enables a good separation of the large number of polyphenolic compounds in a relatively short time. Table 7.1 summarizes UHPLC methods that were used for determining polyphenolic compounds in biological fluids.

The analyses of polyphenolic compounds using UHPLC are primarily conducted in a reversed phase. Silica gel modified by octadecylosilan groups is commonly used as the stationary phase. Furthermore, the separation of polyphenolic compounds using nonpolar sorbents is based on the hydrophobic interaction of individual polyphenols with the stationary phase, whereas the retention times of the individual substances depend on their solubility in water. In most cases aglycones, glycosides, and their derivatives are determined simultaneously in a chromatographic system, and glycosides are eluted followed by their corresponding aglycones.

The determination of polyphenolic compounds in biological fluids uses a short column of small diameter and with a small sorbents particle. Serra et al. (30) used an HSS T3 column of a length of 100 mm for the determination of procyanidins and their metabolites in rat plasma samples. Hence, the proposed method allowed the accurate determination of procyanidins and their metabolites within 5 min. The same column was used to identify twenty-one procyanidins and anthocyanins in rat plasma samples, after due preparation of samples using μSPE. The aforementioned methods were also applied by Marti et al. (29). Used by Wang et al., a system consisting of a BEH C18 column with a length of 100 mm at 30°C and a buffer as the mobile phase allowed the determination of isoflavones in plasma of six rats within 6 min (22).

In the case of simultaneous determination of a great number of polyphenolic compounds such as a mixture of twelve analytes (28), seventeen analytes (26), and twenty-four analytes (27), a gradient elution was most commonly used to obtain good separation of analytes in a relatively short period of time at 3.5, 4.5, and 8.0 min, respectively. Mobile phases consisted of acetonitrile and water more often than of methanol and water. In some cases acetic acid, formic acid, and ammonium acetate were added to the mobile phase to obtain better chromatographic separation and increased sensitivity of the MS (mass spectrometer) detector. The addition of acid

Table 7.1 Ultrahigh Pressure Liquid Chromatography in the Analysis of Polyphenolic Compounds in Biological Fluids

No.	Compound	Column	Mobile Phase	Detection	Type of Sample / Sample Preparation	Linear Dynamic Range	LOD / LOQ	Recovery (%)	Ref.
1.	Isorhamnetin, kaempferol, quercetin, baicalein	Acquity UPLC BEH C18 (50 × 2.1 mm, 1.7 μm)	*Gradient elution* A: methanol B: 0.1% formic acid in water *Flow rate:* 0.4 mL/min	ESI-MS	Plasma / acidic hydrolysis	0.009–1.04 μg/mL	0.78–3.38 ng/mL / 3.63–5.88 ng/mL	57.1–101.8	20
2.	Vitexin-4″-O-glucoside, vitexin-2″-O-rhammoside, rutin, vitexin	Acquity UPLC BEH C18 (50 × 2.1 mm, 1.7 μm)	*Gradient elution* A: acetonitrile B: 0.1% formic acid in water *Flow rate:* 0.2 mL/min	ESI-MS/MS	Plasma / protein precipitation	8–50.000 ng/mL	– / 8–16 ng/mL	72.0–95.0	21
3.	Puerarin, daidzein, baicalin, wogonoside, liquiritin	Acquity UPLC BEH C18 (100 × 2.1 mm, 1.7 μm)	*Gradient elution* A: methanol B: 0.1% formic acid and 5 mM ammonium acetate in water *Flow rate:* 0.2 mL/min	ESI-MS	Plasma / LLE (ethyl acetate)	0.0025–10.2 μg/mL	– / 2.54–10.2 ng/mL	66.9–92.1	22
4.	Rutin, luteolin-7-O-β-D-glucoside, quercetin-3-O-β-D-glucoside, lonicerin, icariin	Zorbax Eclipse XDB-C18 (50 × 4.6 mm, 1.8 μm)	*Gradient elution* A: 0.05% formic acid in water B: acetonitrile *Flow rate:* 0.4 mL/min	ESI-MS/MS	Plasma / protein precipitation	2.5–500 ng/mL	0.5–1.2 ng/mL / 2.5 ng/mL	88.7–92.5	23

No.	Analytes	Column	Mobile phase	Detection	Sample / extraction	Linear range	LOD/LOQ	Recovery (%)	Ref.
5.	Hesperidin, (R)-hesperetin, (S)-hesperetin	Acquity UPLC HSS T3 (100 × 2.1 mm, 1.8 μm) / DAICEL Chiralpak IA-3 (150 × 2.0 mm, 3.0 μm)	*Gradient elution* A: 0.1% formic acid in water B: 0.1% formic acid in acetonitrile *Flow rate:* 0.4 mL/min	ESI-MS/MS	Plasma, urine / enzymatic hydrolysis SPE (column: Oasis MAX)	0.02 –10 μM	25–50 nM (plasma) 50–100 nM (urine) / –	91.9–103.6	24
6.	Procyanidins, anthocyanins	Acquity UPLC HSS T3 (100 × 2.1 mm, 1.8 μm)	*Gradient elution* A: 0.2% acetic acid in water B: acetonitrile *Flow rate:* 0.4 mL/min	ESI-MS/MS	Plasma / μSPE (column: Oasis HLB)	0.01–11.5 μM	0.003–0.7 μM / 0.01–1.38 μM	70.0–100.0	29
7.	Hyperoside, isoquercitrin, quercetin-3'-O-glucoside, hibifolin, quercetin, gossypetin and metabolites	Acquity UPLC HSS T3 (100 × 2.1 mm, 1.8 μm)	*Gradient elution* A: acetonitrile B: 0.05% formic acid in water *Flow rate:* 0.1 mL/min	Q-TOF/MS	Plasma / LLE (ethyl acetate)	–	–	–	31
8.	Narirutin, naringin, hesperidin, neohesperidin and metabolites	Acquity UPLC BEH C18 (100 × 2.1 mm, 1.7 μm)	*Gradient elution* A: 0.1% formic acid in water B: acetonitrile *Flow rate:* 0.4 mL/min	DAD Q-TOF/MS	Plasma / protein precipitation	–	–	–	32
9.	Epimedin A, epimedin B, epimedin C, icariin, sagittatoside B, 2''-O-rhamnosyl icariside II, baohuoside I	Zorbax SB-C18 (50 × 2.1 mm, 1.8 μm)	*Gradient elution* A: 0.3% acetic acid in water B: 0.3% acetic acid in acetonitrile *Flow rate:* 0.4 mL/min	ESI-MS	Plasma / protein precipitation	0.06–21.8 ng/mL	– / 0.065 ng/mL	92.8–114.5	33

(continued)

Table 7.1 Ultrahigh Pressure Liquid Chromatography in the Analysis of Polyphenolic Compounds in Biological Fluids (*Continued*)

No.	Compound	Column	Mobile Phase	Detection	Type of Sample / Sample Preparation	Linear Dynamic Range	LOD / LOQ	Recovery (%)	Ref.
10.	Vitexin-2''-O-rhamnoside, hesperidin	Acquity UPLC BEH C18 (50 × 2.1 mm, 1.7 μm)	*Gradient elution* A: acetonitrile B: 0.1% formic acid in water *Flow rate:* 0.25 mL/min	ESI-MS/MS	Plasma / protein precipitation	10–2500 ng/mL	2 ng/mL / 10 ng/mL	94.6–99.7	34
11.	Procyanidins and metabolites	Acquity UPLC HSS T3 (100 × 2.1 mm, 1.8 μm)	*Gradient elution* A: water:acetic acid (99.8:0.2; *v/v*) B: acetonitrile *Flow rate:* 0.4 mL/min	DAD ESI-MS/MS	Plasma / SPE (column: Oasis HLB)	0.008–5.8 μM	0.003–0.8 μM / 0.01–0.98 μM	65.0–102.0	30
12.	Luteolin-7-O-glucoside, luteolin, apigenin, acacetin, protocatechuic aldehyde, chlorogenic acid, caffeic acid, ferulic acid, rosmarinic acid and metabolites	Acquity UPLC BEH C18 (100 × 2.1 mm, 1.7 μm)	*Gradient elution* A: 0.2% formic acid in water B: acetonitrile *Flow rate:* 0.4 mL/min	DAD Q-TOF/MS	Plasma, urine / protein precipitation	–	–	–	35
13.	Genistein and metabolites	Acquity UPLC BEH C18 (50 × 2.1 mm, 1.7 μm)	*Gradient elution* A: 2.5 mM ammonium acetate pH 7.4 B: acetonitrile *Flow rate:* 0.45 mL/min	ESI-MS/MS	Blood / protein precipitation	1.56–10,000 nM	0.78–6.25 nM /	85.0–115.0	36

No.	Analytes	Column	Elution	Detection	Sample preparation	Linear range	LOD / LOQ	Recovery (%)	Ref.
14.	(±)-catechin, (−)-epicatechin, rutin, quercitrin, hesperidin, neohesperidin, (±)-naringenin, hesperetin and some drugs	Chromolith Fast Gradient Monolithic C18e (50 × 2.0 mm)	Gradient elution A: 0.05% trifluoroacetic acid in water B: acetonitrile Flow: 1.5–2.5 mL/min	UV	Urine / enzymatic hydrolysis SPE (columns: SDB, C18)	0.1–40 µg/mL	23.1–53.3 ng/mL / 70.3–163.3 ng/mL	76.2–101.3	25
15.	(±)-catechin, (−)-epicatechin, rutin, quercitrin, hesperidin, neohesperidin, genistein, daidzein, glycitin, puerarin, biochanin A and some drugs	Hypersil GOLD (50 × 2.1 mm, 1.9 µm)	Gradient elution A: 0.05% trifluoroacetic acid in water B: acetonitrile Flow: 1.0–1.4 mL/min	UV	Urine / enzymatic hydrolysis SPE (column: Oasis HLB)	0.04–20 µg/mL	8.9–66.2 ng/mL / 26.7–198.5 ng/mL	70.3–98.6	26
16.	Genistein, daidzein, glycitin, puerarin, biochanin A and metabolites, some drugs and metabolites	Hypersil GOLD (50 × 2.1 mm, 1.9 µm)	Gradient elution A: 0.05% trifluoroacetic acid in water B: acetonitrile Flow: 0.3–0.8 mL/min	UV	Urine / enzymatic hydrolysis SPE (column: Oasis HLB)	0.08–12 µg/mL	10.7–66.9 ng/mL / 32.2–200.7 ng/mL	68.7–102.2	27

(continued)

Table 7.1 Ultrahigh Pressure Liquid Chromatography in the Analysis of Polyphenolic Compounds in Biological Fluids (*Continued*)

No.	Compound	Column	Mobile Phase	Detection	Type of Sample / Sample Preparation	Linear Dynamic Range	LOD / LOQ	Recovery (%)	Ref.
17.	(±)-catechin, (−)-epicatechin, rutin, quercitin, hesperidin, neohesperidin, hesperetin, genistein, daidzein, glycitin, puerarin, biochanin A	Hypersil GOLD (50 × 2.1 mm, 1.9 μm)	*Gradient elution* A: 0.05% trifluoroacetic acid in water B: acetonitrile *Flow:* 0.4–1.0 mL/min	UV	Urine / enzymatic hydrolysis SPE (column: Oasis HLB)	0.02–10 μg/mL	3.9–107.0 ng/mL / 11.7–321.1 ng/mL	70.0–96.6	28
18.	Scutellarin and metabolites	Acquity UPLC BEH C18 (100 × 2.1 mm, 1.7 μm)	*Gradient elution* A: 0.2% formic acid in water B: acetonitrile *Flow rate:* 0.5 mL/min	TOF/MS	Urine / protein precipitation	1.1–10.6 μg/mL 10.6–63.9 μg/mL	0.365 μg/mL / 1.065 ng/mL	96.5–99.3	37
19.	Anthocyanin derivatives	Acquity UPLC BEH C18 (50 × 2.1 mm, 1.7 μm)	*Gradient elution* A: 1% formic acid in water B: 1% formic acid in acetonitrile *Flow rate:* 0.45 mL/min	ESI-MS/MS	Urine / SPE (column: Oasis HLB)	12.5–500 ng/mL	1.3–3.3 ng/mL / 6.75–10.2 ng/mL	66.0–90.0	38

to the eluent resulted in a significant increase in the selectivity of the separation of compounds, and it also influenced the improvement in peak symmetry.

There is a growing interest on selection of the mobile phase components such as organic solvent, acid, and pH. The results showed that the use of methanol in a mobile phase, induced very slow elution of flavonoids and isoflavones from the chromatography column, and the peaks were not symmetrical. Furthermore, the use of methanol as an organic modifier significantly decreased the efficiency of the chromatographic column, as compared to acetonitrile, which was observed on the basis of the low number of theoretical plates.

The use of a more hydrophobic organic modifier, acetonitrile, resulted in the increased power of elution and thus influenced the retention of the analyzed compounds in a reversed phase. The optimum conditions—in terms of retention time—of the separation of compounds and the symmetry of the peaks were obtained by means of using acetonitrile as the organic component of the mobile phase. Therefore, acetonitrile is commonly used by many researchers as an organic modifier in the separation of the mixtures of polyphenols and their metabolites (21, 23–38).

Studies on various acids (acetic, formic, and trifluoroacetic acid (TFA)) added to the aqueous component of the mobile phase used in the UHPLC-UV technique indicated that the use of acid, preferably TFA, affects the separation of the selected flavonoids and isoflavones. The addition of acid to a mobile phase gave the best shape of peaks that corresponded to aglycone polyphenols. After the addition of acetic acid there was a significant deterioration of resolution as compared both to the results obtained with the use of formic acid and the use of TFA. The use of aqueous acetic acid, as a component of the mobile phase, did not give satisfactory peak symmetry either. The obtained results justify the choice of TFA acid as an additional component of eluent that affects the improvement of the separation of polyphenolic compounds. TFA acid is used as an additive to the aqueous mobile phase component; however, exclusively in the case of coupling liquid chromatography with UV detector.

Detection that uses a mass spectrometer requires TFA acid substitution with another acid because the addition of TFA acid causes the attenuation of the signal and prevents efficient aerosol formation due to the high surface tension of the mobile phase. Frequently, this acid is replaced with formic acid, and less frequently with acetic acid. The addition of acid to the mobile phase used in the UHPLC-MS method or UHPLC-MS/MS improves the shape of the peaks, and it also reduces the influence of matrix effects and improves the ionization, which increases the sensitivity of the method (20–24, 29–35, 37).

Baranowska and Magiera (28) examined the effect of the aqueous component pH of the mobile phase on the separation of polyphenolic compounds. It was observed that there is an impact of pH on the retention time, peak symmetry, and resolution. The low pH of the aqueous component of the mobile phase increased the separation selectivity and method sensitivity and improved peak symmetry, while the increase in mobile phase pH from 2.5 to 6.5 caused a significant deterioration in the separation of flavonoids and isoflavones, increased the width of the peaks, and decreased the number of theoretical plates. Without the addition of acid to the mobile phase, peaks eluted at considerably low retention times. Furthermore, lowering the pH to 2.5 reduced the dissociation of the analytes, thereby increasing their retention on the hydrophobic stationary phase.

Interestingly, an important parameter, which also greatly affects the separation of polyphenolic compounds, is the temperature of the column, which was optimized in the range from 15 to 45°C. It was observed that the retention of compounds decreased with the increase in the temperature of the column. This may be due to a reduced viscosity of the solvent and the acceleration of molecule diffusion. Although the increase of column temperature affected the shortening of the retention times of polyphenols, this did not affect their resolution. The use of higher temperatures of the column decreased the widths of the peaks, increased their height, as well as increased the number of theoretical plates. Column temperatures at 30°C (22), 35°C (31,35), 40°C (20,23,24), and 45°C (36) were used for the separation of polyphenolic compounds.

UHPLC coupled with spectrophotometric detector, diode array detector (DAD) or MS was widely used in the analysis of polyphenolic compounds in biological samples. Polyphenolic compounds are characterized by strong absorption of ultraviolet radiation; therefore, the use of a UV (25–28) detector or DAD (32, 35, 36) allowed their determination after chromatographic separation.

Mass spectrometers with single or triple quadrupole are most commonly used in the determination of polyphenolic compounds where electrospray ionization (ESI) is used as an interface. UHPLC-MS and UHPLC-MS/MS, which work in a selected ion mode (SIM) and multiple reaction monitoring mode (MRM), respectively, offer high sensitivity, selectivity, and high throughput for the determination of polyphenols and their metabolites in biological fluids.

The application of mass spectrometry in the analysis of polyphenolic compounds in biological samples produced excellent results. The coupling of liquid chromatography with mass spectrometry allows for the determination of accurate molecular masses, making it possible to predict the structure of an unknown compound. UHPLC-MS (20, 22) was used to improve the selectivity of methods, and UHPLC-MS/MS (21, 23, 24, 29, 30, 32–34) was used to reduce the interference caused by co-eluted components of the analyzed mixture, to track the mass fragmentation of the analytes, and to aid the elucidation of molecular structures.

When samples are characterized by a complex composition of the matrix (samples of biological fluids), even when using the MRM, there is a certain danger of underreporting (due to the ionization suppression by matrix components) or overstating (the result of insufficient resolution) results. Because of the high sensitivity and selectivity of UHPLC-MS/MS techniques and the ability to identify the structures of polyphenols, this technique is preferred for the quantitative analysis of polyphenols and their metabolites in biological fluids. Procyanidins and their metabolites in rat plasma sample, flavonoids and isoflavones in human urine were determined by UHPLC-ESI-MS/MS (25–28). UHPLC methods were also utilized to study the pharmacokinetics of polyphenolic compounds (biological half-life, clinical functional biological half-life, volume of distribution, and clearance) (20–24, 33, 34, 36).

Furthermore, new UHPLC methodologies can be an indispensable tool to obtain reliable information concerning the interaction between polyphenols and other exogenous compounds (e.g., drugs). For example, effects of polyphenolic compounds in foods and dietary supplements on the bioavailability of drugs were investigated using UHPLC methods (25–28).

UHPLC-TOF/MS (37) and UHPLC-Q-TOF/MS (quadrupole analyzer combined with time-of-flight [TOF] analyzer) (31, 32, 35) methods, as described in specialist literature, are used for the sole identification, but not for the quantitative analysis, of polyphenolic compounds in biological samples. More than that, the abovementioned methods are solely used for detection in biological fluids and their metabolites and for the tracking of the metabolic pathways of polyphenols.

7.2.2 Determination of Polyphenolic Compounds in Foods

Polyphenols are commonly present in fruits, vegetables, nuts, seeds, and flowers. Moreover, polyphenols can be found in such products as cocoa, tea, coffee, wine, jam, and chocolate. Polyphenolic compounds are also often applied in cosmetics, medicines, pharmaceuticals, dietary supplements, and in recent years, for the production of functional foods. The food industry offers a variety of new functional products in which the polyphenol content is usually higher than in products of natural origin. Milk enriched with soy isoflavones, chocolate enriched in procyanidins, drinks with higher amounts of anthocyanins, functional drinks enriched with extracts of tea—these are just a few products that are the result of functional foods revolution. On the other hand, the use of synthetic antioxidants in the food industry is of great concern among consumers who seek to limit their intake. Toxicological and nutritional research indicates adverse effect of some synthetic antioxidants used in food. The use of additives in food products, mainly in edible fats, means that more attention should be paid to antioxidants that are derived from natural plant extracts and used as food additives.

Naturally occurring polyphenols are a large group of compounds with different structures and properties, which only a small part has been discovered so far. Because of the continuous discoveries of new isomers (potential number of isomers increases exponentially), the determination of some compounds (e.g., procyanidins and catechins) is time-consuming work. High complexity and chemical diversity of natural polyphenolic compounds present in food are major challenges for analytical chemistry. Moreover, food is a complex matrix of chemical compounds, which poses certain difficulties and requires the use of specific methods, or very precise separation, of the analyzed compounds from the matrix possibly in unmodified form.

The abovementioned task requires the use of modern analytical techniques. The development of new UHPLC techniques allowed the determination of polyphenolic compounds in food samples in a short time while maintaining good resolution of analytes. Using UHPLC technology enables the identification of unknown compounds in complex matrix of polyphenols, as well as allows to perform study in vivo and in vitro, to increase knowledge and understanding of their beneficial health properties.

Polyphenolic compounds are separated mainly in the reversed phase on the column with C18 sorbent, with inside diameter from 1.7 to 2.1 mm, filled with small size particles (about 1.7 μm). In the event of nonpolar stationary phases the order of elution of polyphenols is consistent with the number of hydroxyl groups in the molecule, the degree of methoxylation, as well as the amount of sugar molecules in the structure of polyphenol.

Columns with different sorbents and of various parameters such as Hypersil Gold C18 (50 × 2.1 mm, 1.9 μm); Acquity UPLC BEH C18 (50 × 2.1 mm, 1.7 μm); Acquity UPLC BEH Shield RP18 (50, 100, 150 × 2.1 mm, 1.7 μm); and Acquity UPLC BEH phenyl (50 × 2.1 mm, 1.7 μm) were examined by Guillarme et al. to determine the eight catechins in tea samples (39). The best selectivity and resolution were obtained by means of the Acquity UPLC BEH column Shield RP18. Depending on the length of the column and gradient elution program used, the run time of an injection was 30 sec (for a column of length 50 mm), 8 min (for a column of length 100 mm), 20 min (for a column of length 150 mm), respectively. Authors used a variety of solvents for the extraction of catechins from tea samples (ether, ethyl acetate, and butanol), and the best recoveries were obtained using ethyl acetate. Klejdus et al. in his numerous works on the determination of isoflavones optimized chromatographic conditions for the simultaneous determination of isoflavones in soy preparations samples. To obtain efficient and rapid separation of isoflavone, columns of different sizes of grains and different column temperatures were investigated. A Zorbax SB C18 column at 80°C and a mobile phase consisting of acidified water and methanol was used to separate thirteen isoflavones within 1 min (40).

Kalili and Villiers have shown that reducing the particle size of sorbent and increasing column temperature can increase the efficiency of chromatographic analysis of polyphenolic compounds. It was shown that the best resolution of anthocyanins was obtained by means of a column with a diameter of 1.7 mm and a column temperature of 50°C. These chromatographic conditions were also used for the analysis of procyanidins in cocoa beans and apples samples (41). Twenty-four anthocyanins in red cabbage were determined in 34 min using a Zorbax SB C18 column and an aqueous solution of formic acid and acetonitrile as the mobile phase (42). Pati et al. have applied similar chromatographic conditions to the determination of 18 monomers and 22 dimers of anthocyanins in grape samples within 15 min. In addition, this method enabled the identity of six new acylated anthocyanins in grape samples (43).

Pongsuwan et al. (44) describe a UHPLC method for the determination of catechins and their metabolites using BEH C18 (150 × 2.1 mm, 1.7 μm) column and a column temperature of 40°C. This method has been applied in metabolic profiling of fifty-six Japanese tea samples. BEH C18 chromatographic column (100 × 2.1 mm, 1.7 μm) at 40°C and a 20-min gradient elution program were used by Trautvetter et al. to identify thirty-seven phenolic acids and flavonoids in honey samples (45).

Monolithic columns are a new type of HPLC column. The use of these columns can reduce analysis time as compared to conventional columns; it allows faster equilibration in the column and higher flow rate of mobile phase. Monolithic columns separated twelve flavonoids in 6 min. Red wine samples were analyzed by a UHPLC method coupled with a monolithic column (46).

Nováková et al. described a UHPLC-DAD method for the simultaneous determination of twenty-nine phenolic acids, flavonoids, and catechins in tea samples using BEH C18 columns and waters and acidified acetonitrile as the mobile phase (47) and for the determination of phenolic acids and twelve flavonols in tea samples (48). Comparatively, in another research similar chromatographic conditions and MS detection were used for the determination of eight catechins in just 2.5 min (49). Furthermore, other studies provided an example of the use of the UHPLC-MS method based on two columns connected in series C18, 70°C, and a ternary mobile phase (water, acetonitrile, and methanol) for the determination of flavonols, anthocyanins, and stilbenes in grape skins and seeds. Thirty compounds were detected in extracts of grape skins, and eleven compounds were found in the extracts of their seeds (50). Gómez-Romero et al. identified 135 polyphenolic compounds (including 21 new, previously unknown compounds) in tomato extracts using a 46-min gradient elution condition (51).

The selection of the detection method depends on the nature of research and the type of sample analyzed. UHPLC coupled with UV detector is commonly used and is an effective tool for analyzing polyphenols in food samples. The identification of polyphenolic compounds is based mainly on a comparison of UV spectra and retention times of polyphenols found in food samples with spectra and retention times obtained for standard solutions. Nonetheless, current requirements to identify new polyphenolic compounds and a large variety of samples where these compounds are detected require a better analytical technique. UHPLC-UV does not provide any information about the structure of the compound, and there is a great danger of analytes' co-elution with the matrix components.

Currently, mass spectrometry is one of the most effective techniques for the analysis of polyphenolic compounds in food samples, because it allows very low limits of detection of polyphenolic compounds to be obtained and allows for the selective detection of individual analytes. Over the past few years there has been considerable development of UHPLC methods combined with mass spectrometry, which contributed to the improvement of methods of identification and determination of new structures of polyphenolic compounds. Detection that uses a mass spectrometer allows an analyte in a sample to be unambiguously identified by comparing the mass spectra and retention time. However, the application of mass spectrometer requires the selection of appropriate work parameters, paying particular attention to the choice of ionization method and mode of formation of ions. In the analysis of polyphenolic compounds in food samples ESI is the most widely use ionization mode, which is more sensitive and provides better stability of polyphenols as compared to atmospheric-pressure chemical ionization (APCI). It was reported that the negative ionization mode achieved significantly greater signal intensity and better accuracy than the positive ionization mode for polyphenols. Tandem mass spectrometry (MS/MS) is particularly useful in the detection and determination of polyphenolic compounds. For the quantitative analysis, the MRM mode is widely used. The MRM scan mode significantly increases the selectivity and sensitivity of the method and thus lowers the limits of detection and quantification of the developed analytical procedure. The combination of UHPLC with tandem mass spectrometry enabled the identification of the test compounds on the basis of mass spectra, which provides high selectivity and allows for the quantitative determination of the test substance in complex mixtures based on selected parent and fragmentation ions without completely separating the analytes.

The mass spectrometer is frequently used to detect polyphenolic compounds in various food products such as twenty-seven anthocyanins and flavonols in drinking milk (analysis time 10 min) (52), seventeen phenolic acids in different beverages (analysis time 10 min) (53), and eleven phenolic acids and flavonols in red wine according to geographical origin and grape variety and vintage year (analysis time 10 min) (54). Xiang et al. also described a method to identify polyphenols and organic acids by UHPLC-MS/MS in tobacco after the prior preparation of samples using the matrix solid phase dispersion (MSPD) (55).

Despite many advantages, MS/MS detection is not sufficient to determine the molecular formula of an unknown. Mass spectrometers with ion trap are able to identify new, unknown polyphenolic compounds in food samples, which are characterized by their ability to detect MS^n experiments. They can generate fragmentation of ions in n generations, which is particularly crucial in the analysis of isomer glycoside flavonoids. The resolution of the masses, achieved by the ion trap mass spectrometer, is comparable to that of the quadrupole mass spectrometers.

Mass spectrometers with the TOF analyzer are characterized by higher resolution; however, it may not be sufficient to confirm or deny the suggested structure of the polyphenols. In this case it is useful to use a quadrupole mass spectrometer with a TOF analyzer (Q-TOF-MS/MS), which provides greater selectivity and provides more information about the detected polyphenols molecule. A UHPLC technique combined with a TOF/MS detector was utilized to identify twenty-three phenolic acids and flavonoids in almonds (56). Taking into account the numerous advantages of UHPLC coupled with various detectors in the analysis of polyphenols in food, it should also be mentioned that there is the possibility of its application in the analysis of the so-called fingerprint technique (an analysis of commercial tea).

In conclusion, UHPLC-MS has a wide range of applications in the determination of polyphenolic compounds in foods. Highly sensitive chromatographic method coupled with mass spectrometry allows the detection and identification of anthocyanins in wine samples (57). Yet another example is the use of the technology to detect polyphenolic compounds. Q-TOF-MS is an excellent tool for the detection and identification of new polyphenols in food. An appropriate selection of chromatographic parameters allows for the quick detection of a large number of polyphenols, including both those with hydrophilic properties as well as hydrophobic properties. The use of small particle size and small diameter of the column represents a major advance for the selectivity, sensitivity, and speed of analysis of polyphenolic compounds in foods. Table 7.2 summarizes the developed UHPLC methods, which were used for the determination of polyphenolic compounds in foods.

Table 7.2 Ultrahigh Pressure Liquid Chromatography in the Analysis of Polyphenolic Compounds in Food

No.	Compound	Column	Mobile Phase	Detector	Type of Sample / Sample Preparation	Linear Dynamic Range	LOD / LOQ	Recovery (%)	Ref.
1.	Anthocyanins (24 compounds)	Acquity UPLC BEH C18 (50 × 2.1 mm, 1.7 μm)	*Gradient elution* A: 0.12% trifluoroacetic acid B: acetonitrile *Flow rate:* 0.5 mL/min	DAD Q-TOF/MS	Wine / SPE (column: Strata SDB)	0.5–100 mg/L	0.05–1 mg/L / 0.1–2.4 mg/L	–	57
2.	(±)-catechin, (−)-epicatechin, rutin, quercitrin, hesperidin, neohesperidin, (±)-naringenin, hesperetin, chrysin	Chromolith® Fast Gradient C18e (50 × 2.0 mm)	*Gradient elution* A: 0.1% formic acid in water B: acetonitrile *Flow rate:* 0.8–2.0 mL/min	UV	Wine / SPE (column: Oasis HLB)	2–40 μg/mL	0.06–0.19 μg/mL / 0.18–0.59 μg/mL	78.0–99.6	46
3.	Catechin, vanillic acid, syringic acid, caffeic acid, *cis*-aconitic acid, protocatechuic acid ethyl ether	Acquity UPLC BEH C18 (100 × 2.1 mm, 1.7 μm)	*Gradient elution* A: 5% formic acid in water B: methanol *Flow rate:* 0.2 mL/min	TOF/MS	Wine / LLE (dichloro-methane)	0.03–172 μg/mL	0.02–0.66 μg/mL / 0.08–2.18 μg/mL	96.0–100.0	58
4.	Trans *p*-coumaric acid, cis *p*-coumaric acid, myricetin, kaempferol, trans-resveratrol, cis-resveratrol, (+)-catechin, (−)-epicatechi, quercetin, gallic acid, caffeic acid, ferulic acid	Rapid Resolution HT C18 (50 × 2.1 mm, 1.8 μm)	*Gradient elution* A: water: acetonitrile: formic acid (99:1:0.1; %/%/%) B: acetonitrile: water:formic acid (99:1:0.1; %/%/%) *Flow rate:* 0.4 mL/min	ESI-MS/MS	Wine / filtration	0.2–7500 μg/L	0.05–30 μg/L / 0.2–90 μg/L	–	54

(continued)

Table 7.2 Ultrahigh Pressure Liquid Chromatography in the Analysis of Polyphenolic Compounds in Food (*Continued*)

No.	Compound	Column	Mobile Phase	Detector	Type of Sample / Sample Preparation	Linear Dynamic Range	LOD / LOQ	Recovery (%)	Ref.
5.	Phenolic acids, flavonoids, catechins, coumarins	Acquity UPLC BEH C18 (100 × 2.1 mm, 1.7 μm)	*Gradient elution* A: 0.1% formic acid in water B: methanol *Flow rate:* 0.45 mL/min	UV	Tea, wine / filtration	–	– / –	–	14
6.	Phenolic acids (17 compounds)	Acquity UPLC BEH C8 (150 × 2.1 mm, 1.7 μm)	*Gradient elution* A: 7.5 mM formic acid in water B: acetonitrile *Flow rate:* 0.25 mL/min	UV ESI-MS/MS	Tea, wine, grapefruit juice / filtration	0. –3000 pmol/inj	0.15–15 pmol/inj / –	85.5–117.5	53
7.	(+)-catechin, (−)-epicatechin, (−)-catechin gallate, (−)-epicatechin gallate, (−)-gallocatechin gallate, (−)-epigallo-catechin gallate, (−)-epigallocatechin, gallic acid	Acquity UPLC BEH Shield RP18 (100 × 2.1 mm, 1.7 μm)	*Gradient elution* A: 0.1% formic acid in water B: acetonitrile *Flow rate:* 0.6 mL/min	UV ESI-MS/MS	Tea / LLE (ether, butanol, ethyl acetate)	–	–	–	39

No.	Analytes	Column	Mobile phase	Detection	Sample / preparation	Linearity range	LOD / LOQ	Recovery (%)	Ref.
8.	Phenolic acids (9), flavonoids (16), coumarins (4) plus caffeine	Acquity UPLC BEH C18 (100 × 2.1 mm, 1.7 μm)	Gradient elution A: 0.1% formic acid in water B: methanol Flow rate: 0.45 mL/min	DAD	Tea / filtration	1–100 μg/mL	0.01–13.9 μg/mL / 0.04–45.8 μg/mL	77.0–110.0	47
9.	(–)-epigallocatechin, (–)-epigallocatechin gallate, (–)-epicatechin gallate	Acquity UPLC BEH C18 (150 × 2.1 mm, 1.7 μm)	Gradient elution A: 0.1% formic acid in water B: acetonitrile Flow rate: 0.3 mL/min	TOF/MS	Tea / extraction (methanol:water:chloroform (2.5:1:1; v/v/v))	–	–	–	44
10.	Phenolic acid relatives, catechnis, purine alkaloids, proanthocyanidins, O-glycosylated flavonols, flavonols, flavone, acylated glycosylated flavonols	Acquity UPLC HSS (100 × 2.1 mm, 1.8 μm)	Gradient elution A: 0.1% formic acid in water B: 0.1% formic acid in acetonitrile Flow rate: 0.5 mL/min	DAD ESI-MS	Tea / extraction (methanol:water (60:40; v/v))	–	–	–	59
11.	(+)-catechin, (–)-catechin gallate, (–)-epicatechin, (–)-epicatechin gallate, (–)-epigallocatechin, (–)-epigallocatechin gallate, (–)-gallocatechin, (–)-gallocatechin gallate	Acquity UPLC BEH C18 (100 × 2.1 mm, 1.7 μm)	Gradient elution A: 0.01% formic acid in water B: methanol Flow rate: 0.45 mL/min	ESI-MS/MS	Tea / filtration	0.02–2.40 μg/mL	3.0–4.8 ng/mL / 9.9–15.8 ng/mL	89.5–113.7	49

(continued)

Table 7.2 Ultrahigh Pressure Liquid Chromatography in the Analysis of Polyphenolic Compounds in Food (*Continued*)

No.	Compound	Column	Mobile Phase	Detector	Type of Sample / Sample Preparation	Linear Dynamic Range	LOD / LOQ	Recovery (%)	Ref.
12.	Chlorogenic acid, caffeic acid, umbelliferone, rutin, quercetin-3-glucoside, apigenin-7-glucoside, quercitrin, quercetin, luteolin, kaempherol, apigenin, isorhamnetin	Acquity UPLC BEH C18 (100 × 2.1 mm, 1.7 μm)	*Gradient elution* A: 0.01% formic acid in water B: methanol *Flow rate:* 0.45 mL/min	ESI-MS/MS	Tea / filtration	0.005–100 μmol/L	0.002–0.015 μmol/L / 0.005–0.02 μmol/L	76.6–126.7	48
13.	Procyanidins, galloylated procyanidins, prodelphinidins, propelargonidins and polymers (22 compounds)	Zorbax SB C18 (50 × 4.6 mm, 1.8 μm)	*Gradient elution* A: 0.1% formic acid in water B: acetonitrile *Flow rate:* 0.8 mL/min	FL DAD ESI-MS/MS	Tea / extraction (70% acetone)	–	–	–	60
14.	Phenolic acids (14 compounds)	Acquity UPLC BEH C18 (100 × 2.1 mm, 1.7 μm)	*Gradient elution* A: acetonitrile: acetic acid:water (3:2:95; %:%:%) B: acetonitrile: acetic acid:water (85:2:13; %:%:%) *Flow rate:* 0.7 mL/min	UV	Brandy / filtration	0.02–117.8 μg/mL	0.009–0.399 μg/mL / 0.031–1.391 μg/mL	95.3–103.2	18

No.	Compounds	Column	Mobile phase	Detection	Sample / Extraction	Linear range	LOD / LOQ	Recovery	Ref.
15.	Catechin, epicatechin, dimer B2, dimer B5, trimer C1, tetramer D	Acquity UPLC BEH C18 (50 × 2.1 mm, 1.7 μm)	Gradient elution A: woda: tetrahydrofurane: trifluoroacetic acid B: 0.1% trifluoroacetic acid in acetonitrile Flow rate: 0.3 mL/min	UV Q-TOF/MS	Chocolate / extraction (hexane, acetone: water:acetic acid (70:28:2; v/v/v))	11.2–2500 μg/g	– / 11.2–62.5 μg/g	~100	17
16.	(–)-epicatechin, (+)-catechin, theobromine, caffeine, dimer B2	Acquity UPLC HSS T3 (100 × 2.1 mm, 1.8 μm)	Gradient elution A: water: acetic acid (99.8:0.2; v/v) B: acetonitrile Flow rate: 0.4 mL/min	ESI-MS/MS	Coca nib / extraction (hexane)	0.01–40 μg/mL	0.001–0.03 μg/mL / 0.003–0.1 μg/mL	–	15
17.	Procyanidins (14 compounds)	Zorbax SB C18 (50 × 4.6 mm, 1.8 μm)	Gradient elution A: acetonitrile B: 0.1% formic acid in water Flow rate: 1.8 mL/min	FL DAD Q-TOF/MS	Cocoa beans, apple / extraction (70% acetone, 70% methanol)	–	–	–	41
18.	Esculetin, chlorogenic acid, caffeic acid, scopoletin, quercitrin, rutin	Acquity UPLC BEH C18 (100 × 2.1 mm, 1.7 μm)	Gradient elution A: methanol B: 0.3% acetic acid in water Flow rate: 0.4 mL/min	ESI-MS/MS	Tobacco / extraction SPE (column: C18, NH$_2$, SAX, Oasis HLB)	0.001–10 μg/mL	0.05–1 ng/mL / 0.17–3.33 ng/mL	88.0–101.0	61
19.	Chlorogenic acid, rutin, scopoletin	Acquity UPLC BEH C18 (100 × 2.1 mm, 1.7 μm)	Gradient elution A: methanol B: 1% acetic acid in water Flow rate: 0.2 mL/min	DAD	Tobacco / UE (60% methanol)	0.5–100 μg/mL	0.39–2.54 μg/g / 1.29–8.47 μg/g	96.5–105.6	62

(continued)

Table 7.2 Ultrahigh Pressure Liquid Chromatography in the Analysis of Polyphenolic Compounds in Food (*Continued*)

No.	Compound	Column	Mobile Phase	Detector	Type of Sample / Sample Preparation	Linear Dynamic Range	LOD / LOQ	Recovery (%)	Ref.
20.	Polyphenols, organic acids	Acquity UPLC BEH C18 (100 × 2.1 mm, 1.7 μm)	*Gradient elution* A: methanol B: 0.3% acetic acid in water *Flow rate:* 0.4 mL/min	ESI-MS/MS	Tobacco / MSPD, UE-SPE, RE	0.002–700 μg/mL	0.15–165.7 ng/g / —	75.4–90.4	55
21.	Phenolic acids, flavonoids	Acquity UPLC BEH C18 (100 × 2.1 mm, 1.7 μm)	*Gradient elution* A: 2% acetic acid in water B: methanol *Flow rate:* 0.3 mL/min	Q-TOF/MS	Honey / homogenization extraction (ethyl acetate)	—	—	—	45
22.	Flavonoids, phenolic acids (32 compounds)	Acclaim RSLC 120 C18 (100 × 2.1 mm, 1.8 μm)	*Gradient elution* A: 0.1% formic acid in water B: 0.1% formic acid in acetonitrile *Flow rate:* 0.6 mL/min	UV TOF/MS	Avocado / extraction (methanol)	0.5–200 ppm	0.5–730.5 ppb / 1.7–2435.1 ppb	97.2–102.0	63
23.	(−)-catechin, (−)-epicatechin, oligomeric proanthocyanidins	Acquity UPLC BEH C18 (100 × 1.0 mm, 1.7 μm)	*Gradient elution* A: 0.1% formic acid in water B: methanol *Flow rate:* 0.2 mL/min	DAD ESI-MS/MS	Berries, strawberries, blueberries, raspberries / extraction (acetone:water: acetic acid (70:29.5:0.5; v/v/v))	—	—	—	64

No.	Analytes	Column	Mobile phase	Detection	Sample / Extraction	Linear range	LOD	Recovery	Ref.
24.	Flavonols, flavanols, antocyanins, stilbenes (33 compounds)	Zorbax Eclipse Plus C18 (100 × 2.1 mm, 1.8 µm)	*Gradient elution* A: acetonitrile:methanol (50:50; v/v) B: 0.1% formic acid in water *Flow rate:* 0.3 mL/min	ESI-MS/MS	Grape berries / extraction (methanol:water:formi acid (70:30:1; v/v/v))	1–2000 µg/g	100–200 ng/g / –	90.0–102.0	50
25.	Anthocyanins (40 compounds)	Zorbax SB C18 (50 × 2.1 mm, 1.8 µm)	*Gradient elution* A: 10% formic acid in water B: acetonitrile *Flow rate:* 0.2 mL/min	DAD ESI-MS/MS	Grape skins / extraction (70% methanol)	–	–	–	43
26.	Phenolic acids, flavonol O-glycosides (21 compounds)	Phenomenex C18 (100 × 3.0 mm, 2.6 µm)	*Gradient elution* A: 0.1% formic acid in water B: 01% formic acid in acetonitrile:water (90:10; v/v) *Flow rate:* 0.75 mL/min	DAD ESI-MS^n	Peels of jocate fruits / extraction (aetone:0.1% formic acid (80:20; v/v)) SPE (column: CC6)	–	–	–	65
27.	Phenolic acids, flavonoids, monoterpen glycosides, galloyl glucoses (23 compounds)	Phenomenex Synergi Fusion RP100A (50 × 2.0 mm, 2.5 µm)	*Gradient elution* A: 0.1% formic acid in water B: 0.1% formic acid in acetonitrile *Flow rate:* 0.5 mL/min	ESI-TOF/MS	Almond skin / extraction (70% methanol)	–	–	–	56
28.	Quercetin, kaempferol, isorhamnetin, myricetin	Acquity UPLC BEH C18 (100 × 2.1 mm, 1.7 µm)	*Gradient elution* A: 0.1% formic acid in water B: methanol *Flow rate:* 0.2 mL/min	DAD Q-TOF/MS	Onion / UE, MAE, PLE	200–400 µg/g	0.9–4.8 µg/g / 20–60 µg/g	98.0–99.0	66

(continued)

Table 7.2 Ultrahigh Pressure Liquid Chromatography in the Analysis of Polyphenolic Compounds in Food (*Continued*)

No.	Compound	Column	Mobile Phase	Detector	Type of Sample / Sample Preparation	Linear Dynamic Range	LOD / LOQ	Recovery (%)	Ref.
29.	Anthocyanins (24 compounds)	Zorbax SB C18 (100 × 2.1 mm, 1.8 µm)	*Gradient elution* A: 5% formic acid in water B: acetonitrile *Flow rate:* 0.2 mL/min	DAD ESI-MS/MS	Red cabbage / extraction (water:ethanol: formic acid (94:5:1; v/v/v))	–	– / –	–	42
30.	Organic acids, amino acids, phenolic acids, phenolic alcohol, flavonoids, fatty acids, triterpenoid glycosides (135 compounds)	Zorbax Eclipse Plus C18 (150 × 4.6 mm, 1.8 µm)	*Gradient elution* A: 0.5% acetic acid in water B: acetonitrile *Flow rate:* 1.8 mL/min	DAD TOF/MS	Tomato / extraction (methanol)	0.5–300 mg/L	0.03–1.5 ppm / –	–	51
31.	Daidzin, glycitin, genistin, ononin, daidzein, glycitein, sissotrin, genistein, formononetin, biochanin A	Zorbax SB C18 (30 × 2.1 mm, 1.8 µm)	*Gradient elution* A: 0.2% acetic acid in water B: methanol *Flow rate:* 1.4 mL/min	UV	Soy / extraction (90% methanol)	100–500 ng/mL	0.5–4.6 ng/mL / 1.6–15.2 ng/mL	–	40
32.	Daidzein, glycitein, genistein	Acquity UPLC BEH C8 (100 × 2.1 mm, 1.7 µm)	*Gradient elution* A: 0.3% formic acid in water B: methanol *Flow rate:* 0.55 mL/min	DAD	Soybean / extraction (80% methanol)	0.125–6 µg/mL	55–94 pg / 184–223 pg	88.0–103.0	67

No.	Compounds	Column	Elution / Flow rate	Detection	Sample / Extraction				Reference
33.	Equol, daidzein, dihydrodaidzein, genistein, dihydrogenistein	Acquity UPLC BEH C18 (50 × 1.0 mm, 1.7 μm)	*Isocratic elution* acetonitrile: 0.1% formic acid (40:60; %:%) *Flow rate:* 0.2 mL/min	ESI-MS/MS	Soy supplement / extraction (methanol)		–	–	13
34.	(–)-epicatechin, (+)-catechin, (–)-epicatechin gallate, 4-hydroxybenzoic acid, vanillic acid, syringic acid, dehydrocaffeic acid, protocatechuic acid, p-coumaric acid, o-coumaric acid, gallic acid	Phenomenex Luna C18 (50 × 2.1 mm, 3.0 μm)	*Gradient elution* A: 0.1% formic acid in water B: 0.1% formic acid in acetonitrile *Flow rate:* 0.8 mL/min	DAD ESI-MS/MS	Supplement (GADF) / extraction (acetone:water (70:30; v/v))		–	–	68
35.	Anthocyanins, flavonols (27 compounds)	Acquity UPLC BEH C18 (50 × 2.1 mm, 1.7 μm)	*Gradient elution* A: water:formic acid (9:1; v/v) B: acetonitrile *Flow rate:* 1.8 mL/min	DAD ESI-MS/MS	Milk-based food products / extraction (water)	0.002–600 μg/mL	0.3–300 ng/mL / 1–1000 ng/mL	16.0–103.0	52

7.2.3 Determination of Polyphenolic Compounds in Plant Extract

Folk medicine has drawn attention abundantly for the healing properties of plants in preventing and eliminating diseases for hundreds of years. This knowledge has not only been preserved by generations but also, for the last few decades, it has been expanded. Currently, there is a great interest in further development of this knowledge. Over recent years, there has been a growing interest in natural drugs and an increasing amount of research that provides the evidence on the importance and mechanisms of action of specific substances contained in plants. The efficacy and health benefits, resulting from the application of biological plant extracts, have been confirmed in numerous clinical trials and research (69, 70).

Apparently, only particular substance among the group of substances contained in plant extracts has an unequivocal biological effect; therefore, there is a need to monitor the qualitative composition of plant extracts. The inherent element in the study of pharmacognostic material is undoubtedly the quantitative assessment of the individual components of plant extracts, because an appropriate amount of a biologically active substance provides therapeutic value of the given extract. Furthermore, it is very important to determine which compounds are endowed with the greatest biological activity.

In the past decade, certain important steps have been taken to improve the quality of herbal product application in medicine. Herbal medicines may contain hundreds of different compounds. Some compounds occur at low concentrations; however, they can be extremely important for the quality, safety, and efficacy of herbal remedies. The great progress of the quality control of plant extracts is mainly due to the emergence of modern separation and spectral techniques. Chromatography offers an extremely important and useful tool to assess the individual profiles of compounds present in samples. Nowadays, chromatographic fingerprinting is the most widely used technique for quality control of herbal products. Depending on the technique, scientists use fingerprints for quality control based on biomarkers to focus on the analysis of a particular group of compounds or to analyze all compounds in a sample.

Various chromatographic techniques were used to study fingerprinting of plant extracts, including UHPLC technology. Toh et al. studied the metabolic profiles of *Panax notoginseng* using UHPLC-TOF/MS (71). Li et al. developed a UHPLC-ESI-MS/MS method for qualitative and quantitative determination of constituents in the flower *Trollius ledebouri* (72). To obtain in-depth information about the composition of the test plants, a nonselectivity solvent such as methanol or ethanol was used to extract a large number of biologically active compounds. An extraction of analytes in the Soxhlet apparatus was preceded by acid hydrolysis. After the extraction of the analytes from the plant samples, four flavonoids in extracts were determined within 3.5 min and more than fifty compounds within 20 min using UHPLC-MRM. Klejdus et al. developed a method for separation of ten isoflavones in less than 2 min, which was then applied to the study of soy products and plant materials (73). The same researchers also described a UHPLC method for the simultaneous determination of phenolic acids and isoflavones in 1.5 min (74). To optimize conditions in separating daidzein, genistein, glycitein, their glycosides, and other derivatives (nineteen

Table 7.3 Ultrahigh Pressure Liquid Chromatography in the Analysis of Polyphenolic Compounds in Plant Extract

No.	Compound	Column	Mobile Phase	Detector	Type of Sample / Sample Preparation	Linear Dynamic Range	LOD / LOQ	Recovery (%)	Ref.
1.	Puerarin-4′-O-glucoside, puerarin-3′-methyoxy-4′-O-glucoside, daidzin-4′,7-O-glucoside, puerarin, mirificin, daidzin, 6″-O-xylosylpuerarin, 3′-methoxypuerarin, genistin, sophoraside A, ononin, daidzein, genistein, formononetin	Zorbax SB C18 (50 × 4.6 mm, 1.8 μm)	*Gradient elution* A: 0.1% formic acid in water B: methanol *Flow rate:* 2.0 mL/min	DAD TOF/MS	*Pueraria lobata* / MAE, UE, RE, PLE	1.18–554.72 μg/mL	0.01–0.36 μg/mL / 0.02–0.65 μg/mL	95.2–102.9	77
2.	Polyphenols (56 compounds)	Acquity HSS T3 (100 × 2.1 mm, 1.8 μm)	*Gradient elution* A: 0.1% formic acid in water B: acetonitrile *Flow rate:* 0.5 mL/min	DAD Q-TOF/MS	*Paeonia lactiflora* / UE (50% methanol)	–	–	–	80
3.	Caffeic acid, chlorogenic acid, rutin, vanillic acid, ferulic acid, benzoic acid, quercetin, cynarin, quercitrin, luteolin, p-coumaric acid, arctiin	Acquity UPLC BEH C18 (150 × 2.1 mm, 1.7 μm)	*Gradient elution* A: 0.1% formic acid in water B: acetonitrile: methanol (20:80; v/v) *Flow rate:* 0.28 mL/min	DAD MS/MS	Burdock leaves / UE/microwave extraction (70% ethanol)	–	–	–	81

(continued)

Table 7.3 Ultrahigh Pressure Liquid Chromatography in the Analysis of Polyphenolic Compounds in Plant Extract (*Continued*)

No.	Compound	Column	Mobile Phase	Detector	Type of Sample / Sample Preparation	Linear Dynamic Range	LOD / LOQ	Recovery (%)	Ref.
4.	Vanillin, *p*-hydroxybenzaldehyde, vanillic acid, pHB acid	Acquity UPLC BEH C18 (50 × 2.1 mm, 1.7 μm)	*Gradient elution* A: water acidified with H_3PO_4 (10^{-2}M) B: acetonitrile *Flow rate:* 0.31 mL/min	DAD	Vanilla beans / Soxhlet extraction	0.12–1835 μg/mL	0.315–0.44 pg / –	–	82
5.	Genistin, genistein, daidzein, daidzin, glycitin, glycitein, ononin, formononetin, sissotrin, biochanin A	Zorbax SB C18 (30 × 2.1 mm, 1.8 μm)	*Gradient elution* A: 0.2% acetic acid in water B: methanol *Flow rate:* 1.4 mL/min	DAD ESI-MS	*Trifolium pratense, Iresine herbstii, Ononis spinosa* / Soxhlet extraction	100–500 ng/mL	0.49–4.62 ng/mL / 1.62–15.25 ng/mL	–	73
6.	Genistin, genistein, daidzin, daidzein, glycitin, glycitein, ononin, sissotrin, formononetin, and biochanin A, gallic, protocatechuic, *p*-hydroxybenzoic, vanillic, caffeic, syringic, *p*-coumaric, ferulic, sinapic acid	Zorbax SB C18 (50 × 2.1 mm, 1.8 μm)	*Gradient elution* A: 0.3% acetic acid in water B: methanol *Flow rate:* 0.9 mL/min	DAD	*Trifolium pratense, Glycine max, Pisum sativum, Ononis spinosa* / Soxhlet extraction	–	0.19–0.57 ng/mL / 0.63–1.89 ng/mL	–	74

No.	Analytes	Column	Mobile phase	Detection	Sample / Extraction	Linear range	Recovery (%)	LOD / LOQ	Ref.
7.	Daidzein, genistein, glycitein, coumestrol	Acquity UPLC BEH C18 (50 × 2.1 mm, 1.7 μm)	*Gradient elution* A: 0.3% acetic acid in water B: methanol *Flow rate:* 0.7 mL/min	DAD	Different plants / UE (50% ethanol)	2–400 ng/mL	–	– / 1.97–4.08 ng/mL	83
8.	Polyphenols (59 compounds)	Acquity UPLC BEH Shield C18 (150 × 2.1 mm, 1.7 μm)	*Gradient elution* A: acetonitrile B: 0.1% formic acid in water *Flow rate:* 0.4 mL/min	DAD MS/MS	*Salvia officinalis* / filtration	1.5–750 mg/L	–	–	84
9.	Genistin, genistein, daidzein, daidzin, glycitin, glycitein, ononin, formononetin, sissotrin, biochanin A	Atlantis dC18 (20 × 2.1 mm, 3 μm)	*Gradient elution* A: 0.2% formic acid in water B: acetonitrile *Flow rate:* 0.45 mL/min	DAD	Soy bits / Soxhlet extraction, sonication/PLE	–	97–104	1.1–9.4 ng/mL / 3.8–31.5 ng/mL	75
10.	Daidzin, glycitin, genistin, ononin, daidzein, glycitein, genistein	Atlantis dC18 (20 × 2.1 mm, 3 μm)	*Gradient elution* A: 0.1% acetic acid in water B: methanol *Flow rate:* 0.35 mL/min	DAD	Soy plants / SPE (column: HLB)	–	96–106	1.1–9.4 ng/mL / –	85
11.	Ginsenosides, notoginsenoside	Acquity UPLC BEH C18 (100 × 2.1 mm, 1.7 μm)	*Gradient elution* A: 0.1% formic acid in water B: 0.1% formic acid in acetonitrile *Flow rate:* 0.5 mL/min	TOF/MS	*Panax notoginseng* / UE (70% methanol)	–	–	–	71

(continued)

Table 7.3 Ultrahigh Pressure Liquid Chromatography in the Analysis of Polyphenolic Compounds in Plant Extract (*Continued*)

No.	Compound	Column	Mobile Phase	Detector	Type of Sample / Sample Preparation	Linear Dynamic Range	LOD / LOQ	Recovery (%)	Ref.
12.	Flavonoid conjugates (38)	Acquity HSS T3 (100 × 2.1 mm, 1.8 μm)	*Gradient elution* A: water: acetonitrile:formic acid (95:4.5:0.5; v/v/v) B: acetonitrile: water:formic acid (95:4.5:0.5; v/v/v) *Flow rate:* 0.6 mL/min	DAD ESI-MS/MS	*Lupinus angustifolius* / UE (80% methanol)	–	–	–	86
13.	Coumaric acid, rutin	Acquity UPLC BEH C18 (150 × 2.1 mm, 1.7 μm)	*Gradient elution* A: 0.1% formic acid in water B: 0.1% formic acid in acetonitrile *Flow rate:* 0.4 – 0.7 mL/min	DAD TOF/MS	*Arabidopsis thaliana* / UE (85% methanol)	–	–	–	87
14.	Steviol, steviolbioside, stevioside, rebaudioside A, rebaudioside B, rebaudioside C, dulcoside A, apigenin-7-*O*-glucoside, quercetin-3-*O*-glucoside, quercetin-3-*O*-rhamnoside, luteolin-7-*O*-glucoside and quercetin-3-*O*-rutinoside	Agilent Zorbax RRHD SB-C18 UHPLC (30 × 2.1 mm, 1.8 μm)	*Gradient elution* A: water B: acetonitrile *Flow rate:* 20 μL/min	DAD	*Stevia rebaudiana* / extraction (acetonitrile: water (80:20; v/v))	–	–	–	88

No.	Analytes	Column	Mobile phase	Detection	Sample/extraction	Linear range	LOD	Recovery (%)	Ref.
15.	Glycosides of quercetin and kaempferol (13 compounds)	Acquity UPLC BEH C18 (50 × 2.1 mm, 1.7 μm)	*Gradient elution* A: 0.1% acetic acid in water B: 40% acetonitrile *Flow rate:* 0.35 mL/min	DAD MS/MS	*Aesculus hippocastanum* / extraction (80% ethanol)	–	–	–	89
16.	Isoflavones and saponins (13 compounds)	Agilent ZorBax SB-C18 (50 × 4.6 mm, 1.8 μm)	*Gradient elution* A: 0.2% formic acid in water B: acetonitrile *Flow rate:* 0.6 mL/min	DAD TOF/MS	*Radix Astragali* / extraction (methanol)	0.08–82.4 μg/mL	0.001–0.45 ng / 0.05–2.25 ng	90–110	90
17.	Phenoloc acids, phthalides, saponins, isoflavones (39 compounds)	Agilent ZorBax SB-C18 (50 × 4.6 mm, 1.8 μm)	*Gradient elution* A: 0.2% formic acid in water B: acetonitrile *Flow rate:* 0.6 mL/min	DAD TOF/MS	*Danggui Buxue Tang* / extraction	–	–	–	79
18.	Chlorogenic acid, caffeic acid, umbelliferone, rutin, quercetin-3-glucoside, apigenin-7-glucoside, quercitrin, quercetin, luteolin, kaempherol, apigenin, isorhamnetin	Acquity UPLC BEH C18 (100 × 2.1 mm, 1.7 μm)	*Gradient elution* A: 0.1% formic acid in water B: methanol *Flow rate:* 0.45 mL/min	MS/MS	*Matricaria recutita* / extraction (60% methanol)	0.005–100 μmol/L	0.002–0.015 μmol/L / 0.005–0.05 μmol/L	77–127	91
19.	Glucosinolates and flavonoids	Acquity UPLC BEH Shield C18 (150 × 1.0 mm, 1.7 μm)	*Gradient elution* A: 0.1% formic acid:acetonitrile (95:5; v/v) B: 0.1% formic acid:acetonitrile (40:60; v/v) *Flow rate:* 0.45 mL/min	DAD MS/MS	*Brassica oleracea* / extraction (methanol)				92

(continued)

Table 7.3 Ultrahigh Pressure Liquid Chromatography in the Analysis of Polyphenolic Compounds in Plant Extract (*Continued*)

No.	Compound	Column	Mobile Phase	Detector	Type of Sample / Sample Preparation	Linear Dynamic Range	LOD / LOQ	Recovery (%)	Ref.
20.	Isovitexin, vitexin, luteolin-7-O-glucoside, hyperoside, luteolin, apigenin	Acquity BEH C18 (100 × 2.1 mm, 1.7 μm)	*Gradient elution* different mixtures of solvents (water, methanol, THF, acetonitrile)	DAD	*Passiflora incarnata* / extraction (60% ethanol)	–	–	–	78
21.	Phenolic compounds (21 compounds)	Acquity UPLC BEH C18 (100 × 2.1 mm, 1.7 μm)	*Gradient elution* A: 0.05% trifluoroacetic acid in water B: acetonitrile *Flow rate:* 0.3 mL/min	Q-TOF/ MS/MS	*Rosa bourboniana, Rora brunonii, Rosa damascene* / extraction (methanol)	–	–	–	93
22.	Chlorogenic acid, theobromine, caffeine, rutin, fructose, glucose and sucrose	Acquity UPLC BEH C18 (50 × 2.1 mm, 1.7 μm)	*Gradient elution* A: 1% acetic acid in water B: 1% acetic acid acetonitrile *Flow rate:* 0.3 mL/min	ELSD DAD ESI-MS	*Ilex paraguariensis* / extraction (ethanol)	–	–	–	94
23.	Flavonoids (18 compounds)	Acquity UPLC BEH C18 (150 × 2.1 mm, 1.7 μm)	*Gradient elution* A: 0.1% acetic acid in water B: 0.1% acetic acid acetonitrile *Flow rate:* 0.3 mL/min	DAD ESI-MS	*Glycyrrhiza glabra* / extraction (70% ethanol)	–	–	–	95

No.	Compounds	Column	Mobile phase	Detection	Sample / Pretreatment	Linear range	LOD / LOQ	Recovery (%)	Ref.
24.	Phenolic acid, clovamide, flavonoids, isoflavones	Acquity UPLC BEH C18 (50 × 2.1 mm, 1.7 µm)	Gradient elution A: 0.1% acetic acid in water B: 40% acetonitrile Flow rate: 0.35 mL/min	DAD	Trifolium tops / SPE (column: C18)	–	–	–	76
25.	Flavonoids (15 compounds)	Acquity UPLC BEH C18 (50 × 2.1 mm, 1.7 µm)	Gradient elution A: 50 mM acetic acid in water B: acetonitrile Flow rate: 0.25 mL/min	DAD	Epimedium / PLE (70% ethanol)	1.76–134 µg/mL	0.05–0.13 ng / 0.22–0.52 ng	95.0–103.7	96
26.	Isoorientin, isoorientin 3''-O-glucopyranoside, isovitexin, isovitexin 3''-O-glucopyranoside, isoscoparin, isoscoparin 3''-O-glucopyranoside, isovitexin 6''-O-glucopyranoside and saponarin	Acquity UPLC BEH C18 (100 × 2.1 mm, 1.7 µm)	Gradient elution A: 0.1% formic acid in water B: methanol Flow rate: 0.25 mL/min	DAD ESI-MS/MS	Isatis indigotica / extraction (methanol:water (50:50; v/v))	0.04–8.2 µg/mL	0.8–4.5 ng/mL	96.6–100.2	97
27.	2''-O-β-L-galactopyranosyl-orientin, 2''-O-β-arabinopyranosyl-orientin, orientin, vitexin	Acquity UPLC BEH C18 (100 × 2.1 mm, 1.7 µm)	Gradient elution A: 0.1% acetic acid in water B: acetonitrile Flow rate: 0.4 mL/min	MS/MS	Trollius ledibouri / extraction (60% ethanol)		6–43 µg/g / 220–540 µg/g	96–100.3	72

compounds), they investigated stationary phases modified by octadecyl, phenyl, and cyanopropyl groups. In another work, Klejdus et al. compared the different techniques of plant sample preparation (pressurized liquid extraction (PLE), sonication, Soxhlet extraction, sonication/PLE) (75). An accelerated solvent extraction has been applied for the isolation of isoflavones from soybean samples. In this paper the authors described the very extensive study on the impact of the components of the mobile phase, pH, program of gradient elution, flow rate, and column temperature on resolution and symmetry of peaks. Other researchers used a UHPLC technique for the determination of various groups of polyphenols in the aerial parts of *Trifolium* in less than 8.5 min (76). Du et al. reported the use of UHPLC-DAD-TOF/MS in the determination of fourteen polyphenols in *Pueraria lobata* (77) and extensively discussed the plant sample preparation, which was based on the use of microwave-assisted extraction (MAE). A central composite design (CCD) was used to optimize temperature and time of extraction and type and volume of solvent. Furthermore, the authors compared the MAE with other types of extraction such as pressurized liquid extraction, reflux extraction, and ultrasonic extraction. Pietrogrande et al. investigated new stationary phases used by UHPLC methods for the determination of the components of *Passiflora incarnata* extracts (78). They compared relevant separation parameters such as total number of chemical components, separation efficiency, peak capacity, overlap degree of peaks, and peak purity autocovariance function (ACVF). Qi et al. performed simultaneous identification of thirty-nine analytes in *Danggui Buxue Tang* samples (79) by UHPLC. The work not only showed the significant reduction in analysis time as compared to the analysis using HPLC, but also a large improvement in selectivity. Miniaturization of the UHPLC apparatus, new stationary phases, and use of multidimensional liquid chromatography were reported in the analysis of polyphenolic compounds. Table 7.3 summarizes UHPLC methods developed for the determination of polyphenolic compounds in plant extract.

7.3 SUMMARY

It can be concluded that applications of UHPLC techniques has been increasing each year. Impressive technological advances in the UHPLC have made it possible to simultaneously determinate many polyphenolic compounds. UHPLC provides high speed and high sensitivity and consumes less mobile phase. Moreover, the combination of ultrahigh pressure liquid chromatography with mass spectrometry greatly increases the sensitivity and selectivity of the quantitative analysis of polyphenolic compounds in complex matrixes. UHPLC-Q-TOF/MS and TOF/MS techniques enable the identification of new polyphenolic compounds in plants.

REFERENCES

1. Tsao, R. *Nutrients*. 2010, 2: 1231–1246.
2. Han, X., Shen, T., Lou, H. *Int. J. Mol. Sci.* 2007, 8: 950–988.

3. HAVSTEEN, B.H. *Pharm. Therap.* 2002, 96: 67–202.
4. AHERNE, S.A., O'BRIEN, N.M. *Nutrition.* 2002, 18: 75–81.
5. XING, J., XIE, Ch., LOU, H. *J. Pharm. Biomed. Anal.* 2007, 44: 368–378.
6. VACEK, J., ULRICHOVÁ, J., KLEJDUS, B., ŠIMÁNEK, V. *Anal. Methods.* 2010, 2: 604–613.
7. STALIKAS, C.D. *J. Sep. Sci.* 2007, 30: 3268–3295.
8. VALLS, J., MILLÁN, S., MARTÍ, M.P., BORRÀS, E., AROLA, L. *J. Chromatogr. A.* 2009, 1216: 7143–7172.
9. ROSTAGNOA, M.A., VILLARES, A., GUILLAMÓN, E., GARCÍA-LAFUENTE, A., MARTÍNEZ, J.A. *J. Chromatogr. A.* 2009, 1216: 2–29.
10. VACEK, J., KLEJDUS, B., LOJKOVÁ, L.L., KUBÁN, V. *J. Sep. Sci.* 2008, 31: 2054–2067.
11. IGNAT, I., VOLF, I., POPA, M.I. *Food Chem.* 2011, 126: 1821–1835.
12. KALILI, K.M., DE VILLIERS, A. *J. Sep. Sci.* 2011, 34: 854–876.
13. CHURCHWELL, M.I., TWADDLE, N.C., MEEKER, L.R., DOERGE, D.R. *J. Chromatogr. B.* 2005, 825: 134–143.
14. SPÁČIL, Z., NOVÁKOVÁ, L., SOLICH, P. *Talanta.* 2008, 76: 189–199.
15. ORTEGA, N., ROMERO, M.P., MACIÀ, A., REGUANT, J., ANGLÈS, N., MORELLÓ, J.R., MOTILVA, M.J. *J. Food Comp. Anal.* 2010, 23: 298–305.
16. VILLIERS, A., KALILI, K.M., MALAN, M., ROODMAN, J. *LC-GC Eur.* 2010, 23: 466–478.
17. COOPER, K.A., CAMPOS-GIMÉNEZ, E., ALVAREZ, D.J., NAGY, K., DONOVAN, J.L., WILLAMSON, G. *J. Agric. Food Chem.* 2007, 55: 2841–2847.
18. SCHWARZ, M., RODRÍGUEZ, M.C., GUILLÉN, D.A., BARROSO, C.G. *J. Sep. Sci.* 2009, 32: 1782–1790.
19. KLEJDUS, B., LOJKOVÁ, L.L., LAPČIK, O., KOBLOVSKÁ, R., MORAVCOVÁ, J., KUBÁN, V. *J. Sep. Sci.* 2005, 28: 1334–1346.
20. LI, G., ZENG, X., XIE, Y., CAI, Z., MOORE, J.C., YUAN, X., CHENG, Z., JI, G. *Fitoterapia.* 2012, 83: 182–191.
21. ZHANG, W., XU, M., YU, Ch., ZHANG, G., TANG, X. *J. Chromatogr. B.* 2010, 878: 1837–1844.
22. WANG, Y., YAO, Y., AN, R., YOU, L., WANG, X. *J. Chromatogr. B.* 2009, 877: 1820–1826.
23. LUO, Y., CHEN, J., LI, P. *Chromatographia.* [Online] March 21, 2009. DOI: 10.1365/s10337-009-1048-5.
24. LÉVÈQUES, A., ACTIS-GORETTA, L., REIN, M.J., WILLIAMSON, G., DIOISI, F., GIUFFRIDA, F. *J. Pharm. Biomed. Anal.* 2012, 57: 1–6.
25. BARANOWSKA, I., MAGIERA, S., BARANOWSKI, J. *J. Liq. Chromatogr. Rel. Tehnol.* 2011, 34: 421–435.
26. BARANOWSKA, I., MAGIERA, S., BARANOWSKI, J. *J. Chromatogr. Sci.* 2011, 49: 764–773.
27. BARANOWSKA, I., MAGIERA, S., BARANOWSKI, J. *J. Chromatogr. B.* 2011, 879: 615–626.
28. BARANOWSKA, I., MAGIERA, S. *Anal. Bioanal. Chem.* 2011, 399: 3211–3219.
29. MARTÍ, M.P., PANTALEÓN, A., ROZEK, A., SOLER, A., VALLS, J., MARIÀ, A., ROMERO, M.P., MOTILVA, M.J. *J. Sep. Sci.* 2010, 33: 2841–2853.
30. SERRA, A., MACIÀ, A., ROMERO, M.P., SALVADÓ, M.J., BUSTOS, M., FERNÁNDEZ-LARREA, J., MOTILVA, M.J. *J. Chromatogr. B.* 2009, 877: 1169–1176.
31. GUO, J., SHANG, E., DUAN, J., TANG, Y., QIAN, D., SU, S. *Rapid Commun. Mass Spectrom.* 2010, 24: 443–453.
32. MA, Ch., GAO, W., GAO, Y., MAN, S., HUANG, L., LIU, Ch. *Phytochem. Anal.* 2011, 22: 112–118.
33. WU, C., ZHANG, J., ZHOU, T., GUO, B., WANG, Y., HOU, J. *J. Pharm. Biomed. Anal.* 2011, 54: 186–191.
34. YING, X., LU, X., SUN, X., LI, X., LI, F. *Talanta.* 2007, 72: 1500–1506.
35. NI, S., QIAN, D., DUAN, J., GUO, J., SHANG, E., SHU, Y., XUE, C. *J. Chromatogr. B.* 2010, 878: 2741–2750.
36. YANG, Z., ZHU, W., GAO, S., XU, H., WU, B., KULKARNI, K., SINGH, R., TANG, L., HU, M. *J. Pharm. Biomed. Anal.* 2010, 53: 81–89.
37. LIU, Q., SHI, Y., WANG, Y., LU, J., CONG, W., LUO, G., WANG, Y. *Talanta.* 2009, 80: 84–91.
38. LEHTENEN, H.M., RANTALA, M., SUOMELA, J.P., VIITANEN, M., KALLIO, H. *J. Agric. Food Chem.* 2009, 57: 4447–4451.
39. GUILLARME, D., CASETTA, C., BICCHI, C., VEUTHEY, J.L. *J. Chromatogr. A.* 2010, 1217: 6882–6890.

40. KLEJDUS, B., VACEK, J., BENEŠOVÁ, L., KOPECKÝ, J., LAPČÍK, O., KUBÁŇ, V. *Anal. Bioanal. Chem.* 2007, 389: 2277–2285.
41. KALILI, K.M., VILLIERS, A. *J. Chromatogr. A.* 2009, 1216: 6274–6284.
42. ARAPITSAS, P., SJÖBERG, P.J.R., TURNER, Ch. *Food Chem.* 2008, 109: 219–226.
43. PATI, S., LIBERATORE, M.T., GAMBACORTA, G., ANTONACCI, D., NOTTE, E.L. *J. Chromatogr. A.* 2009, 1216: 3864–3868.
44. PONGSUWAN, W., BAMBA, T., HARADA, K., YONETANI, T., KOBAYASHI, A., FUKUSAKI, E. *J. Agric. Food Chem.* 2008, 56: 10705–10708.
45. TRAUTVETTER, S., KOELLING-SPEER, I., SPEER, K. *Apidologie.* 2009, 40: 140–150.
46. BARANOWSKA, I., MAGIERA, S. *J. AOAC Internat.* 2011, 94: 786–794.
47. NOVÁKOVÁ, L., SPÁČIL, Z., SEIFRTOVÁ, M., OPLETAL, L., SOLICH, P. *Talanta.* 2010, 80: 1970–1979.
48. NOVÁKOVÁ, L., VILDOVA, A., MATEUS, J.P., GONÇALVES, T., SOLICH, P. *Talanta.* 2010, 82: 1271–1280.
49. SPÁČIL, Z., NOVÁKOVÁ, L., SOLICH, P. *Food Chem.* 2010, 123: 535–541.
50. CAVALIERE, Ch., FOGLIA, P., GUBBIOTTI, R., SACCHETTI, P., SAMPERI, R., LAGANA, A. *Rapid Commun. Mass Spectrom.* 2008, 22: 3089–3099.
51. GÓMEZ-ROMERO, M., SEGURA-CARRETERO, A., FERNÁNDEZ-GUTIÉRREZ, A. *Phytochem.* 2010, 71: 1848–1864.
52. NAGY, K., REDEUIL, K., BERTHOLET, R., STEILING, H., KUSSMANN, M. *Anal. Chem.* 2009, 81: 6347–6356.
53. GRUZ, J., NOVAK, O., STRNAD, M. *Food Chem.* 2008, 111: 789–794.
54. JAITZ, L., SIEGL, K., EDER, R., RAK, G., ABRANKO, L., KOELLENSPERGER, G., HANN, S. *Food Chem.* 2010, 122: 366–372.
55. XIANG, G., YANG, L., ZHANG, X., YANG, H., REN, Z., MIAO, M. *Chromatographia.* 2009, 70: 1007–1010.
56. ARRÁEZ-ROMÁN, D., FU, S., SAWALHA, S.M.S., SEGURA-CARRETERO, A., FERNÁNDEZ-GUTIÉRREZ, A. *Electrophoresis.* 2010, 31: 2289–2296.
57. PAPOUŠKOVA, B., BEDNÁŘ, P., HRON, K., STÁVEK, J., BALÍK, J., MYJAVCOVÁ, R., BARTÁK, P., TOMÁNKOVÁ, E., LEMR, K. *J. Chromatogr. A.* 2011, 1218: 7581–7591.
58. SÁENZ-NAVAJAS, M.P., FERREIRA, V., DIZY, M., FERNÁNDEZ-ZURBANO, P. *Anal Chim Acta.* 2010, 673: 151–159.
59. ZHAO, Y., CHEN, P., LIN, L., HARNLY, J.M., YU, L., LI, Z. *Food Chem.* 2011, 126: 1269–1277.
60. KALILI, K.M., VILLIERS, A. *J. Sep. Sci.* 2010, 33: 853–863.
61. XIA, Z., WEI, L., YONG, X., LIU, Y., WEI-SONG, K., XIAO-DONG, R., SHUAI, Y., YONG-KUAN, Ch., MING-MING, M. *Chem. Res. Chinese Universities.* 2011, 27: 550–556.
62. RUI-FENG, Z., FENG, L., JING, H. *Anal. Methods.* 2011, 3: 2421–2424.
63. HURTADO-FERNÁNDEZ, E., PACCHIAROTTA, T., GÓMEZ-ROMERO, M., SCHOENMAKER, B., DERKS, R., DEELDER, A.M., MAYBORODA, O.A., CARRASCO-PANCORBO, A., FERNÁNDEZ-GUTIÉRREZ, A. *J. Chromatogr. A.* 2011, 1218: 7723–7738.
64. HOSSEINIAN, F.S., LI, W., HYDAMAKA, A.W., TSOPMO, A., LOWRY, L., FRIEL, J., BETA, T. *J. Agric. Food Chem.* 2007, 55: 6970–6976.
65. ENGELS, Ch., GRÄTER, D., ESQUIVEL, P., JIMÉNEZ, V.M., GÄNZLE, M.G., SCHIEBER, A. *Food Res. Internat.* [Online] April 13, 2011. DOI: 10.1016/j.foodres.2011.04.003.
66. SØLTOFT, M., CHRISTENSEN, J.H., NIELSEN, J., KNUTHSEN, P. *Talanta.* 2009, 80: 269–278.
67. FIECHTER, G., OPACAK, I., RABA, B., MAYER, H.K. *Food Res. Internat.* [Online] March 25, 2011. DOI: 10.1016/j.foodres.2011.03.038.
68. TOURINO, S., FUGUET, E., JÁUREGUI, O., SAURA-CALIXTO, F., CASCANTE, M., TORRES, J.L. *Rapid Commun. Mass Spectrom.* 2008, 22: 3489–3500.
69. TISTAERT, Ch., DEJAEGHER, B., HEYDENMM, Y.V. *Anal Chim Acta.* 2011, 690: 148–161.
70. LIANG, X., JIN, Y., WANG, Y., JIN, G., FU, Q., XIAO, Y. *J. Chromatogr. A.* 2009, 1216: 2033–2044.
71. TOH, D.F., NEW, L.S., KOH, H.L., CHAN, E.Ch.Y. *J. Pharm. Biomed. Anal.* 2010, 52: 43–50.
72. LI, X., XIONG, Z., YING, X., CUI, L., ZHU, W., LI, F. *Anal Chim Acta.* 2006, 580: 170–180.
73. KLEJDUS, B., VACEK, J., BENEŠOVÁ, L., KOPECKÝ, J., LAPČÍK, O., KUBÁŇ, V. *Anal. Bioanal. Chem.* 2007, 389: 2277–2285.

74. KLEJDUS, B., VACEK, J., LOJKOVÁ, L., BENEŠOVÁ, L., KUBÁŇ, V. *J. Chromatogr. A.* 2008, 1195: 52–59.
75. KLEJDUS, B., MIKELOVÁ, R., PETRLOVÁ, J., POTĚŠI, D., ADAM, V., STIBOROVÁ, M., HODEK, P., VACEK, J., KIZEK, R., KUBÁŇ, V. *J. Chromatogr. A.* 2011, 1084: 71–79.
76. OLESZEK, W., STOCHMAL, A., JANDA, B. *J. Agric. Food Chem.* 2007, 55: 8095–8100.
77. DU, G., ZHAO, H.Y., ZHANG, Q.W., LI, G.H., YANG, F.Q., WANG, Y., LI, Y.C., WANG, Y.T. *J. Chromatogr. A.* 2010, 1217: 705–714.
78. PIETROGRANDE, M.Ch., DONDI, F., CIOGLI, A., GASPARRINI, F., PICCIN, A., SERAFINI, M. *J. Chromatogr. A.* 2010, 1217: 4355–4364.
79. QI, L.W., WEN, X.D., CAO, J., LI, Ch.Y., LI, P., YI, L., WANG, Y.X., CHENG, X.L., GE, X.X. *Rapid Commun. Mass Spectrom.* 2008, 22: 2493–2509.
80. LI, S.L., SONG, J.Z., CHOI, F.F.K., QIAO, Ch.F., ZHOU, Y., HAN, Q.B., XU, H.X. *J. Pharm. Biomed. Anal.* 2009, 49: 253–266.
81. LOU, Z., WANG, H., ZHU, S., ZHANG, M., GAO, Y., MA, Ch., WANG, Z. *J. Chromatogr. A.* 2010, 1217: 2441–2446.
82. CICCHETTI, E., CHAINTREAU, A. *J. Sep. Sci.* 2009, 32: 3043–3052.
83. KISS, B., POPA, D.S., HANGANU, D., POP, A., LOGHIN, F. *Rev. Roum. Chim.* 2010, 55: 459–465.
84. ZIMMERMANN, B.F., WALCH, S.G., TINZOH, L.N., STÜHLINGER, W., LACHENMEIER, D.W. *J. Chromatogr. B.* 2011, 879: 2459–2464.
85. KLEJDUS, B., MIKELOVÁ, R., PETRLOVÁ, J., POTĚŠI, D., ADAM, V., STIBOROVÁ, M., HODEK, P., VACEK, J., KIZEK, R., KUBÁŇ, V. *J. Agric. Food Chem.* 2005, 53: 5848–5852.
86. MUTH, D., MARSDEN-EDWARDS, E., KACHLICKI, P., STOBIECKI, M. *Phytochem. Anal.* 2008, 19: 444–452.
87. GRATA, E., GUILLARME, D., GLAUSER, G., BOCCARD, J., CARRUPT, P.A., VEUTHEY, J.L., RUDAZ, S., WOLFENDER, J.L. *J. Chromatogr. A.* 2009, 1216: 5660–5668.
88. CACCIOLA, F., DELMONTE, P., JAWORSKA, K., DUGO, P., MONDELLO, L., RADER, J.I. *J. Chromatogr. A.* 2011, 1218: 2012–2018.
89. KAPUSTA, I., JANDA, B., SZAJWAJ, B., STOCHMAL, A., SIACENTE, S., PIZZA, C., FRANCESCHI, F., FRANZ, Ch., OLESZEK, W. *J. Agric. Food Chem.* 2007, 55: 8485–8490.
90. QI, L.W., CAO, J., LI, P., YU, Q.T., WEN, X.D., WANG, Y.X., LI, Ch.Y., BAO, K.D., GE, X.X., CHENG, X.L. *J. Chromatogr. A.* 2008, 1203: 27–35.
91. NOVÁKOVÁ, L., VILDOVÁ A., MATEUS, J.P., GONÇALVES, T., SOLICH, P. *Talanta.* 2010, 82: 1271–1280.
92. GRATACÓS-CUBARSÍ, M., RIBAS-AGUSTÍ, A., GARCÍA-REGUEIRO, J.A., CASTELLARI, M. *Food Chem.* 2010, 121: 257–263.
93. KUMAR, N., BHANDARI, P., SINGH, B., BARI, S.S. *Food Chem. Toxicol.* 2009, 47: 361–367.
94. DARTORA, N., SOUZA, L.M., SANTANA-FILHO, A.P., IACOMINI, M., VALDUGA, A.T., GORIN, P.A.J., SASSAKI, G.L. *Food Chem.* 2011, 129: 1453–1461.
95. SIMONS, R., VINCKEN, J.P., BAKX, E.J., VERBRUGGEN, M.A., GRUPPEN, H. *Rapid Commun. Mass Spectrom.* 2009, 23: 3083–3093.
96. CHEN, X.J., JI, H., ZHANG, Q.W., TU, P.F., WANG, Y.T., GUO, B.L., LI, S.P. *J. Pharm. Biomed. Anal.* 2008, 46: 226–235.
97. DENG, X., GAO, G., ZHENG, S., LI, F. *J. Pharm. Biomed. Anal.* 2008, 48: 562–567.

Chapter 8

UHPLC for Characterization of Protein Therapeutics

Jennifer C. Rea, Yajun Jennifer Wang, and Taylor Zhang

8.1 INTRODUCTION

Protein-based therapeutics represent a significant portion of the pharmaceutical market, with over 130 proteins currently approved for clinical use by the FDA for a wide variety of indications (1). Many protein therapeutics are manufactured in bacterial or mammalian cells in large bioreactors, which is in contrast to the laboratory synthesis of small molecule drugs (2).

Protein products are often heterogeneous, containing mixtures of product and product variants that vary in molecular structure, including molecular size and charge. Protein variants can be produced during protein synthesis or can result from post-translational modifications or degradation during the manufacturing and storage of the material. Molecular size, charge, and other product attributes are often critical quality attributes that can impact the safety and efficacy of the therapeutics; thus, detailed characterization of biopharmaceuticals, both the product and its variants, is needed for regulatory approval. To provide a comprehensive assessment of a complex protein biopharmaceutical product, orthogonal analytical techniques are required for characterization studies. Among an array of analytical techniques, high performance liquid chromatography (HPLC) is the most widely used chromatographic technique in the biopharmaceutical industry to characterize protein hydrophobicity, charge and size heterogeneities, sequence variants, glycan profile, and other protein modifications.

While HPLC has been the standard analytical tool for protein characterization for many years, ultrahigh performance liquid chromatography (UHPLC), sometimes referred to as ultrahigh pressure liquid chromatography, has the potential to increase resolution and efficiency compared to conventional HPLC and, thus, facilitate improved protein characterization. Like conventional HPLC, UHPLC can provide

Ultra-High Performance Liquid Chromatography and Its Applications, First Edition. Edited by Quanyun Alan Xu.
© 2013 John Wiley & Sons, Inc. Published 2013 by John Wiley & Sons, Inc.

qualitative and quantitative profiles of the product and product variants and can be used for monitoring lot-to-lot comparability among different production batches, thus meeting requirements for regulatory submissions. UHPLC can also be used for comparing biosimilars and proprietary drug substances (3, 4). This chapter will review development and applications of UHPLC for protein drug characterization. Various modes of separation will be discussed: reversed-phase chromatography (including reversed-phase liquid chromatography/mass spectrometry (LC/MS) and liquid chromatography tandem mass spectrometry (LC/MS/MS)), hydrophilic interaction chromatography, hydrophobic interaction chromatography, ion exchange chromatography, and size exclusion chromatography.

8.2 PROTEIN CHARACTERIZATION AND LOT RELEASE TESTING

Protein therapeutics range in protein type (e.g., antibody, enzyme, etc.), size, and molecular structure and are used to treat a variety of indications (Table 8.1) (1). Proteins often have highly complex structures; changes in physicochemical properties of the protein therapeutic can affect product safety and efficacy. Protein characterization involves determining physicochemical properties such as amino acid sequence, molecular size, charge, hydrophobicity, electrophoretic mobility, isoelectric point (pI), sedimentation velocity, glycosylation, spectral properties, and primary, secondary, tertiary, or quaternary structures. Protein modification can occur throughout the manufacturing process at various steps, including cell culture, harvest, purification, formulation, filling, and during shelf life, resulting in a heterogeneous mixture of product and structurally related product variants. Therefore, product characterization and quality control testing are required throughout clinical development and at critical points during manufacturing to identify and control for these product variants (5).

Protein therapeutics are characterized by physicochemical, immunochemical, and biological methods. Guidance documents have been issued by regulatory agencies and industry representatives recommending approaches for protein characterization

Table 8.1 Examples of Commercially Available Protein Therapeutics

Therapeutic	Indications	Approximate Molecular Mass	Reference
Deoxyribonuclease I	Cystic fibrosis	29 kDa	6
Ranibizumab	Neovascular age-related macular degeneration	48 kDa	7
Tissue plasminogen activator	Pulmonary embolism, acute ischemic stroke, myocardial infraction	59 kDa	8
Trastuzumab	Breast cancer	150 kDa	9

Table 8.2 Commonly Used HPLC-based Tests for Characterization and Lot Release Testing of Biotherapeutics

Attribute	Test Name
Identity	Peptide mapping by reversed-phase HPLC (RP-HPLC)
Purity	Size exclusion chromatography (SEC)
	Ion exchange chromatography (IEC)
	Glycosylation profile by hydrophilic interaction chromatography (HILIC)
	Hydrophobic interaction chromatography (HIC)

(4, 10–12). Assays for protein characterization, such as identity and purity assays, are often based on HPLC methods. These assays evaluate critical quality attributes, such as primary structure, size, and charge heterogeneities, which are part of the overall quality target product profile. The extent of characterization is linked to the level of risk associated with each phase of drug development. For example, while there may not be sufficient time or resources for extensive characterization of a protein therapeutic during early stage development, it is expected that the product is well characterized before the license application is submitted to regulatory agencies.

Once the protein therapeutic is purified and formulated, it must be tested prior to lot release. A set of tests and acceptance criteria are established based on protein characterization and regulatory requirements to ensure product quality. HPLC is one of the most frequently used techniques for characterization of biotherapeutics (Table 8.2) (10). These tests are often also used for lot release testing of biotherapeutics. Once these tests are performed and the results meet the established acceptance criteria, a certificate of analysis (COA) is generated. Adequate stability studies should be performed on the protein drug substance (e.g., frozen bulk for storage) and drug product (e.g., final vial) according to regulatory guidelines (13, 14).

8.3 HPLC AND UHPLC

As mentioned earlier, HPLC is one of the most widely used techniques in the biopharmaceutical industry for characterizing proteins. The diverse modes of chromatography separate molecules based on different attributes, such as molecular size, surface charge, and hydrophobic and affinity interactions. In some chromatography modes, subtle differences in the physicochemical properties of the product could result in significant changes in elution profiles and elution time, thus providing valuable information regarding the various product variants present in the protein sample. The high-resolution separation of protein variants provides an effective means to study product profiles, which can be used for lot-to-lot comparisons.

UHPLC has recently been adopted for protein characterization (15). Conventional HPLC and UHPLC attributes and chromatographic equipment are shown in Table 8.3. UHPLC methods typically use small particle (sub-2 μm) columns for improved efficiency, resolution, speed, and peak capacity (number of peaks resolved

Table 8.3   Comparison of HPLC and UHPLC Attributes and Equipment

Attribute	HPLC	UHPLC
Pressure	Up to 600 bar	Up to 1200 bar
Column particle size	>2 μm	<2 μm

Instrumentation examples	Agilent
	    1100, 1200, 1260
	Waters
	    Alliance 2690, 2695, 2796
	Dionex
	    UltiMate 3000
	Thermo Scientific
	    Surveyor Plus
		Agilent
		    1290
		Waters
		    Acquity H-Class
		Dionex
		    UltiMate 3000 RSLC
		Thermo Scientific
		    Accela

per unit time) (15). As the particle size decreases, there is a significant gain in efficiency, and the efficiency does not diminish at increased linear velocities (Figure 8.1) (16). Therefore, high flow rates are typically used for UHPLC methods, resulting in shorter run times than conventional HPLC methods (17).

Since the commercialization of UHPLC technologies by Waters (Milford, Massachusetts) and Agilent (Santa Clara, California) in the early 2000s (18), most of the top chromatographic equipment vendors have introduced competitive UHPLC instruments. These UHPLC chromatographic systems have low system volumes, column heaters capable of heating to 90°C, and pressure limits exceeding the conventional HPLC pressure limits of 400–600 bar (13). In addition, UHPLC

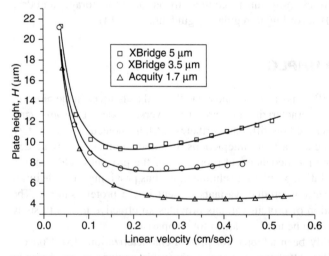

Figure 8.1   Van Deemter plots obtained for acetophenone on Acquity and XBridge columns. Columns: Acquity BEH C_{18}, 1.7 μm, 2.1 mm × 10 cm; XBridge C_{18}, 3.5 μm, 4.6 mm × 15 cm; XBridge C_{18}, 5 μm, 4.6 mm × 25 cm. Reproduced with permission from de Villiers, A. et al. *J. Chromatogr. A.* 2006.

instrumentation includes detectors with rapid scan rates to accommodate the high flow rates, high system pressures, and narrow peak widths typically associated with UHPLC methods.

Several biocompatible HPLC and UHPLC systems are also available, engineered with a bio-inert flow path made of non-stainless steel materials (such as titanium or PEEK) for protein analysis. Stainless steel equipment and tubing should be avoided for protein analysis by aqueous/salt-based methods because protein chelation with metals can negatively affect separation (19). These interactions can either occur with metal contaminating the column or with corroded surfaces within the instrument. In addition to affecting separation, corrosion can result in physical damage to the system, such as pump seal failure and compromised performance of the detector cells. Stainless steel systems may require periodic passivation for reliable usage (20).

Converting HPLC methods into UHPLC methods, as well as transferring HPLC and UHPLC methods to different UHPLC instruments, can be challenging. Factors such as geometry of the pump, gradient delay volumes, gradient mixing, system pressure limits, and column compartment temperature ranges must be taken into consideration for successful method transfer between different instruments. In some instances, the UHPLC can be "detuned" by adjusting instrument settings and chromatographic conditions to give equal performance to its HPLC predecessor (21). Some UHPLC manufacturers have introduced emulation software to facilitate method transfer. Alternatively, some manufacturers make UHPLCs that can perform both HPLC and UHPLC methods on a single instrument. It is up to the analyst to determine instrument settings that would be optimal for a particular application.

Although the advantages of columns with small particles for protein analysis have been long recognized (22), modern commercially available columns designed for UHPLC were only recently introduced, accompanied by the debut of UHPLC systems. Agilent's 1.8-μm particle "rapid resolution high throughput" (RRHT) column and Waters' 1.7-μm bridged ethane-silicon hybrid (BEH) particles were some of the first commercial columns used for protein separation by UHPLC (23). Surface chemistries, which give columns their selectivity, are somewhat more limited for UHPLC compared with HPLC. However, UHPLC column selection is quickly expanding, with numerous vendors offering a full range of column chemistries.

UHPLC systems can be used for performing analysis to facilitate real-time decision making, such as for pooling of a process chromatography column (24); the UHPLC method in this study compared favorably with the traditional off-line HPLC method in terms of accuracy and variability. Also, the UHPLC method had significantly faster separations while retaining the separation efficiency, making it feasible for process analytical technology (PAT) applications. The authors concluded that implementation of PAT based on UHPLC can allow for control schemes that rely on measurement of product quality attributes and enable real-time decisions, which may result in more consistent product quality and higher operational efficiency.

The following sections will cover various separation modes used for the characterization of proteins by UHPLC: reversed-phase chromatography (RP-UHPLC) of proteins and peptide maps, RP-UHPLC for LC/MS and LC/MS/MS applications, hydrophilic interaction chromatography (HILIC) for glycan profiling, hydrophobic interaction chromatography (HIC), ion exchange chromatography (IEC), and size exclusion chromatography (SEC).

8.4 REVERSED-PHASE UHPLC FOR PROTEIN ANALYSIS

UHPLC is becoming increasingly utilized for reversed-phase analysis of proteins. UHPLC may prove particularly attractive for protein separations due to decreased carryover compared to conventional HPLC. By employing RP-HPLC columns with 1.5-μm particles and using a pressure range from 160–1600 bar on an in-house UHPLC, Eschelbach and Jorgenson found that, as the pressure was increased, carryover was diminished (25). The recovery was enhanced for each of the proteins studied, approaching 100% for certain proteins.

In addition to decreased carryover, another advantage that UHPLC offers for protein analysis is speed. In one study, three columns with different stationary phases were evaluated for antibody-free thiol variant separation: Waters Acquity BEH C_8 (1.7 μm), Agilent diphenyl (3.5 μm), and Agilent Eclipse C_{18} (1.8 μm) (26). The results demonstrated that free thiol variants could be separated in less than 3 min as shown in Figure 8.2, with the C_{18} column demonstrating better resolution compared to the C_8 and diphenyl columns.

In addition to analyzing intact antibodies, UHPLC has been used for characterizing reduced monoclonal antibodies. Stackhouse and colleagues developed a

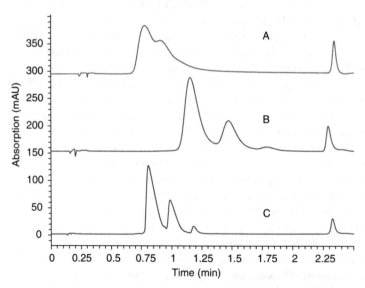

Figure 8.2 UHPLC reversed-phase separation of monoclonal antibody-free thiol variants using three different columns: Waters Acquity BEH C_8 (1.7 μm) (A), Agilent diphenyl (3.5 μm) (B), and Agilent Eclipse C_{18} (1.8 μm) (C). These chromatograms show the different selectivity for the free thiol variants with the different columns. Reproduced with permission from Jeong, J. et al. *Am. Pharmaceut. Rev.* 2011.

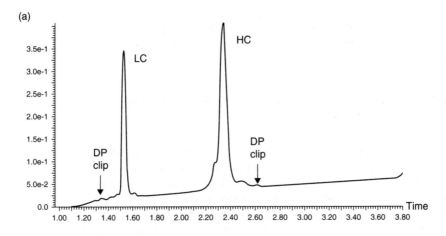

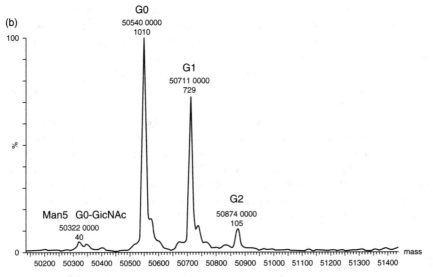

Figure 8.3   Phenyl UHPLC reversed-phase separation of a reduced IgG2 molecule. (a) UHPLC chromatogram of the phenyl UHPLC separation of the reduced IgG2 molecule; LC represents light chain and HC represents heavy chain; DP clip represents acid-induced aspartic acid/proline clipping. (b) The deconvoluted mass spectrum of the HC. Each peak in the spectrum is labeled with nominal mass and intensity, and peak assignments were based on mass. G0, G1, G2, G0–GlcNAc, and Man5 are glycoforms typically observed in IgG molecules and their assignment is based on the mass of the peak. Reproduced with permission from Stackhouse, N., et al. *J. Pharm. Sci.* 2011.

high-throughput method with a short cycle time of 5 min, achieved using a UHPLC system with a 1.7-μm phenyl column (27). This UHPLC method allowed quantitation of acid-induced aspartic acid/proline (D/P) clipping in an IgG2 molecule and peak assignments of glycoforms (Figure 8.3). The results from the UHPLC method were comparable to those obtained with a reduced capillary electrophoresis sodium dodecyl sulfate (rCE–SDS) method (27). Additionally, the phenyl column offered partial resolution of oxidation and other chemical modifications (data not shown), demonstrating UHPLC as a tool for high-throughput process characterization and formulation screens.

8.5 REVERSED-PHASE UHPLC FOR PEPTIDE MAP ANALYSIS

Peptide mapping is one of the most important and widely used techniques in protein characterization and is used to assess product quality and process consistency. To produce a peptide map, proteins are cleaved into peptide fragments by enzymes with different specificities, such as trypsin, which selectively cleaves the C-terminus of the basic residues Lys and Arg. After the protein is digested, the peptide fragments are typically separated on an RP-HPLC column using an acetonitrile (ACN)/water gradient with added trifluoroacetic acid (TFA), and the elution profile is monitored via UV detection at 214 nm, and/or mass spectrometry. Because peptide mapping can detect small changes in the protein amino acid sequence, peptide mapping is valuable for identity testing and monitoring site-specific modifications including expression errors, mutations, location of glycosylation, disulfide linkages, posttranslational modifications, and degradation such as deamidation, oxidation, and a host of other covalent modifications.

With the advent of UHPLC, peptide mapping analysis using smaller sized particles has been developed (28). Liu and colleagues studied the effects of column length, particle size, gradient length, and flow rate of a UHPLC system on peptide peak capacity (29). Columns packed with 1.7 and 3 μm C_{18} materials were tested with different gradient lengths and flow rates. In most cases, an increase in peak capacity of more than 45% was obtained with the reduction in particle size from 3 to 1.7 μm (Figure 8.4). In a flow rate range of 100–700 nL/min, increasing the flow rate

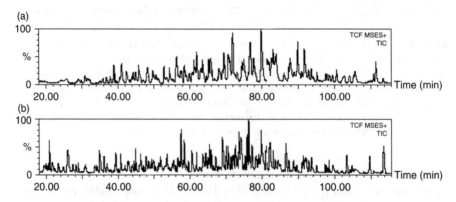

Figure 8.4 Total ion current (TIC) traces of a 5-protein digest mixture. (a) 75 μm × 15 cm column packed with 3 μm particles; (b) 75 μm × 15 cm column packed with 1.7 μm particles. Reproduced with permission from Liu, H., et al. *J. Chromatogr. A.* 2007.

improved peak capacity for columns packed with 1.7 and 3 μm materials compared to slower flow rates. Increasing the flow rate increased back pressure, and back pressure was higher for the 1.7 μm particles compared to the 3 μm particles. With the same back pressure generated, a shorter 1.7 μm particle column outperformed a longer column packed with the 3 μm material. The authors concluded that for improved peptide analysis, the use of a 1.7 μm particle column is generally advantageous over the use of a longer 3 μm particle column, with higher peak capacity using the same gradient length and flow rate.

In addition to particle size, the porosity of particles has been studied for protein and peptide analysis. Staub and colleagues evaluated columns with sub-2 μm totally porous particles and sub-3 μm superficially porous particles for peptide and protein analysis (30). The columns employed in this study were three UHPLC fully porous particles, namely Acquity C_{18} BEH 120, Acquity C_{18} BEH 300, and Acquity C_4 BEH 300, all made by Waters, and three superficially porous columns: a Halo C_{18} by Advanced Materials Technology, a Kinetex C_{18} by Phenomenex, and a Poroshell 120 EC-C_{18} by Agilent Technologies. This study showed that good resolution of small proteins with molecular weights lower than 40,000 Da could be achieved with chromatographic conditions similar to those used for peptide analysis. Larger pore size columns may be more appropriate for large proteins.

8.6 UHPLC FOR LC-MS AND LC-MS/MS APPLICATIONS

With the continued maturation of UHPLC technology, major advances in the field of UHPLC-MS and UHPLC-MS/MS have been made, with applications in biomolecule analysis, drug metabolism, and targeted screening assays, achieving performance in qualitative and quantitative analysis equivalent to HPLC-MS and HPLC-MS/MS (31). Throughput increases and resolution enhancements afforded by UHPLC can be applied to mass spectrometry, so long as the impact of UHPLC conditions on MS detection capabilities is taken into account (e.g., acquisition rate, limits of detection, and matrix effects). Most UHPLC-MS applications use reversed-phased chromatography, thus the examples cited herein are for reversed-phase applications.

Tolley and colleagues demonstrated a very high pressure system for the analysis of tryptic digests of proteins by LC-MS/MS using ESI-MS (electrospray ionization-mass spectrometry) (32). Columns were 150 μm × 22 cm and were slurry packed with 1.5 μm C_{18} bonded nonporous silica particles. The results of this analysis were compared to an analysis performed on the same tryptic digest using a nanoelectrospray-MS/MS direct infusion technique. Although both techniques were able to identify the unknown protein by analysis of the peptides from the tryptic digest, the UHPLC method gave twice as many sequenced peptides as nanoelectrospray and improved the signal-to-noise ratio of the spectra by at least a factor of ten (32).

Kocher and colleagues used RP-UHPLC to demonstrate a linear relationship between peak capacity and number of identified peptides, providing a rationale for using UHPLC technology for LC-MS/MS (33). Using two different column lengths and gradient times between 1 and 10 h, Kocher identified on average 2516 proteins based on 14,292 peptides in the tryptic digest of 1 μg of HeLa lysate using an 8-h gradient on a 50-cm column packed with 2 μm C_{18} reversed-phase chromatographic material. Currently, this is the highest number of proteins reported for a single-run LC-MS/MS experiment, which was possible due to the increased peak capacity afforded by UHPLC (33).

Everley and Croley used a combination of high temperatures (up to 65°C), a strong organic modifier (i.e., isopropanol), and columns packed with sub-2 μm particles at very high pressure to yield enhanced resolution, sensitivity, and a threefold increase in throughput using UHPLC-MS as compared to conventional HPLC-MS (34). Ten protein standards ranging in mass from 6 to 66 kDa were chosen to examine the performance of UHPLC relative to HPLC. Intact proteins were separated by reversed-phase chromatography, which was directly interfaced to a mass spectrometer. By using UHPLC, the run time was reduced by two-thirds without affecting MS data quality (Figure 8.5). Furthermore, data analysis time, for example, deconvolution time, was also reduced using the UHPLC method. For the 75-min HPLC chromatogram, the time required to perform automated charge state deconvolution was approximately 30 min. In contrast, the 25-min UHPLC chromatogram required only 10 min to perform deconvolution per chromatogram. Reducing the deconvolution time minimized the gap between data collection and interpretation, allowing data-directed decisions to be made in less time.

In another study, a selective, rapid, and sensitive 12.7-min ultrahigh performance liquid chromatography—isotope dilution tandem mass spectrometry (UHPLC–ID-MS/MS) method was developed and compared to a conventional 74-min

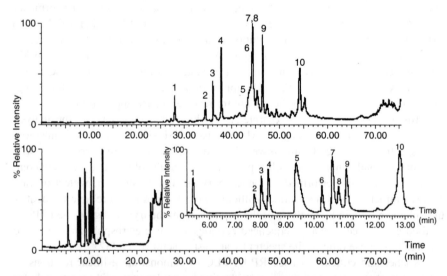

Figure 8.5   Comparison of chromatograms obtained using a 3.5 μm particle column (top) and 1.7 μm particle column (bottom). The 1.7 μm particle method is completed prior to the elution of the first protein in the 3.5 μm particle method. The inset more clearly reveals the resolution advantage afforded by the 1.7 μm particle method. Reproduced with permission from Everley, R.A., et al. *J. Chromatogr. A.* 2008.

high-performance liquid chromatography—isotope dilution tandem mass spectrometry (HPLC—ID-MS/MS) method for the absolute quantitative determination of multiple proteins from complex matrixes (35). Detection was performed on a mass spectrometer operated in the multiple reaction monitoring (MRM) mode. Relative standard deviation values equal to or less than 6.5% were obtained by the UHPLC—ID-MS/MS method, thus demonstrating performance equivalent to conventional HPLC—ID-MS/MS for isotope dilution quantification of peptides and proteins. UHPLC also provided the additional advantages of rapid analysis time and high sample throughput, which expand laboratory capabilities over conventional HPLC. The conventional HPLC method allows for a maximum of 19 runs in a 24-h period, whereas the UHPLC method provides the capability of 4–5 runs per hour and up to 113 runs per day.

Kay and colleagues demonstrated the first large scale and high-throughput UHPLC-MS/MS-based quantitation of a medium abundance protein in human serum (36). Insulin-like growth factor-I (IGF-I) is a known biomarker of recombinant human growth hormone (rhGH) abuse and is also used clinically to confirm acromegaly. The protein leucine-rich a-2-glycoprotein (LRG) was recently identified as a putative biomarker of rhGH administration. The use of a 5-min UHPLC-MS/MS-based selected reaction monitoring (SRM) assay detected both IGF-I and LRG at endogenous concentrations. Serum samples from two rhGH administrations were extracted, and their UHPLC-MS/MS-derived IGF-I concentrations correlated well against immunochemistry-derived values (36).

In another application, UHPLC-MS technology was developed for rapid comparison of a candidate biosimilar to an innovator monoclonal antibody (mAb) (37). In this study, UHPLC-MS was developed for rapid verification of identity and characterization of sequence variants and posttranslational modifications (PTMs) for mAb products. Although the biosimilar product is expected to have the same amino acid sequence and modifications as the innovator's product, the observed intact mass by UHPLC-MS was different for the biosimilar compared to the innovator protein. Peptide mapping using UHPLC-MS/MS (38) revealed that the mass difference between the biosimilar and the innovator's product was due to a two amino acid residue variance in the heavy chain sequence of the biosimilar (Figure 8.6).

8.7 HYDROPHILIC INTERACTION CHROMATOGRAPHY (HILIC) FOR GLYCAN PROFILING

Many protein-based biopharmaceuticals are glycosylated proteins. Protein glycosylation has implications for a variety of biological functions, including cell–cell signaling, protein stability, solubility, and affinity to a target molecule (39). The glycan profile is particularly relevant for some therapeutic proteins because it can impact efficacy and safety of the product (39, 40). Thus, glycans of biopharmaceuticals are often characterized, and in many cases, the relative amounts of the individual glycan structures are monitored during production.

For glycan analysis, the glycans are released by chemical or enzymatic methods. The prevailing analytical methods for glycan profiling involve some form of labeling

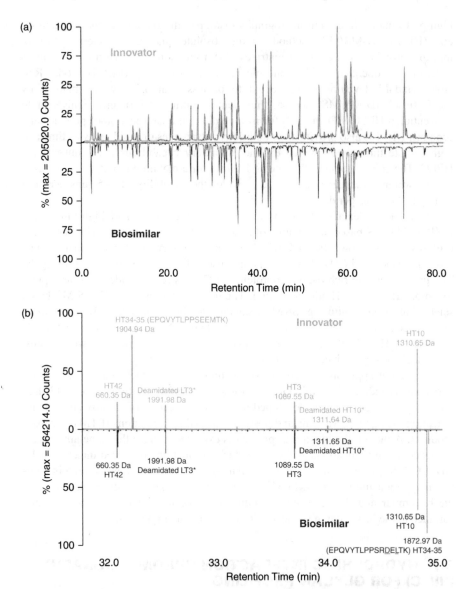

Figure 8.6 Mirror plots of UHPLC/MS peptide maps from tryptic digests of an innovator mAb and a biosimilar mAb. Peptide mixture was separated on a 2.1 × 150 mm BEH300 1.7 μm column at 65°C using a 90-min gradient (1%–40%B). The total ion chromatograms (TIC) obtained by the mass spectrometer (A) and a zoom view of charge-reduced, isotope-deconvoluted UHPLC-MS data from 32–35 min (B) are shown. Reproduced with permission from Xie, H., et al. *MAbs*. 2010.

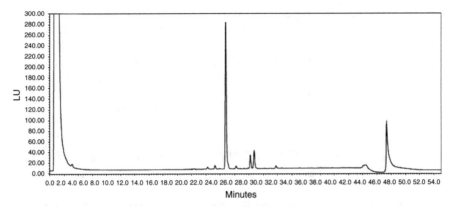

Figure 8.7   Chromatographic profile of released glycans from a monoclonal antibody following fluorescent labeling. Sample was analyzed with a Waters BEH Glycan column (1.7 μm, 1.5 × 150 mm) and fluorescent detection. Reproduced with permission from Cook, K.S., et al. *Biologicals*. 2012.

the released glycans with UV- or fluorescence-active reagents, followed by appropriate HPLC or UHPLC separations (41). Perhaps the most common reagent today is 2-aminobenzamide, or 2-AB. 2-AB and other commonly used reagents are compatible with fluorescence detection for best sensitivity, which is one reason why UHPLC with fluorescence detection is rapidly becoming a standard method for glycoprofiling. There are commercially available UHPLC columns for glycan analysis that are optimized for use with UHPLC systems with fluorescence detection, separating the released glycans of biopharmaceuticals as their 2-AB derivatives, including high mannose, complex, hybrid, and sialylated glycans. These columns for glycan analysis are based on HILIC rather than reversed-phase chromatography. HILIC refers to liquid chromatography utilizing a polar stationary phase (e.g., amide, diol, or cyano phases) and mobile phases using a high percentage of organic, such as acetonitrile. The strongly retained polar analytes are eluted with an aqueous gradient. The high selectivities, as well as the commercial availability of UHPLC columns in different phases, render HILIC as increasingly popular for the analysis of glycans (42). In one application, the use of a BEH Glycan column reduced the glycan analysis time from 3 h to 1 h (43), thus increasing throughput. The analysis time could potentially be reduced further, as the peaks of interest elute within a timeframe of approximately 10 min (from 22–32 min) (Figure 8.7).

8.8 HYDROPHOBIC INTERACTION CHROMATOGRAPHY FOR HYDROPHOBICITY ANALYSIS

Hydrophobic interaction chromatography (HIC) is known as an effective means of resolving proteins under mild conditions by employing a descending salt gradient to elute the protein (44). Although HIC exhibits favorable features with regard to preservation of biological activity of proteins, as a separation technique it suffers from low peak capacities and long analysis times compared with RP-HPLC. Early work by Janzen and colleagues demonstrated HIC separation of proteins using nonporous 1.5 μm silica columns (45). With UHPLC systems and nonporous 1.5 μm silica as a stationary phase for protein separation, further improvements in resolution, peak capacity, and analysis time appear to be attainable. However, characterization of protein therapeutics by HIC using UHPLC has yet to be widely published, partially due to the lack of commercially available HIC columns for UHPLC.

8.9 ION EXCHANGE CHROMATOGRAPHY FOR CHARGE VARIANT ANALYSIS

Analytical biochemists routinely use IEC for resolving charge variants of proteins. IEC is a popular method due to the fact that it preserves the native conformation and maintains bioactivity of the protein, is relatively easy to use, is supported by the maturity of the HPLC equipment and consumables market, and has widespread use in the biopharmaceutical industry (46). Charge-based methods are an integral component of protein characterization studies and quality control strategies because they are sensitive to many types of modifications. Charge profiling of intact proteins can resolve species related to protein conformation, size, sequence variant species, glycosylation, and posttranslational modifications (47–49).

To date, UHPLC packing materials suitable for protein separations by IEC are limited commercially, with only few manufacturers producing sub-2 μm particles. Sepax Technologies (Newark, Delaware) has demonstrated improved resolution as particle size decreased from 10 to 1.7 μm, enabling detection of a minor peak (Peak N, Figure 8.8).

8.10 SIZE EXCLUSION CHROMATOGRAPHY FOR SIZE HETEROGENEITY ANALYSIS

SEC has become a very important technique for analysis of protein therapeutics, in part because it is able to resolve high-molecular-weight aggregates of proteins. Aggregates are generally dimers and larger non-covalent aggregated forms, which are usually formed in equilibrium with the protein monomer as a function of temperature, time, solvent conditions, and pressure. Aggregates are product-related impurities, and those larger than dimers may be immunogenic (50). Regulatory agencies typically require quantitation of aggregates present in the drug substance and drug product for lot release and regulatory approval.

UHPLC separation by SEC was initially regarded as challenging, as the appropriate pore volume for size-based separations of proteins may compromise the physical strength of such a particle at high pressure. Nevertheless, SEC packings with mechanical strengths able to withstand high back pressures, higher temperatures, and higher flow rates have been introduced for UHPLC separation. Although SEC has traditionally been a low-resolution technique because of the size and slow mass transfer of

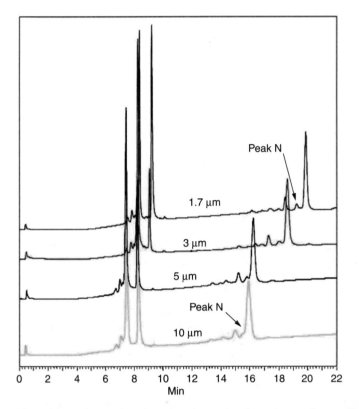

Figure 8.8   IEC separation of proteins using weak cation exchange columns with particle size ranging from 1.7–10 μm. Column dimensions were 4.6 × 50 mm. Sample was a mixture of ribonuclease A, cytochrome C, and lysozyme. Reproduced with permission from Sepax Technologies (Newark, Delaware).

these analytes, more recent size-exclusion UHPLC has been shown to achieve good resolution in shorter run times than conventional HPLC methods with reproducible results (Figure 8.9) (3, 26, 51).

8.11 CONCLUSION

UHPLC is rapidly evolving and is an exciting new area of liquid chromatography, particularly for protein separations. As a natural extension of HPLC, UHPLC requires minimal training and has benefited from growing support from instrument and column manufacturers. Despite being more expensive than traditional HPLC, UHPLC equipment and consumables costs will likely come down due to improved unit volume and value engineering. UHPLC equipment will continue to improve with the potential for even higher pressure limits, peak capacities, and throughput. In addition,

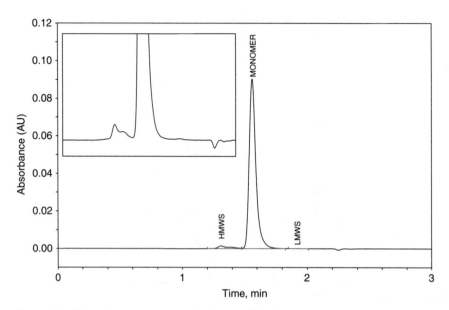

Figure 8.9 Size exclusion chromatography (SEC) analysis of a monoclonal antibody using UHPLC. Column dimensions were 4.6 × 150 mm, with 1.7 μm particles. HMWS indicate high molecular weight species (i.e., dimers and higher order aggregates); LMWS indicate low molecular weight species (i.e., fragments). Reproduced with permission from Rea, J.C., et al. *Biopharm International*. 2010.

UHPLC column technology will continue to develop, such that in addition to RP-UHPLC, HILIC, and SEC applications, increasing UHPLC applications involving HIC and IEC may be developed. UHPLC is in position to play an important role in high-resolution and high-throughput characterization of protein therapeutics.

ACKNOWLEDGMENT

The authors would like to acknowledge Marian Eng, Justin Jeong, Tony Moreno, Yun Lou, Yung-Hsiang Kao and Dell Farnan at Genentech for contributions to this work.

REFERENCES

1. LEADER, B., BACA, Q.J., GOLAN, D.E. *Nat. Rev. Drug. Discov.* 2008, 7(1): 21–39.
2. WURM, F.M. *Nat. Biotechnol.* 2004, 22(11): 1393–1398.
3. KRULL, I.S., RATHORE, A.S., WHEAT, T.E. *LCGC N. America.* 2011, 29(12): 1052–1062.
4. WOODCOCK, J., GRIFFIN, J., BEHRMAN, R., CHERNEY, B., CRESCENZI, T., FRASER, B., HIXON, D., JONECKIS, C., KOZLOWSKI, S., ROSENBERG, A., SCHRAGER, L., SHACTER, E., TEMPLE, R., WEBBER, K., WINKLE, H. *Nat. Rev. Drug Discov.* 2007, 6(6): 437–442.
5. HARRIS, R.J., SHIRE, S.J., WINTER, C. *Drug Dev. Res.* 2004, 61(3): 137–154.

6. FUCHS, H.J., BOROWITZ, D.S., CHRISTIANSEN, D.H., MORRIS, E.M., NASH, M.L., RAMSEY, B.W., ROSENSTEIN, B.J., SMITH, A.L., WOHL, M.E. *N. Engl. J. Med.* 1994, 331(10): 637–642.

7. ROSENFELD, P.J., BROWN, D.M., HEIER, J.S., BOYER, D.S., KAISER, P.K., CHUNG, C.Y., KIM, R.Y. *N. Engl. J. Med.* 2006, 355(14): 1419–1431.

8. ALBERS, G.W., BATES, V.E., CLARK, W.M., BELL, R., VERRO, P., HAMILTON, S.A. *JAMA.* 2000, 283(9): 1145–1150.

9. VOGEL, C.L., COBLEIGH, M.A., TRIPATHY, D., GUTHEIL, J.C., HARRIS, L.N., FEHRENBACHER, L., SLAMON, D.J., MURPHY, M., NOVOTNY, W.F., BURCHMORE, M., SHAK, S., STEWART, S.J., PRESS, M. *J. Clin. Oncol.* 2002, 20(3): 719–726.

10. SCHENERMAN, M., SUNDAY, B.R., KOZLOWSKI, S., WEBBER, K., GAZZANO-SANOTORA, H., MIRE-SLUIS, A. *BioProcess Int.* 2004, 2: 42–52.

11. *ICH Harmonised Tripartite Guideline (ICH Q5E): Comparability of Biotechnological/Biological Products Subject to Changes in their Manufacturing Process Q5E.* Federal Register. 2004, 70(125): 37861–37862.

12. *ICH Harmonised Tripartite Guideline (ICH Q6B): Specifications: Test Procedures and Acceptable Criteria for Biotechnological/Biological Products Q6B.* Federal Register. 1999, 64: 44928.

13. *ICH Harmonised Tripartite Guideline (ICH Q1A): Stability Testing of New Drug Substances and Products Q1A(R2).* Federal Register. 2003, 68(225): 65717–65718.

14. *ICH Harmonised Tripartite Guideline (ICH Q5C): Quality of Biotechnologica/Biological Products: Stability Testing of Biotechnological/Biological Products Q5C.* Federal Register. 1995, 61: 36466.

15. SWARTZ, M.E. *J. Liquid Chromatogr. Rel. Technol.* 2005, 28(7–8): 1253–1263.

16. DE VILLIERS, A., LESTREMAU, F., SZUCS, R., GELEBART, S., DAVID, F., SANDRA, P. *J. Chromatogr. A.* 2006, 1127(1–2): 60–69.

17. CHESNUT, S.M., SALISBURY, J.J. *J. Sep. Sci.* 2007, 30(8): 1183–1190.

18. DAWSON, M. *Am. Lab.* 2011, 43(5): 46–49.

19. RAO, S., POHL, C. *Anal. Biochem.* 2011, 409(2): 293–295.

20. COLLINS, K.E., COLLINS, C.H., BERTRAN, C.A. *LC-GC.* 2000, 18(6): 688–692.

21. STEVENSON, R.L. *Am. Lab.* 2011, 43(5): 50–51.

22. DANIELSON, N.D., KIRKLAND, J.J. *Anal. Chem.* 1987, 59(20): 2501–2506.

23. MAZZEO, J.R., NEUE, U.D., KELE, M., PLUMB, R.S. *Anal. Chem.* 2005, 77(23): 460A–467A.

24. RATHORE, A.S., WOOD, R., SHARMA, A., DERMAWAN, S. *Biotechnol. Bioeng.* 2008, 101(6): 1366–1374.

25. ESCHELBACH, J.W., JORGENSON, J.W. *Anal. Chem.* 2006, 78(5): 1697–1706.

26. JEONG, J., ZHANG, T., ZHANG, J., KAO, Y. *Am. Pharmaceut. Rev.* 2011, 14(2): 44–51.

27. STACKHOUSE, N., MILLER, A.K., GADGIL, H.S. *J. Pharm. Sci.* 2011, 100(12): 5115–5125.

28. KRULL, I.S., RATHORE, A.S., WHEAT, T.E. *LCGC N. Am.* 2011, 29(9): 838–852.

29. LIU, H., FINCH, J.W., LAVALLEE, M.J., COLLAMATI, R.A., BENEVIDES, C.C., GEBLER, J.C. *J. Chromatogr. A.* 2007, 1147(1): 30–36.

30. STAUB, A., ZURLINO, D., RUDAZ, S., VEUTHEY, J.-L., GUILLARME, D. *J. Chromatogr. A.* 2011, 1218(49): 8903–8914.

31. GUILLARME, D., SCHAPPLER, J., RUDAZ, S., VEUTHEY, J.-L. *Trend. Anal. Chem.* 2010, 29(1): 15–27.

32. TOLLEY, L., JORGENSON, J.W., MOSELEY, M.A. *Anal. Chem.* 2001, 73(13): 2985–2991.

33. KÖCHER, T., SWART, R., MECHTLER, K. *Anal. Chem.* 2011, 83(7): 2699–2704.

34. EVERLEY, R.A., CROLEY, T.R. *J. Chromatogr. A.* 2008, 1192(2): 239–247.

35. LUNA, L.G., WILLIAMS, T.L., PIRKLE, J.L., BARR, J.R. *Anal. Chem.* 2008, 80(8): 2688–2693.

36. KAY, R.G., BARTON, C., VELLOSO, C.P., BROWN, P.R., BARTLETT, C., BLAZEVICH, A.J., GODFREY, R.J., GOLDSPINK, G., REES, R., BALL, G.R., COWAN, D.A., HARRIDGE, S.D., ROBERTS, J., TEALE, P., CREASER, C.S. *Rapid Commun. Mass Spectrom.* 2009, 23(19): 3173–3182.

37. XIE, H., CHAKRABORTY, A., AHN, J., YU, Y.Q., DAKSHINAMOORTHY, D.P., GILAR, M., CHEN, W., SKILTON, S.J., MAZZEO, J.R. *MAbs.* 2010, 2(4): 379–394.

38. PLUMB, R.S., JOHNSON, K.A., RAINVILLE, P., SMITH, B.W., WILSON, I.D., CASTRO-PEREZ, J.M., NICHOLSON, J.K. *Rapid Commun. Mass Spectrom.* 2006, 20(13): 1989–1994.

252   Chapter 8   UHPLC for Characterization of Protein Therapeutics

39. JEFFERIS, R. *Nat. Rev. Drug. Discov.* 2009, 8(3): 226–234.
40. LI, H., D'ANJOU, M. *Curr. Opin. Biotechnol.* 2009, 20(6): 678–684.
41. AHN, J., BONES, J., YU, Y.Q., RUDD, P.M., GILAR, M. *J. Chromatogr. B.* 2010, 878(3–4): 403–408.
42. MELMER, M., STANGLER, T., PREMSTALLER, A., LINDNER, W. *J. Chromatogr. A.* 2011, 1218(1): 118–123.
43. COOK, K.S., BULLOCK, K., SULLIVAN, T. *Biologicals.* 2012, 40: 109–117.
44. ROETTGER, B.F., LADISCH, M.R. *Biotechnol. Adv.* 1989, 7(1): 15–29.
45. JANZEN, R., UNGER, K.K., GIESCHE, H., KINKEL, J.N., HEARN, M.T.W. *J. Chromatogr. A.* 1987, 397: 91–97.
46. REA, J.C., MORENO, G.T., LOU, Y., FARNAN, D. *J. Pharm. Biomed. Anal.* 2011, 54(2): 317–323.
47. HE, Y., LACHER, N.A., HOU, W., WANG, Q., ISELE, C., STARKEY, J., RUESCH, M. *Anal. Chem.* 2010, 82(8): 3222–3230.
48. HARRIS, R.J., SHIRE, S.J., WINTER, C. *J. Chromatogr. B.* 2001, 752(2): 233–245.
49. GAZA-BULSECO, G., BULSECO, A., CHUMSAE, C., LIU, H. *J. Chromatogr. B.* 2008, 862(1–2): 155–160.
50. HERMELING, S., CROMMELIN, D.J., SCHELLEKENS, H., JISKOOT, W. *Pharm. Res.* 2004, 21(6): 897–903.
51. REA, J.C., MORENO, G.T., LOU, Y., FARNAN, D. *BioPharm International.* 2010, 23: 44–51.

Chapter 9

UHPLC/MS Analysis of Illicit Drugs

Guifeng Jiang, Jason R. Stenzel, Ray Chen, and Diab Elmashni

9.1 INTRODUCTION

Despite decades of national, regional, and international efforts to stem the scourge of illicit drugs, the prevalence of illegal drug use persists throughout the world. The United Nations Office of Drugs and Crime (UNODC) estimates that between 155 and 250 million people worldwide (representing 3.5%–5.7% of the population aged 15–64) used illicit substances at least once in 2008, with 16–38 million estimated to be heavy drug users (1). Cannabis, by far the most widely produced and used illicit drug in the world, was consumed by an estimated 129–191 million people in 2008, and seizures of herbal cannabis and cannabis resin have increased globally. The number of cocaine users in 2008 was estimated at 15.0–19.3 million people worldwide. While cocaine use has been steadily declining in North America—the world's largest cocaine market with an estimated 5.3 million users—this has been partially offset by increasing cocaine demand in Europe. Of the world's 12.8–21.8 million users of opiates (opium, heroin, and morphine), the majority consume heroin, a highly addictive and lethal form that is associated with high incidences of over-dose, mortality, and blood-borne diseases such as HIV/AIDS and hepatitis B and C. Recently, increasing abuse of prescription drugs has contributed to a rise in opiate addiction in North America. Globally, however, drug consumption appears to be shifting away from opiates and cocaine and moving toward newer synthetic drugs like the amphetamine-type stimulants. In 2008, between 13.7 and 52.9 million people worldwide were estimated to have consumed amphetamine-group substances (pre-dominantly methamphetamine and amphetamine), while the number of ecstasy-group (3,4-methylenedioxymethamphetamine, or 3,4-MDMA, and its analogs) users was estimated to be in the range of 10.5–25.8 million people. Clandestine manufacture of amphetamine-type substances is on the rise and seizures have increased globally.

Ultra-High Performance Liquid Chromatography and Its Applications, First Edition. Edited by Quanyun Alan Xu.
© 2013 John Wiley & Sons, Inc. Published 2013 by John Wiley & Sons, Inc.

The UNODC projects that the number of users of synthetic drugs will soon exceed the combined number of cocaine and opiate users. Inappropriate and extended use of any controlled substance can lead to devastating health outcomes. Moreover, the illicit drug trade fuels crime and violence and can destabilize whole communities. Analytical techniques to accurately identify illegal drugs, precursor chemicals, and new drug derivatives are crucial to law enforcement efforts to curb the production, trafficking, and consumption of controlled substances.

In accordance with guidelines established by the Scientific Working Group for the Analysis of Seized Drugs (SWGDRUG), multiple uncorrelated analytical techniques are usually required for the detection and identification of illicit drugs (2). Suspected materials are typically first examined for the presence of controlled substances using simple and rapid presumptive tests, such as color tests, which provide preliminary identification of drug classes. These tests may be conducted by law enforcement officers in the field or by forensic chemists in laboratories. A positive presumptive test is sufficient to establish probable cause and make an arrest. To minimize false positives and wrongful convictions, all presumptive positives are usually followed by a confirmatory test to conclusively verify the identity of the illicit substance. Confirmation of the presence of a controlled substance is typically performed by gas chromatography-mass spectrometry (GC/MS), the established "gold standard" of identification and quantitation in forensic analysis.

GC/MS is highly selective, provides confirmation of multiple classes of drugs in a single analytical run, and is relatively inexpensive and easy to operate. However, GC/MS lacks sensitivity for certain drugs and cannot detect polar, thermally labile, or nonvolatile compounds without chemical derivatization, which is costly, laborious, and time-consuming and increases the likelihood for sample contamination. Furthermore, the rapid emergence of new and more potent synthetic drugs has fueled the need for accurate and reliable analytical techniques that can quickly identify and quantitate a wider range of compounds at lower limits of detection with minimal sample preparation.

Recent improvements in liquid chromatography (LC) throughput and mass spectrometry (MS) detection capabilities have led to a surge in the use of LC/MS-based techniques for screening, confirmation and quantitation of a diverse range of drug compounds in pharmaceutical applications. LC/MS methods eliminate the need to derivatize and often simplify sample preparation. While long run times, poor resolution, and low separation efficiencies can limit the utility of conventional HPLC, ultrahigh performance liquid chromatography (UHPLC) enables faster separations and higher resolution through the use of sub-2 μm diameter particles. UHPLC coupled with sensitive MS detection provides a powerful alternative to GC/MS for drug analysis. In forensics, however, concerns over resources and training have prevented the widespread adoption of newer and more powerful analytical techniques. Specifically, the comparatively high cost and perceived complexity of LC/MS have limited its use in forensic drug analysis. To address these concerns, LC/MS instrument manufacturers have introduced systems that are cost-effective and easy to operate. Furthermore, gains in workflow efficiency and laboratory productivity through the use of UHPLC/MS methods lead to further reductions in total cost of ownership.

In this chapter, the utility of UHPLC/MS in forensic drug analysis is demonstrated. The following sections describe how UHPLC/MS methods can be optimized and used to provide highly efficient separation, identification, and quantitation of illicit drugs and drug precursors in a variety of matrices, including tablets and food items. Several UHPLC/MS analyses of real case samples are presented.

9.2 APPLICATIONS OF UHPLC/MS IN ILLICIT DRUG ANALYSIS

9.2.1 Development of a UHPLC/MS Method for Simultaneous Identification of Multiple Drugs of Abuse

Sensitive and selective methods that can rapidly analyze multiple drugs of abuse simultaneously from a single sample are essential for forensic applications such as toxicology screening and the analysis of multiple-unit exhibits. GC/MS is commonly employed for the separation and identification of drugs and metabolites in forensic toxicology. This methodology has been the gold standard in terms of admissibility and defensibility in court because of its good sensitivity, excellent selectivity, and high degree of standardization. However, laborious and time-consuming derivatization/labeling procedures and sample cleanup are mandatory in most cases. UHPLC/MS is an attractive alternative method for illicit drug analysis, enabling fast and efficient separations and eliminating the need to derivatize. UHPLC performs separations 5–10 times faster than conventional HPLC by employing sub-2 μm diameter particles, and the 1–2 second peak widths and relatively high separation efficiency of UHPLC are more competitive than capillary GC. To fully maximize the benefits of UHPLC, method parameters such as column selectivity and mobile phase composition must be optimized.

Three stationary phases were evaluated for the UHPLC separation of fourteen illicit and licit drugs (Figure 9.1). A polar endcapped C18 phase did not resolve the early eluting compounds, including amphetamine, methamphetamine, hydrocodone, oxycodone, caffeine, and MDMA, due to secondary interactions with the polar analytes. A C18 phase showed improved selectivity for all analytes except caffeine (peak 1) and oxycodone (peak 7). A pentafluorophenyl (PFP) phase produced the optimal separation of all fourteen drugs, dramatically improving the resolution of the earlier eluting compounds.

Trifluoroacetic acid (TFA), formic acid, and acetic acid can be added into the mobile phase to generate differences in selectivity. UHPLC separation of the fourteen drugs on a PFP column was evaluated by using TFA or acetic acid as eluent modifier (Figure 9.2). The separation method with 0.02% TFA (Figure 9.2a) provided fast separation performance with good resolution and sharp peaks. However, the use of TFA is generally not recommended with MS detection due to its effect on signal suppression. Most of the analytes are well separated with adequate resolution using 0.06% acetic acid as an eluent modifier (Figure 9.2b). However, under such

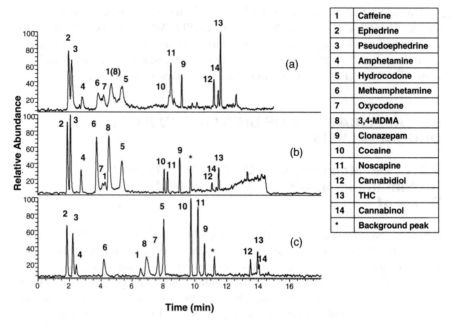

1	Caffeine
2	Ephedrine
3	Pseudoephedrine
4	Amphetamine
5	Hydrocodone
6	Methamphetamine
7	Oxycodone
8	3,4-MDMA
9	Clonazepam
10	Cocaine
11	Noscapine
12	Cannabidiol
13	THC
14	Cannabinol
*	Background peak

Figure 9.1   Comparison of stationary phases for the UHPLC separation of fourteen illicit and licit drugs. (a) Column: 1.9 μm Thermo Fisher Scientific Hypersil GOLD™ aQ (100 × 2.1 mm); (b) Column: 1.9 μm Thermo Fisher Hypersil GOLD™ (100 × 2.1 mm); (c) Thermo Fisher Scientific Hypersil GOLD PFP™ (100 × 2.1 mm). Mobile phase: A = Water, 0.1% formic acid, B = Acetonitrile, 0.1% formic acid. Flow rate = 1 mL/min. Instrument: Thermo Fisher Scientific Accela™ UHPLC system coupled to Thermo Fisher Scientific MSQ Plus™ single quadrupole MS detector.

conditions, a few pairs of compounds, such as oxycodone and methamphetamine (peaks 7 and 6), hydrocodone and 3,4-MDMA (peaks 5 and 8), cocaine and noscapine (peaks 10 and 11), are not baseline resolved.

When separating complex mixtures, a simple binary solvent gradient will sometimes fail to provide good resolution of all analytes in the short time enabled by UHPLC methods. One of the best techniques to improve the resolution is to manipulate the separation factor (α) with a third solvent. The separation of the drug mixture was dramatically improved by using three solvents: water, acetonitrile, and methanol (Figure 9.3). Using this optimized UHPLC/MS method, the fourteen drugs were baseline separated within 12 min. Methanol, a weaker eluent compared with acetonitrile, provided better resolution for most of the analytes. However, the flow rate had to be reduced to accommodate high column back pressure caused by the high viscosity of methanol. Adding acetonitrile reduced the column back pressure so as to maintain the same separation speed.

The limit of quantitation (LOQ) and the limit of detection (LOD) of the drug compounds were determined based on the calibration curve of signal-to-noise ratio versus concentration and the definitions of LOQ and LOD using S/N = 10 and 3,

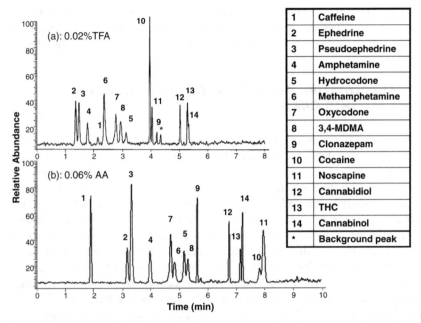

1	Caffeine
2	Ephedrine
3	Pseudoephedrine
4	Amphetamine
5	Hydrocodone
6	Methamphetamine
7	Oxycodone
8	3,4-MDMA
9	Clonazepam
10	Cocaine
11	Noscapine
12	Cannabidiol
13	THC
14	Cannabinol
*	Background peak

Figure 9.2   UHPLC/MS chromatograms of fourteen drugs with acidic solvent modifiers. (a) 0.02%TFA; (b) 0.06% acetic acid. Column: 1.9 μm Thermo Fisher Scientific Hypersil GOLD™ PFP (100 × 2.1 mm). Mobile phase: A = Water with acidic modifier, B = Acetonitrile with acidic modifier. Flow rate = 1 mL/min. Instrument: Thermo Fisher Scientific Accela™ UHPLC system coupled to a Thermo Fisher Scientific MSQ Plus™ single quadrupole MS detector.

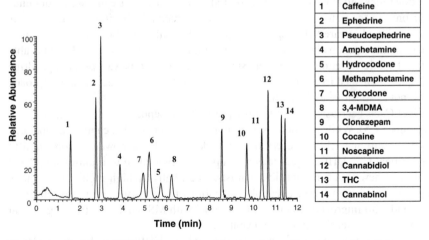

1	Caffeine
2	Ephedrine
3	Pseudoephedrine
4	Amphetamine
5	Hydrocodone
6	Methamphetamine
7	Oxycodone
8	3,4-MDMA
9	Clonazepam
10	Cocaine
11	Noscapine
12	Cannabidiol
13	THC
14	Cannabinol

Figure 9.3   Optimized UHPLC/MS separation of fourteen drugs with ternary gradient. Column: 1.9 μm Thermo Fisher Scientific Hypersil GOLD™ PFP (100 × 2.1 mm). Mobile phase: A = Water, 0.06% acetic acid, B = Acetonitrile, 0.06% acetic acid, C = Methanol, 0.06% acetic acid. Flow rate = 1 mL/min. Instrument: Thermo Fisher Scientific Accela™ UHPLC system coupled to a Thermo Fisher Scientific MSQ Plus™ single quadrupole MS detector.

respectively. LOQs for all drugs were in the range of 0.96–300 ng/mL, while LODs were from 0.29–90.0 ng/mL. Detection by single quadrupole MS at the ppb (ng/mL) level is more than sufficient to identify and quantitate illicit drugs in real samples.

9.2.2 Quantitative Analysis of Pseudoephedrine Tablets by UHPLC/MS

Pseudoephedrine is commonly used as a decongestant in over-the-counter cough, cold, and allergy medicines. Pseudoephedrine hydrochloride and sulfate salts are found in those products either as single-ingredient preparations, or more commonly in combination with antihistamines, naproxen, paracetamol (acetaminophen), and/or ibuprofen. Pseudoephedrine, a U.S. Drug Enforcement Agency (DEA) List I chemical, is highly coveted by drug traffickers who use it to manufacture methamphetamine (commonly known as meth), a DEA Schedule II controlled substance, for the illicit market. The diversion of over-the-counter pseudoephedrine-containing products is one of the major contributing factors to the methamphetamine situation in the United States (3). The separation and identification of pseudoephedrine from illicit drug mixtures that include amphetamine, methamphetamine, and 3,4-methylenedioxy-N-methamphetamine (3,4-MDMA) will help to identify the sources and the manufacturing pathways of the methamphetamine seized in the illicit market.

While GC/MS has traditionally been the screening method of choice for controlled substances, this approach does not provide molecular structural information and cannot be used as a principal means of identification for pseudoephedrine. GC/MS cannot distinguish pseudoephedrine from its sister compound ephedrine; their GC retention times and mass spectra are essentially the same because the two compounds differ only in their stereochemistry. To achieve structural discrimination, scientists have combined infrared (IR) spectroscopy and GC/MS. The IR radiation provides structural information about the intact form of the molecules under study and is able to detect subtle variations in different structural forms. In this way, pseudoephedrine can be distinguished from ephedrine. However, while highly selective, this approach is still not as sensitive as GC/MS.

HPLC has been used to successfully profile pseudoephedrine content in methamphetamine (4,5). While useful for separating out compounds that cannot be separated by GC, traditional HPLC suffers from a lack of resolving power as compared to conventional GC. UHPLC provides a competitive alternative to traditional GC and a powerful new tool for forensic scientists. The relatively high mobile phase flow rates associated with UHPLC enable separations to be performed up to ten times faster and with improved resolution. Combined with MS, UHPLC offers significant improvements in sensitivity over conventional HPLC.

A simple, fast, and reliable method based on UHPLC/MS has been developed to effectively separate and identify pseudoephedrine, ephedrine, amphetamine, methamphetamine, and 3,4-MDMA (Figure 9.4a). The analytes are baseline separated within 4 min with excellent peak efficiency and resolution. The MS spectra of the drug standards show both $[M+ACN+H]^+$ and $[M+H]^+$ ion signals. Full scans

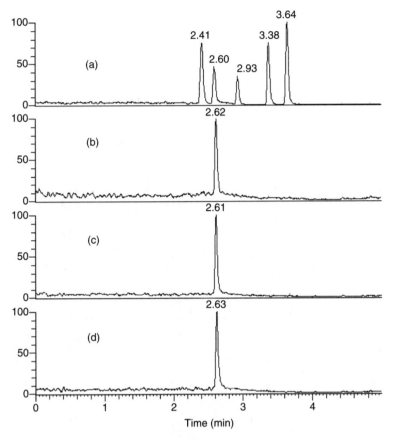

Figure 9.4   Chromatograms of drug standards (a), with the elution order of ephedrine, pseudoephedrine, amphetamine, methamphetamine, and 3,4-MDMA and assay samples (b–d), extracted from over-the-counter cold products with brand names A, B, and C. Column: 1.9 μm Thermo Fisher Scientific Hypersil GOLD[TM] PFP (100 × 2.1 mm). Mobile phase: A = Water, 0.06% acetic acid, B = Acetonitrile, 0.06% acetic acid. Flow rate = 1 mL/min. Instrument: Thermo Fisher Scientific Accela[TM] UHPLC system coupled to a Thermo Fisher Scientific MSQ Plus[TM] single quadrupole MS detector.

(100–200 m/z) were employed for the confirmation of the five compounds, and selected ion monitoring (SIM) modes were used for sensitivity and quantitation studies. Correlation coefficients with $R^2 = 0.996$ or better were achieved for the five drug standards over the concentration range of 1.25–1667 ng/mL (equivalent to 1.25–1667 pg into column). LOQs for all five drugs were in the range of 0.96–1.7 ng/mL, while LODs were from 0.29–0.53 ng/mL. Using this UHPLC/MS method, pseudoephedrine was identified as the major active ingredient for three common over-the-counter cold products (Figure 9.4b–d). The peak retention time of 2.62 min for all three samples matched very well with the retention time of the pseudoephedrine standard at 2.60 min. The confirmation of pseudoephedrine at 2.6 min was further

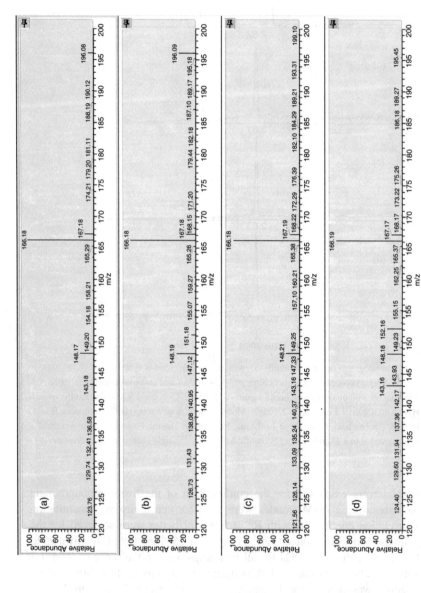

Figure 9.5 MS spectra of pseudoephedrine standard (a) and assay samples (b–d) from over-the-counter drugs with brand names A, B, and C. Molecular ion of pseudoephedrine at m/z 166.18 was the major MS peak. Fragmentation ion of pseudoephedrine at m/z 148.17, created by the in-source CID of the MSQ Plus with a cone voltage of 55 volts, provided the additional MS confirmation of the assay samples.

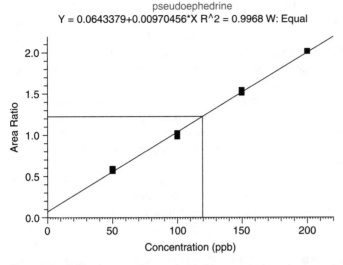

Figure 9.6   Calibration curve for pseudoephedrine quantitation using internal standard method.
Excellent linearity with a correlation coefficient of $R^2 = 0.997$ was obtained.

assured by the match of the MS spectra of the three samples (Figure 9.5b–d) with the
pseudoephedrine standard (Figure 9.5a). An internal standard method was employed
for the quantitative determination of pseudoephedrine in tablet form (Figure 9.6 and
Table 9.1).

9.2.3 Identification of Cannabinoids in Baked Goods by UHPLC/MS

Marijuana (herbal cannabis) is the most common illegal drug in the world. Seizures
have increased globally with the Americas accounting for the highest share (1). U.S.
seizures of marijuana alone exceed 1 million kilograms each year (6). Evidence submitted to forensic laboratories is screened for marijuana by microscopic inspection
and simple chemical tests such as the Duquenois-Levine test. Presumptive positive results are confirmed by GC/MS to positively identify cannabinoids, including

Table 9.1   Pseudoephedrine Quantitation Using Amphetamine as Internal Standard

	Experimental Value (mg/tablet)	Reported Value (mg/tablet)	% Recovery	%RSD (n = 6)
Brand name A	120.09	120	100.1	1.9
Brand name B	112.33	120	93.6	5.6
Brand name C	104.49	120	87.1	2.3

Δ9-tetrahydrocannabinol (THC, the main psychoactive component), cannabinol (the main degradation product of THC), and cannabidiol. This traditional approach works fairly well for leaf marijuana, hashish, hash oil, and residue collected from smoking paraphernalia. GC/MS is less useful, however, for confirming the presence of marijuana in complex food matrices such as baked goods. Simple sample preparation procedures using methanol or methylene chloride co-extract many small molecules found in baked goods that can co-elute with the target cannabinoids. Cholesterol, fatty acids, and caffeine can contaminate the gas chromatography, forcing the analyst to clean the instrument and re-run all subsequent samples. More extensive sample preparation methods are time-consuming and often require greater amounts of the controlled substance than are present in the evidence.

An alternative analytical methodology utilizes UHPLC/MS, which enables fast, selective, and sensitive confirmatory analysis of cannabinoids of forensic importance. THC, cannabinol, and cannabidiol standards were baseline separated and detected within 6 min using an integrated UHPLC/MS platform (Figures 9.7a and 9.8a–c). Good chromatographic peak shapes were obtained, ensuring maximum detection sensitivity. Using a simple methanol extraction and UHPLC/MS, baked goods that had been submitted as evidence were analyzed for cannabinoids. A brownie sample, taken from an adjudicated case and known to contain THC, tested positive for THC (Figures 9.7b and 9.8d). Compared to the GC/MS method employed for the original casework, UHPLC/MS minimized sample preparation, obviated the need for derivatization, required less sample material for analysis, and reduced instrument cleanup. THC was also detected in a 10-year-old cookie sample from an adjudicated case with a good signal-to-noise ratio by increasing the sample injection volume (from 2 to 10 μL) (Figures 9.7c and 9.8e).

9.2.4 Identification of Psychotropic Substances in Mushrooms and Chocolate by UHPLC/MS

The identification of controlled substances in food matrices is a challenge for forensic laboratories. Classical techniques, such as color tests and thin-layer chromatography, do not provide molecular structural information and cannot be used as a principal means of identification. IR spectroscopy and GC/MS can identify controlled substances, but suffer from several shortcomings. IR analysis requires that the substances be pure for identification, and clean-up methods are cumbersome—if they work at all—especially in light of the requirement to leave more than half of the exhibit available for subsequent testing. GC/MS methods, the typical screening approach for controlled substances, suffer from co-elution of small molecules present in food and beverage matrices. Worse yet, a sample with a high concentration of cholesterol, fatty acids, or caffeine can contaminate the GC/MS instrument, forcing the analyst to clean the instrument and reanalyze all subsequent samples. Some compounds, such as psilocybin, a component of the "magic mushrooms" commonly found in food matrices, are thermally labile and do not survive the conditions of GC/MS intact. To ameliorate these shortcomings, extensive wet-chemistry preparation methods have

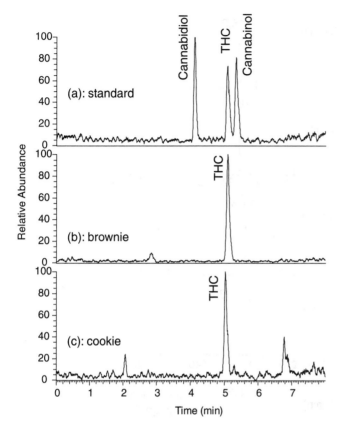

Figure 9.7   Extracted ion chromatograms (m/z; 310.5–311.5, 314.5–315.5) of cannabinoid standards (a) and extracts from brownie (b) and cookie (c) by UHPLC/MS. Column: 1.9 μm Thermo Fisher Scientific Hypersil GOLD™ PFP (100 × 2.1 mm). Mobile phase: A = Water, 0.06% acetic acid, B = Acetonitrile, 0.06% acetic acid, C = Methanol, 0.06% acetic acid. Flow rate = 1 mL/min. Instrument: Thermo Fisher Scientific Accela™ UHPLC system coupled to a Thermo Fisher Scientific MSQ Plus™ single quadrupole MS detector.

been developed; unfortunately, they are time-consuming and often require greater amounts of the controlled substance than are present in the evidence. These preparation schemes also do not eliminate some of the more problematic small molecules and can also exclude controlled substances during the course of the separation.

LC/MS holds several advantages over the traditional methods of analysis. For example, psilocybin does not decompose at the lower temperatures used in HPLC. The low concentration of psilocybin and psilocin are not an issue due to increased sensitivity of this technique. Many matrix components that interfere in GC/MS methods do not interfere in HPLC methods because of much greater differences in analyte solubility as compared to analyte volatility. However, the employment of traditional HPLC has been limited by lack of resolving power compared to capillary GC.

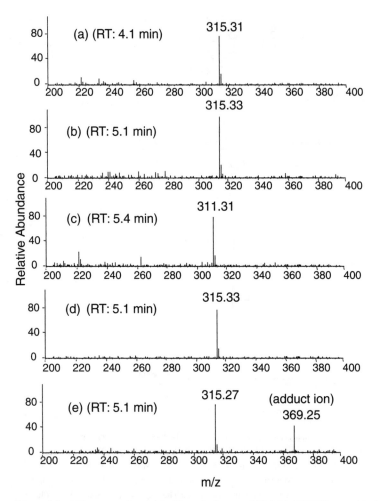

Figure 9.8   MS spectra of cannabinoid standards, cannabidiol (a), THC (b), cannabinol (c), eluted at 4.1 min, 5.1 min, and 5.4 min, respectively, and extracts from brownie (d) and cookie (e), eluted at 5.1 min.

UHPLC/MS is an excellent means of identifying psilocin and psilocybin in both raw mushrooms and chocolates. Psilocin and psilocybin standards eluted rapidly at 0.65 min and 2.25 min, respectively, with excellent resolution (Figure 9.9). Authentic mushroom samples, taken from a training sample known to contain psilocybin and psilocin, were prepared by methanol extraction and a series of simple filtrations. These mushroom samples were analyzed using the same method developed for the standards and showed peaks with retention times consistent with the psilocybin and psilocin standards (Figure 9.10). The identities of these peaks were also confirmed as psilocybin and psilocin by MS detection. The chocolate samples, which were taken from an adjudicated case and were known to contain psilocybin and/or psilocin, also

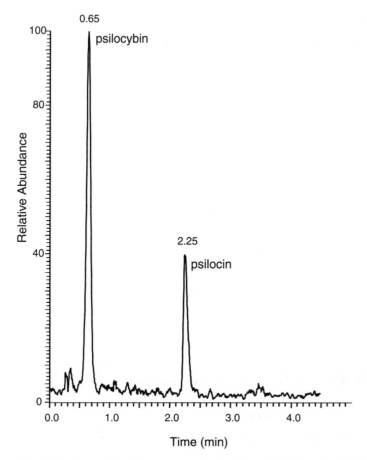

Figure 9.9   Extracted ion chromatograms of psilocybin and psilocin standards by UHPLC/MS. Column: 1.9 μm Thermo Fisher Scientific Hypersil GOLD™ PFP (100 × 2.1 mm). Mobile phase: A = Water, 0.06% acetic acid, B = Acetonitrile, 0.06% acetic acid, C = Methanol, 0.06% acetic acid. Flow rate = 1 mL/min. Instrument: Thermo Fisher Scientific Accela™ UHPLC system coupled to a Thermo Fisher Scientific MSQ Plus™ single quadrupole MS detector.

showed the presence of both psilocybin and psilocin (Figure 9.11). Blanks were run successively after each run, with no apparent carryover from one run to the next.

9.2.5 Identification of Lysergic Acid Diethylamide in Candy by UHPLC/MS

Lysergic acid diethylamide (LSD) is notorious in forensic chemistry as a difficult controlled substance to identify. Its myriad evidentiary forms include paper tabs, eye drops, sugar cubes, and small sugary candies such as sweet tarts, valentine hearts, or mints. Because it is such a potent hallucinogen, typical street doses require only

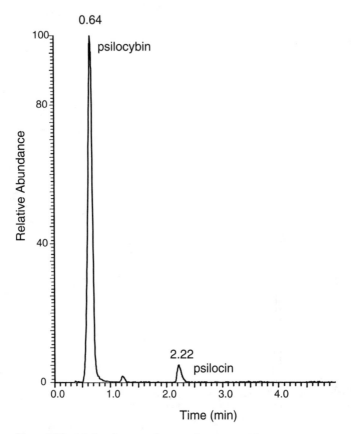

Figure 9.10   Methanol extract of raw mushroom material.

40–120 μg of LSD. The small personal-use amounts seized by state and local law enforcement often lack sufficient drug to allow both forensic analysis by traditional means and archiving of some of the evidence for follow-up testing.

Most forensic laboratories confirm the presence of LSD by using GC/MS. LSD is extracted from the evidence with an organic solvent, derivatized, and determined by GC/MS. GC/MS resolves LSD from other compounds and provides structural information that can be compared to reference spectra in a searchable library. The disadvantages of GC/MS are its requirements for extensive sample preparation, including chemical derivatization of LSD to a more volatile form, and its impaired performance with analytes that are polar, thermally labile, or nonvolatile. LSD has a high affinity for active sites in liners that can spoil chromatographic resolution. LSD-doped sugar cubes or candy can foul the GC with sugars, increasing the burden of instrument maintenance and hindering throughput.

A more efficient way to positively identify LSD in complex food matrices is to use UHPLC/MS. LSD elutes at 4.49 min and is detected by using full scans (m/z = 125–425) of a single quadrupole mass spectrometer. The extracted ion chromatograms from m/z 324 ± 0.5 are shown in Figure 9.12. The MS spectrum

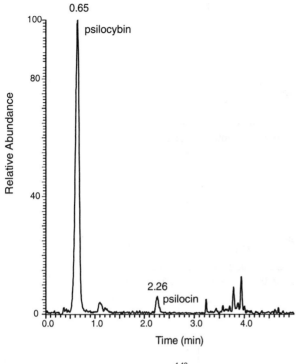

Figure 9.11   Methanol extract of chocolates containing psilocybin and psilocin.

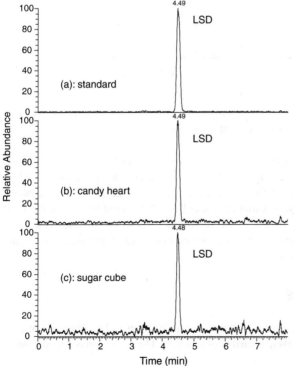

Figure 9.12   Extracted ion chromatograms at m/z = 324 ± 0.5 amu obtained by UHPLC/MS. (a) LSD standard; (b) methanol extract of LSD-doped candy heart; (c) methanol extract of LSD-doped sugar cube. Column: 1.9 μm Thermo Fisher Scientific Hypersil GOLD[TM] PFP (100 × 2.1 mm). Mobile phase: A = Water, 0.06% acetic acid, B = Acetonitrile, 0.06% acetic acid, C = Methanol, 0.06% acetic acid. Flow rate = 1 mL/min. Instrument: Thermo Fisher Scientific Accela[TM] UHPLC system coupled to a Thermo Fisher Scientific MSQ Plus[TM] single quadrupole MS detector.

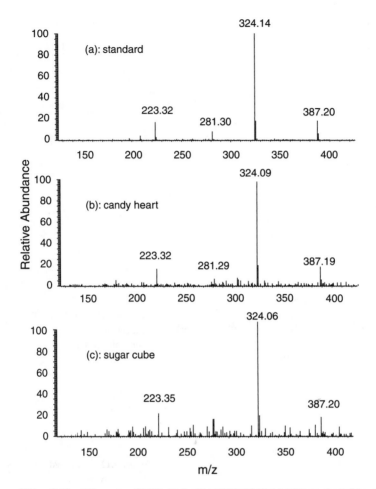

Figure 9.13 MS spectra of LSD obtained by UHPLC/MS. (a) LSD standard; (b) methanol extract of LSD-doped candy heart; (c) methanol extract of LSD-doped sugar cube.

of the LSD standard shows two molecular ion signals: $[M+H]^+$ at m/z 324.1 and $[M+ACN+Na]^+$ at m/z 387.2 (Figure 9.13). Two fragment ion signals at m/z 223.3 and 281.3 are also observed. Methanol extracts from candy hearts and sugar cubes, doped with LSD, were analyzed using the same UHPLC/MS method as for the LSD standard (Figure 9.12b–c). Positive confirmation of LSD in the samples is assured both by retention-time matching and MS spectra matching of the samples (Figure 9.13b–c) with the LSD standard.

9.3 CONCLUSIONS

UHPLC/MS enables fast, accurate, and robust identification and quantitation of controlled substances at ppb levels in a variety of matrices. Confirmation is accomplished with two uncorrelated parameters: retention time and mass spectral signature. Compared to GC/MS, UHPLC/MS quickly and efficiently analyzes a wider range of drugs and drug precursors without the need for costly and laborious derivatization steps. UHPLC/MS is a powerful analytical tool that has the potential to increase the efficiency and productivity of forensic drug analysis laboratories through faster analyses and simplified sample preparation.

REFERENCES

1. United Nations Office on Drugs and Crime. *World Drug Report 2010*. 2010.
2. Scientific Working Group for the Analysis of Seized Drugs (SWGDRUG) Recommendations. Available at http://www.swgdrug.org/approved.htm.
3. National Drug Intelligence Center, U.S. Dept. of Justice. *Methamphetamine National Drug Threat Assessment 2009*, 2009. Available at http://www.justice.gov/ndic/pubs31/31379/meth.htm.
4. MAKINO, Y., URANO, Y., NAGANO, T. *J. Chrom. A.* 2002, 947: 151–154.
5. IWANICKI, R.M., MAIER, K., ZLOTNICK, J.A., LIU, R.H., KUO, T.L., TAGLIARO, F. *J. Forensic Sci.* 1999, 44: 470–474.
6. National Drug Intelligence Center, U.S. Dept of Justice. *National Drug Threat Assessment 2010*, 2010. Available at http://www.justice.gov/ndic/pubs38/38661/index.htm.

Chapter 10

Ultra-High Performance Liquid Chromatography – Mass Spectrometry and Its Application

Zhili Xiong, Ying Deng, and Famei Li

10.1 INTRODUCTION

High performance liquid chromatography (HPLC), a well-known technique, has been widely used in laboratories worldwide over the past 30-plus years, but it forces scientists to consistently choose between speed and resolution. With the growing demand for high-throughput and highly efficient separations, scientists have tried several approaches to shorten the analysis time and maintain good efficiency, including the following:

1. High temperatures (up to 200°C) to decrease mobile-phase viscosity and polarity
2. Monolithic supports instead of columns packed with silica particles
3. Small particles (1, 2)

High temperature liquid chromatography (HTLC) can be recognized as a valuable tool to decrease analysis time (2). Raising the temperature (from 80 to 200°C) can reduce five- to tenfold in the mobile phase viscosity, resulting in a significant reduction in column back pressure with a constant flow rate. So the speed of separation was increased due to the high flow rates (3, 4). Longer columns can also be employed at elevated temperatures, although it was difficult for them to work under optimal flow rate conditions (5). Besides, the proportion of organic solvent is decreased

Ultra-High Performance Liquid Chromatography and Its Applications, First Edition. Edited by Quanyun Alan Xu.
© 2013 John Wiley & Sons, Inc. Published 2013 by John Wiley & Sons, Inc.

significantly due to the decreasing polarity, and the peak shape of basic compounds is also improved (6). However, when using the HTLC technique, the possible thermal degradation of compounds should be considered, so the stability of compounds becomes the main limitation of this approach (7). In addition, the limited choices of stable stationary phase compatible with an elevated temperature also restrict the use of this technique (8).

Monolithic columns are porous rod structures characterized by mesopores and macropores, and these pores provide monolithic column with high permeability, a large number of channels, and a high surface area available for reactivity. The chromatographic performance can be controlled by modifying the size of mesopores and macropores (9). The low column back pressure due to high permeability facilitates high flow rates of mobile phase, leading to the reduction of analysis time. However, monoliths also suffer from several drawbacks, such as no available narrow bore column, no straightforward transfer method, and the limited compatibility with mass spectrometry due to large mobile phase flow rates and the large consumption of organic solvent (10). These obvious disadvantages make monoliths an undesirable tool in the pharmaceutical industry (11).

The most straightforward way to improve resolution and gain speed on HPLC columns is to reduce the particle size of the packing material, as demonstrated by Knox, Giddings, and other authors (12–14). Therefore, shortening the column can also achieve fast separations (15). In 1997, MacNair et al. (16) applied the columns packed with sub-2 μm particles while operating at high pressures. This method can significantly decrease the analysis time.

However, the use of smaller particle packing materials increases the resistance to flow so that the operating pressure required becomes very high, therefore, the whole chromatographic system should not only have the capacity of withstanding high pressure but also possess low dead volume. Yet conventional silica-based materials do not possess the mechanical strength or efficiency necessary to meet these requirements (17). This presents a great challenge to the pressure limitations of a conventional HPLC system. The appearance of UHPLC makes this challenge a reality. Coupling UHPLC with mass spectrometry (MS) can maximize the superiority of UHPLC. On the basis of complementary advantages of UHPLC and MS, UHPLC-MS has become a powerful analytical tool in many fields. This chapter will mainly introduce instrumentation of UHPLC and its applications in analysis of traditional Chinese medicines, metabonomics, drug metabolism and pharmacokinetics.

10.2 INSTRUMENTATION OF UHPLC-MS

10.2.1 UHPLC System

A new tool designed to meet the challenges which conventional HPLC system encountered became generally available in 2004. In this year, an ACQUITY UPLC system was introduced (18). It is the first commercially available UHPLC system solving the problem of elevated pressure and breaking through the bottleneck of chromato-graphic science. UHPLC refers to ultra-high performance liquid chromatography used to separate, identify, and quantify compounds both quickly and effectively. The UHPLC system removed the barrier of traditional chromatographic packing material by inventing a new, highly efficient, mechanically strong, hybrid material called ethylene bridged hybrid (BEH) (19). These particles can also provide a broader range of chemical stability, especially the pH operating range (pH 1–12) (20, 21), while the possibility of the interactions between the matrix and analyte functional groups was also decreased. With the new packing materials, the particle size was successfully decreased to sub-2 μm compared to 5 μm of the conventional HPLC column, while the pressure limit elevated up to 15,000 psi from a typical 4000 psi.

Indeed, the 30-plus year progress of liquid chromatography is a history of the development of packing materials. The particle size directly influences the column efficiency and thus further affects the separation results. The underlying theory is the van Deemter equation, which is the empirical formula that shows the relationship between linear velocity (flow rate) and plate height (column efficiency). From the van Deemter equation we can know that, as the particle diameter decreases, there is a significant gain in efficiency even when the flow rates are increased. When the particle size was reduced to sub-2 μm, the analytical process was speeded up by a factor of nine without compromise of efficiency, or in other words, the efficiency was increased by a theoretical ninefold for a similar run time (22).

Compared to HPLC, a UHPLC system can take full advantages of smaller particles (23, 24). Firstly, the UHPLC system can greatly increase efficiency and improve separation. Besides, small particles can work at higher linear velocities without com-promising resolution; so by using shorter columns packed with small particles, we can obtain superior separation speed and reduce analytical time noticeably. The less analysis time may facilitate in analysis of complex mixtures and reduce solvent con-sumption at the same time. In addition, small particles will generate narrower and taller peaks, reduce the noise, and further improve signal-to-noise ratio, which is related to sensitivity. Therefore, the sensitivity has been enhanced markedly.

However, with sub-2 μm particles, the width of half-height peak obtained is less than a second, posing a significant challenge to the detector. To integrate an analyte peak accurately and reproducibly, the detector sampling rate must be high enough to capture enough data points across the peak. Additionally, the detector cell must have minimal dispersion to maintain separation efficiency. Mass spectrometer is a suitable detector satisfying all these requirements.

10.2.2 Mass Spectrometer

Mass spectrometry (MS) is a method involving the production of gas-phase ions from a sample and then measuring charged particles according to their mass-to-charge ratio (m/z). MS is a powerful analytical technique used for qualitative and quanti-tative determination, and this particular spectroscopic technique has been placed in the forefront of research capability giving more information than any other single technique.

The performance of the MS detector is dramatically enhanced by UHPLC, because UHPLC reduces the analysis time, strengthens chromatographic resolution overall, and reduces co-elution, and that, in turn, leads to a decrease of ion suppression, improving MS sensitivity and reliability. Therefore, the combination of UHPLC with an MS detector appears to be a suitable approach that satisfies key requirements in respect of resolution, sensitivity, selectivity, and peak-assignment certainty (23, 25).

10.2.2.1 Interface

MS works by using magnetic and electric fields to exert forces on the charged ions under a 10^{-5}–10^{-7} torr vacuum. Thus, the interface must connect the atmospheric environment of UHPLC to the high-vacuum environment of mass spectrometer perfectly. Besides, the interface must be capable of ionizing the target compounds so that they can be drawn into a mass spectrometer to be detected. For the reasons mentioned, the interface becomes an indispensable and important part of a mass spectrometer.

The most common ionization sources interfaced to UHPLC are electrospray ionization (ESI) and atmospheric chemical ionization (APCI) in both positive and negative ion modes, because they can both be operated at atmospheric pressure and offer a user-friendly way to couple UHPLC with MS (26). Most commercial MS systems are equipped with these two interfaces, allowing the easy switch between them.

10.2.2.2 Electrospray Ionization

ESI has been widely used for highly polar and ionized materials. It is a soft ionization technique that results in little fragmentation. In the ESI interface, ionization process requires three steps: droplet formation, solvent evaporation, and gas ion formation. The sample solution is led to the ion source through a stainless steel capillary with high voltage. High electrostatic fields on the tip of the capillary will lead the positive ions accumulated at the liquid surface and drift downfield toward the liquid front (when positive ions are analyzed). Therefore a cone-shaped flow is formed on the tip of the capillary. Gradually the tip of the cone-shaped flow starts to break up and small positive charged droplets are formed. As the solvent evaporates from the droplets, they gradually become smaller. The droplet contracts, but the charge density at the surface increases. When the coulomb repulsion on the surface exceeds the surface tension of the droplet, a set of charged smaller droplets is formed. This process is repeated until the size of the droplet is small enough to emit the gas ion (27–29).

ESI can produce not only single charged ions but also multiple charged ions, so large molecular-weight peptides and proteins can be detected easily by ESI. ESI can also analyze large thermolabile molecules because there is no heating process during the ionization. However, due to the nature of ESI, it is best suited to ionization of polar compounds that can already be charged in the solution. Moreover, ion suppression, which is caused by the competition of the charge between the analytes, additives, and impurities, is another disadvantage of ESI. Thus, nonvolatile buffers cannot be used, and sample preparation should be careful enough to remove most of the contaminants.

10.2.2.3 Atmospheric Pressure Chemical Ionization

APCI is also a soft ionization. The ionization mechanism is less complicated than that in ESI. Ionization of APCI takes place mainly in the gas phase, whereas ionization of ESI occurs mostly in the solution. In APCI, the ionization process is as follows: liquid sample evaporated, charged by a corona discharge needle, and ionized through gas-phase reactions in the atmospheric pressure region. The sample solution is nebulized into droplets by high nitrogen flow after eluted from the tip of a capillary coupled with a nebulizer. Then the fine droplets are carried to a heated chamber where the solvent is evaporated. After that, the eluent components and analytes are ionized via gas-phase ion reactions initiated by a corona discharge needle. The most important gas phase reactions are charge transfer, charge exchange, and adduct formation (27–29).

APCI is more convenient for using, which is the main advantage of APCI over ESI. For example, polar and nonpolar solvents can be used, and higher concentrations of salt and additive can be tolerated in the mobile phase. Because the compounds are vaporized to the gas phase by thermal energy, APCI is well suited for the analysis of compounds with certain volatility and thermostability, but not suitable for large biomolecules such as protein or oligonucleotides. Besides, APCI can produce only single-charged ions.

ESI and APCI are complementary ionization methods, and they are both excellent UHPLC-MS interfaces. Users can choose the appropriate interface depending on the nature of the sample.

10.2.2.4 Mass Analyzer

Mass analyzer is the core part of a mass spectrometer. It separates the ionized compound based on their mass-to-charge ratio (m/z). The main purpose of mass analyzer is to select specific mass ions and then take these ions into the detector for counting. There are several kinds of mass analyzers commercially available, and each analyzer has its own special benefits and limitations. However, no ideal mass analyzer available is suited for all applications. We will focus on the commonly used mass analyzers.

Quadrupole Mass Analyzer   The quadrupole consists of four parallel hyperbolic rods, with adjacent rods having opposite charges. A potential is applied to one pare of diagonally opposite rods consisting of a DC voltage and an RF voltage. To the other pair of rods, a DC voltage of opposite polarity and an RF voltage with a 180° phase shift are applied. The quadrupole mass analyzer uses oscillating electrical fields to selectively stabilize or destabilize the paths of ions passing through a radio frequency quadrupole field. Only a single selected ion is passed through the system at any time. All other ions do not have a stable trajectory through the quadrupole mass analyzer and will collide with the quadrupole rods, never reaching the detector. Because the single quadrupole mass analyzer has some disadvantages, the triple quadrupole mass analyzer has been implemented through the pharmaceutical studies. The triple quadrupole mass analyzer has three quadrupoles arranged in series to incoming ions. The first quadrupole acts as a mass filter. The second quadrupole acts as a collision cell where selected ions are broken into fragments. The third quadrupole scans these resulting fragment ions before they reach the mass detector (30).

As a classical mass analyzer, the quadrupole has several advantages including relatively small and low-cost systems, good reproducibility, and a linear mass scale (27). However, the lower sensitivity and resolution at high masses are the limitations of the quadrupole.

Ion Trap Mass Analyzer   The ion trap is relatively inexpensive and compatible with a wide range of inlet and ionization systems. The ion trap is formed from a ring electrode and two endcap electrodes. Ions are transmitted into the ion trap through an aperture on an endcap electrode. In the ion trap, the motion of the ions may be stable or unstable like that in quadrupole mass analyzer. The DC voltage and RF voltage can be changed either in combination or singly so that trapped ions with consecutive m/z values become unstable (31,32). The unstable ions are ejected from the aperture and a mass spectrum is generated. A low pressure of helium (10^{-3} torr) reduces the kinetic energy of the ions, focuses them toward the center of the trap, and thus allows the ion traps to measure all ions retained until the trapping stops. Consequently, sensitivity and resolution are enhanced significantly (10–1000 times).

Other ion traps such as the quadrupole ion trap and linear quadrupole ion trap have garnered a lot interest (33). The quadrupole ion trap works on the same physical principles as the quadrupole mass analyzer, but ions are trapped and sequentially ejected. A linear quadrupole ion trap is similar to a quadrupole ion trap, but it traps ions in a two-dimensional quadrupole field, instead of a three-dimensional quadrupole field as in a quadrupole ion trap.

Because of the low cost, high sensitivity, large mass range, and ability to perform many sequential fragmentation experiments (MS^n), ion traps played an important role in pharmaceutical studies, especially in drug metabolites identification.

Time-of-Flight Mass Analyzer   The time-of-flight (TOF) mass analyzer measures the mass-dependent time it takes ions of different masses to move from the ion source to the detector. The TOF analyzer uses an electric field to accelerate the ions through the same potential. Ions of different m/z values will have the same kinetic energy and travel at different velocities after acceleration out of the ion source through a fixed potential into the TOF drift tube. The time they reach the detector after traveling the same distance is different. Lighter ions will reach the detector first. The m/z value of the ion can be determined by measuring the time it takes to reach the detector after it is formed (34,35). The TOF is well suited for pulsed ionization methods such as matrix assisted laser desorption ionization (MALDI) mass spectrometer system, and they can be used for analysis of large biological molecules (proteins, peptides, and polynucleotides).

Recently, UHPLC-MS has become the mainstream of development of liquid chromatography, which has been successfully applied to many fields, including rapid qualification and quantification of traditional Chinese medicine, metabonomics, drug metabolism, and pharmacokinetics study.

10.3 UHPLC-MS APPLICATIONS

10.3.1 UHPLC-MS Applications in Analysis of Traditional Chinese Medicines

Traditional Chinese medicine (TCM) is the oldest and most documented traditional medicine in the world, which has a long history dating back to about 2700 BC. It includes a huge number of medicine plants, composite patent products, and prescriptions. TCM has been used for the treatment of diseases for thousands of years. Throughout China's long history, Chinese people have accumulated a rich empirical knowledge of the properties and use of natural medicines as well as enormous amounts of human clinical data on their efficacy and toxicity. The specific therapeutic effects and minimum side effects of many herbal remedies have recently been demonstrated or verified by modern scientific investigations. These therapeutic effects are often complementary to those of western drugs, which makes TCM research attract the attention of more scientists, medical doctors, and the pharmaceutical industry.

Although many TCMs have a good efficacy in the treatment of many types of diseases, their mechanisms of action are not well understood. In general, TCM works depending on the combined action of a number of bioactive constituents in a selected treatment. The analysis of the active constituents in Chinese medicinal herbs is a key to help unlock the secret of their effectiveness. Because the constituents of TCM are various and structurally complex, it is difficult to develop a rapid identification and analysis method using conventional HPLC. However, UHPLC-MS can solve this problem, due to its enhanced selectivity, sensitivity, high resolution, and structure identification, which have made it the most important technique for both quantitative and qualitative analysis of complex samples such as TCM.

Li et al. (36) developed a rapid UHPLC-ESI-MS/MS method to determine the constituents of the flower of *Trollius ledibouri Reichb* quantitatively and qualitatively. The dried flower of *T. ledibouri Reichb* recorded in the medicinal literature in *Zhong Hua's Herbal Classic* (37), possesses antimicrobial and antiviral actions and has been used in the treatment of colds, fever, chronic tonsillitis, and acute tympanitis for a long time. The analysis was carried out on an ACQUITY UPLC^{TM} BEH C$_{18}$ column using gradient elution with a mobile phase of 0.1% acetic acid and acetonitrile. The triple quadrupole mass spectrometer was operated in either full scan mode or in MS/MS mode for the qualitative and quantitative determination of the constituents, respectively. Water was chosen as the extraction solvent for the quantitative experiment because the four investigated flavanoids are mainly polar compounds and the peak sizes of these early eluted compounds are big, while 95% EtOH was chosen for the qualitative analysis due to its number of peaks and bigger

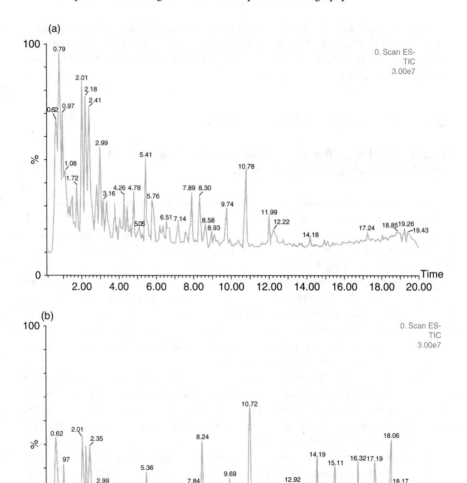

Figure 10.1   The total ion chromatograms of the (a) water and (b) 95% EtOH extract of the flower of *T. ledibouri* Reichb in negative ionization mode.

peak sizes of late eluents. Figure 10.1 shows the total ion chromatograms of aqueous exact and 95% EtOH extract.

During qualitative analysis, a good chromatographic separation of constituents in 95% EtOH extracts of the flower of *T. ledibouri Reichb* was achieved on a reversed-phase column. The peak capacity was increased significantly by UHPLC. Within

20 min, over fifty peaks were detected in the MS-TIC chromatogram of the 95% EtOH extract. The molecular weights of thirty-three constituents were deduced according to their positive and negative ion spectra. Based on the mass spectrometric fragmentation mechanism and data from UHPLC-ESI-MS/MS, the chemical structures of fifteen constituents were identified (shown in Figure 10.2).

During quantitative analysis, a BEH C_{18} column (100×2.1 mm i.d., 1.7 µm) provided good separation of the four flavanoids, $2''$-O-β-galactopyranosylorientin, $2''$-O-β-arabinopyranosylorientin, orientin, and vitexin. Figure 10.3 shows the UHPLC-ESI-MS/MS chromatograms of these four constituents. The benefits of using UHPLC technology involves improved peak resolution and significant reductions in analysis times, resulting in reductions in solvents and argon consumption. In this experiment, the total run time was less than 7.5 min per sample, which was much shorter than that reported in the literature (38, 39). Moreover, the chromatographic efficiency and sensitivity was enhanced dramatically on account of the very narrow peaks generated by UHPLC. Due to using the MRM mode for peak detection, no interference from other constituents and the matrix was observed at the retention time of each analyte.

The method was fully validated for quantitative determination and successfully applied to evaluate the quality of the flower of *T. ledibouri Reichb*. In summary, the powerful new technique UHPLC-ESI-MS/MS improved the assay performance and increased the sample throughput for the analysis of TCM.

UHPLC together with ESI-MS/MS can also be used for direct online identification of known constituents present in complex plant extracts in the absence of authentic samples of these compounds. Li et al. (40) established a UHPLC-ESI-MS/MS and principal component analysis (PCA) method to investigate the chemical constituents and variation of the flower buds of the *Lonicera* species. This method enabled the simultaneous identification of the major constituents present in the flower of the *Lonicera* species without time-consuming and pre-purification steps and can form the basis for the successful quality control of the *Lonicera* species. Among the thirty-three constituents detected, six caffeoylquinic acids (including caffeic acid), eight flavonoids, and eight iridoid glycosides were characterized based on their fragmentation patterns in collision-induced dissociation (CID) experiments. Furthermore, the seven *Lonicera* species studied were divided into well-defined groups directly based on PCA in terms of the log transformed relative contents of the major caffeoylquinic acids (including caffeic acid) as the variables. This method can reveal characteristic details of the chemical constituents of different *Lonicera* species and classify them.

In recent years, some pharmaceutical manufacturers, driven by economic profits, illegally add synthetic antidiabetic drugs into their Chinese proprietary medicines (CPMs) to compensate the slow pharmacological action of Chinese medicines. The consequences of adulteration are very serious, and sometimes life-threatening, especially when taken by patients unwittingly and continually. Moreover, adulteration of CPMs and dietary supplements with synthetic drugs has had a bad influence on Chinese medicinal products. Therefore, it is of significant importance to analyze the

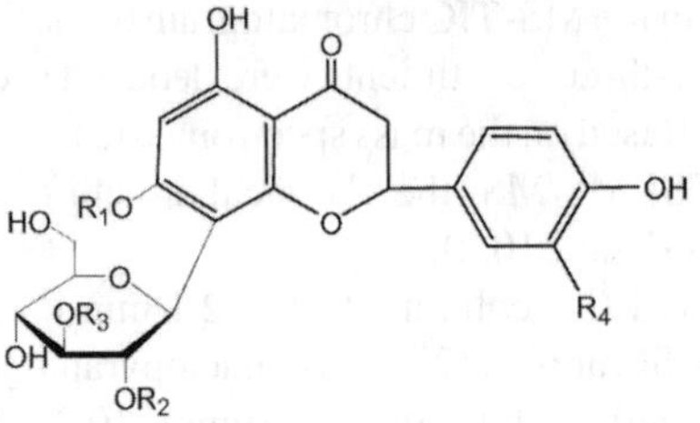

compounds no.	R_1	R_2	R_3	R_4
2	H	β-L-galactopyranosyl	H	OH
3	H	β-D-pyranarabolose	H	OH
4	H	H	H	OH
5	H	β-D-pyranxylose	H	H
6	H	H	H	H
7	CH₃	H	H	OH
8	CH₃	H	H	H
10	H	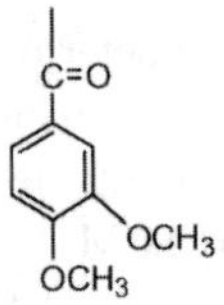 (structure C=O with OCH₃, OCH₃)	H	OH
11	H	Et(Me)CHCO	H	OH
12	H	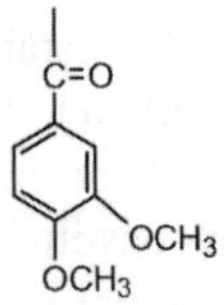 (structure C=O with OCH₃, OCH₃)	H	H
13	H	Et(Me)CHCO	H	H
14	CH₃	Et(Me)CHCO	H	H
14	CH₃	H	Et(Me)CHCO	H

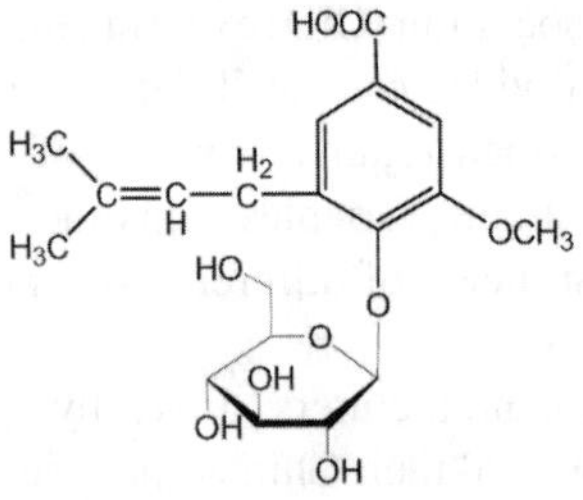

Figure 10.2   The chemical structures of constituents identified in an extract of *T. ledibouri Reichb*. 2″-O-β-l-galactopyranosylorientin (2), 2″-O-β-arabinopyranosylorientin (3), orientin (4), 2″-O-β-d-pyranxylvitexin (5), vitexin (6), isoswertiajaponin (7), isoswertisin (8), trollioside (9), 2″-O-(3‴,4‴-dimethoxybenzoyl) orientin (10), 2″-O-(2‴-methylbutyryl) orientin (11), 2″-O-(3‴,4‴-dimethoxybenzoyl) vitexin (12), 2″-O-(2‴-methylbutyryl) vitexin (13), 2″-O-(2‴-methylbutyryl)isoswertisin or 3″-O-(2‴-methylbutyryl)isoswertisin (14).

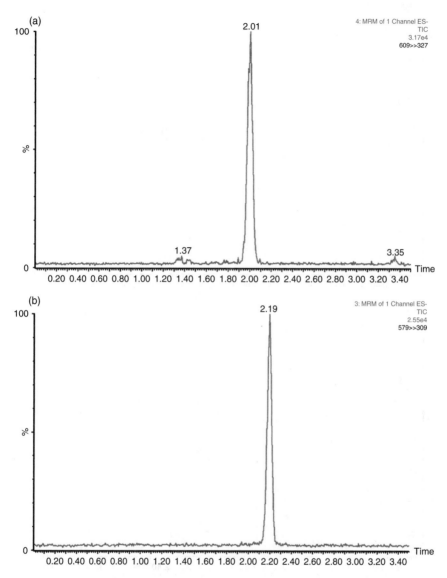

Figure 10.3   UPLC–ESI-MS/MS chromatogram of (a) 2''-O-β-l-galactopyranosylorientin (1467 ng mL^{-1}), (b) 2''-O-β-arabinopyranosylorientin (699 ng mL^{-1}), (c) orientin (444 ng mL^{-1}), and (d) vitexin (102 ng mL^{-1}) in an extract of *T. ledibouri Reichb*.

illegal synthetic drugs in CPMs and dietary supplements to ensure patient safety and protect the reputation of Chinese medicines. Li et al. (41) reported a rapid and reliable UHPLC-MS/MS method for the identification and quantification of fourteen synthetic antidiabetic drugs illegally added in CPMs and dietary supplements. These fourteen synthetic antidiabetic drugs with different chemical groups varied

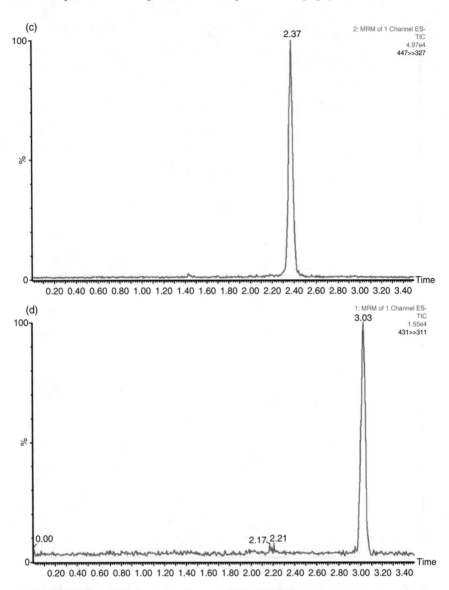

Figure 10.3   (*Continued*)

significantly in polarity, which poses a great challenge for simultaneous qualification and quantification of all these drugs in the complex matrix of CPMs and dietary supplements by conventional HPLC method. However, UHPLC-MS/MS resolved this difficult problem well, because UHPLC using a narrow-bore column packed with very small particles (sub-2 μm) and a mobile phase delivery system operating at high back pressure could offer satisfactory resolution between analytes with similar polarities, reduce elution windows between analytes with very different polarities, and increase sensitivity. The separation was performed on an ACQUITY UPLC™ BEH C_{18} column using gradient elution with a mobile phase of acetonitrile containing 0.10% formic acid (FA) and water containing 0.10% FA. For MS/MS detection, electrospray was operated in positive ion mode. Two transitions from protonated molecules were monitored for each synthetic antidiabetic drug under their respective optimal collision energy (CE). The more intensive transition was used for quantification (called quantifier), and the other transition was used for identification confirmation (called qualifier). The total analysis time for each sample was 6 min. This method provided a reliable identification and satisfactory linearity, accuracy, precision, and sensitivity in quantification and also showed advantages in simplicity and saving time. It can be used as a routine tool for monitoring illegal synthetic drugs added to CPMs and dietary supplements.

10.3.2   UHPLC-MS Applications in Metabonomics

Metabonomics, one of the main branches of system biology, is defined as the quantitative measurement of time-related multiparametric metabolic response of multicellular systems to pathophysiological stimuli or generic modification (42). It mainly investigates the change of endogenous metabolism of the biological system that is stimulated or disturbed and studies the metabolic pathways of the biological system. Recently, metabonomics has attracted increasing attention from scientists around the globe.

Traditional Chinese medical theory favors a holistic approach and believes that human is an organic entirety. Metabonomics focuses on the investigation of the recognition and measurement of the entire metabolic reaction of an organism in response to an internal or external influence. The strategy of metabonomics is consistent with the overall concept of TCM. Therapeutic effects of TCM are usually the combined results of multiple constituents present in TCM. However, the analysis of one or two major constituents cannot reveal their complexity of ingredients and the aspects of opposing, combinatory, and synergistic bioactivities. Metabonomics is trying to study interaction and changes of the low molecular weight (<1500 Daltons) metabolites found in a system and to seek the dynamic markers of biological status. Furthermore, it provides abundant information that can reflect the integrated function of the organism. Therefore, this would be very helpful to explore the therapeutic basis and to clarify possible pharmacological function mechanisms of the complicated TCM.

In metabonomic study, there are always a large amount of biological samples that should be processed and analyzed. Metabolites present in the samples are generally labile species, chemically diverse and in a wide range of concentrations. So modern analysis technology that provides high throughput and sensitivity should be employed to meet the requirements of metabonomic study. Among the analytical techniques in metabonomic research, nuclear magnetic resonance (NMR) is the earliest and most popular method. Much of the original work of metabonomics was performed by NMR because of its high speed, nondiscrimination in sensitivity for all hydrogen-containing compounds, and no requirement for complicated sample preprocessing. More recently, a combination of mass spectrometry and chromatographic techniques has been used for metabonomic analysis, because this approach could provide higher sensitivity and resolution and detect more subtle fluctuations of the metabolite composition. Compared with LC-MS, as mentioned previously, UHPLC-MS can provide higher resolution and peak capacity and can also perform rapid and high-throughput analysis. All of these advantages can meet the requirements for determination of complex metabonomic samples even at low concentration, and have made UHPLC-MS a potentially powerful tool to investigate metabonomics.

Li et al. (43) employed a metabonomic approach to investigate the therapeutic basis and metabonomic effects of *Epimedium brevicornum* Maxim on an animal model induced by a high dose of hydrocortisone using UHPLC-MS. *Epimedium brevicornum* Maxim is one of the most frequently used TCM, which can replenish the kidney and strengthen the bone. In this study, a high dose of hydrocortisone was administered to rats to induce a pathological condition similar to the kidney deficiency syndromes in TCM. Then *Epimedium brevicornum* Maxim extract was administrated to rats showing kidney deficiency syndrome. UHPLC-MS technique was used to analyze the endogenous metabolite profiles in urine (serum) of controlled group (rats before hydrocortisone intervention), model group (hydrocortisone intervention rats), and treated group (*Epimedium brevicornum* Maxim treated rats). The restoration of abnormalities of the metabolic pathway in treated rats was investigated by comparing the endogenous metabolite profile using PCA. Significant difference in endogenous metabolite profiles was observed from the model group compared with the control group, and the abnormality of metabolism which recovered toward the normal level after administration with *Epimedium brevicornum* Maxim extract was also observed. The chemical constituents and their in vivo metabolites of this herbal drug were investigated at the same time. Four active constituents of *Epimedium brevicornum* Maxim,—epimedin C, icariin, icariside II and 2″-O-rhamnosoyl icaride—were found in blood circulation of kidney-deficient rats exerting pharmacology action, and two metabolites of icariin and epimedin C were found in the urine.

Action mechanisms of drugs have been studied extensively in the medicinal field. However, routine pharmacological studies always focus on investigating a specific target instead of changes of the whole metabolic spectrum. Metabonomics is a platform identifying and measuring metabolic profile dynamics of host changes after administration, reflecting changes of biochemical process and state in vivo directly, and exploring the cause of these changes to clarify the action mechanisms of drugs. The UHPLC-MS/MS-based metabonomics method is also a powerful tool to investigate the action mechanisms of drugs.

Metformin hydrochloride is an oral biguanide antihyperglycemic drug, which has been used for over 40 years in the treatment of type 2 diabetes. However, the exact mechanism of metformin is not very clear yet. Our research group employs

UHPLC-MS/MS technology in combination with metabonomics methodology to investigate the hypoglycemic mechanism of metformin (44). Based on this technology, an analytical platform was developed to reveal the metabolic spectrum in serum and urine samples, which can present efficient separation of complex components in serum and urine samples and also provide abundant information of metabolites with small molecular weight. This method was fully validated, and the results were approved in accordance to analytical requirements of biological samples. The platform was applied to generate a metabolic fingerprint for the metabonomic analysis of serum samples obtained from type 2 diabetes mellitus patients without any drug treatment and type 2 diabetes mellitus patients treated with metformin hydrochloride. The possible biomarkers (tryptophan, phenylalanine, and lysophosphatidylcholine) related to action mechanism of metformin were indicated through pattern recognition techniques and analyzing fragments in the second-order mass spectrum. Through analyzing biomarkers and clinical parameters, the possible metformin action mechanism was that metformin can inhibit hepatic gluconeogenesis, regulate lipid metabolism, and reduce oxidative stress. These results illustrate the UHPLC-MS/MS method as a very powerful tool for metabonomic research.

10.3.3 UHPLC-MS Applications in Drug Metabolism

Absorption, distribution, metabolism, and excretion (ADME) studies are widely used in drug discovery and development to determine the metabolism and pharmacokinetic properties of drugs and drug candidates. Among ADME properties, metabolic characterization is a key issue that has become one of the main drivers in drug discovery and development process. Drug metabolism is the biochemical modification of pharmaceutical substances by living organisms, usually through specialized enzymatic systems. Liver is the primary metabolic organ in the body due to its various kinds of enzymatic systems, while some drugs can also be metabolized by other tissues. Through metabolism, lipophilic chemical compounds can be converted into more polar products and readily excreted from the body. Generally, metabolic reactions can be classified into phase I reactions and phase II reactions. Phase I reactions, also termed functionalization reactions, often convert the parent drug to a more polar metabolite by introducing or unmasking a functional group, including oxidation, reduction, hydrolysis, and isomerization. Many drugs undergo a number of these reactions, and the main function of phase I reactions is to prepare the compound for phase II reactions. Phase II reactions, usually known as conjugation reactions, produce water soluble and easily excreted products, including glucuronic acid conjugation, sulfation, methylation, acetylation, amino acid conjugation and glutathione conjugation. The metabolism reactions, especially phase I reactions, usually result in the formation of pharmacologically active or toxic metabolites, so studies on the metabolic fate of drug are of great importance to drug safety.

Finding and identifying trace, even ultra-trace, amounts of metabolites is a difficult and time-consuming process. UHPLC represents a new era of separation science that is characterized by superior speed, sensitivity, and resolution. When coupled to tandem mass spectrometry, it has significant advantages in metabolism study. UHPLC-MS/MS has become an indispensable tool in the rapid detection and characterization of metabolites.

Sun et al. (45) reported a selective UHPLC-ESI-MS/MS method for the identification and structure elucidation of metabolites of mosapride, a selective gastroprokinetic agent that enhances the gastrointestinal (GI) motility and gastric emptying by accelerating acetylcholine release in the GI tract. Eighteen metabolites in all were found. Among these eighteen metabolites, the identification of two new phase I metabolites (M17 4-amino-5-chloro–N-(2-hydroxypropylamine)-2-ethoxybenzamide and M18 4-glucosidemosapride) and one known metabolite (M5 4-amino-5-chloro-2-ethoxy-N-[3-(4-fluorobenzylamino)-2-hydroxypropyl] benzamide) were achieved with available reference standards, while the structures of fifteen metabolites (nine phase I metabolites and six phase II metabolites) of mosapride were elucidated based on the characteristics of their protonated molecular ions, product ions, and chromatographic retention times. UHPLC-MS/MS chromatograms of mosapride and its metabolites by daughter scan mode are shown in Figure 10.4. The analysis time was only 15 min per run. Four main metabolism pathways may be responsible for the transformation of the phase I metabolites, including dealkylation, N-oxidation, morpholine ring cleavage, and hydroxylation, with dealkylation as the predominant metabolic pathway, while glucuronidation was the main metabolism pathway for phase II metabolites.

In clinic, combination therapy is the coadministration of more than one kind of drug to enhance effectiveness through synergic action between drugs. However, inappropriate coadministration may lead to severe drug-drug interaction. Many drug-drug interactions that occurred in the past were due to alterations in drug metabolism. One notable metabolism enzyme system involved in drug-drug interaction is cytochrome P450 (CYP450), which is a large and diverse group of enzymes responsible for approximately 70%–80% of the rate-limiting phase I metabolism of drugs. Medicines inducing or inhibiting CYP450 activities may lead to drug-drug interaction. Therefore, evaluation of the effect of new drugs or drug candidates on CYP450 enzyme activities is of significant importance in pharmaceutical development. Jiang et al. (46) assessed the interference of trantinterol with human liver CYP450 enzyme activities. Trantinterol is a novel β_2-adrenoceptor agonist used for the treatment of asthma, and it is necessary to investigate the effects of trantinterol on the activities of human liver CYP450 enzyme to evaluate the potential of drug-drug interaction. The ability of trantinterol to inhibit and induce CYP450 activities were evaluated in vitro with human liver microsome and human cryohepatocytes, respectively. The analysis was performed on an ACQUITY UPLC^{TM} BEH C_{18} column with gradient elution using mobile phase of 0.1% formic acid and methanol at a flow rate of 0.25 ml/min. The detection was performed on a triple quadrupole tandem mass spectrometer using ESI technique under selected ion reaction (SIR) and multiple reaction monitor (MRM). The results revealed that trantinterol would not produce CYP450 enzyme inhibition or induction at clinical relevant concentrations.

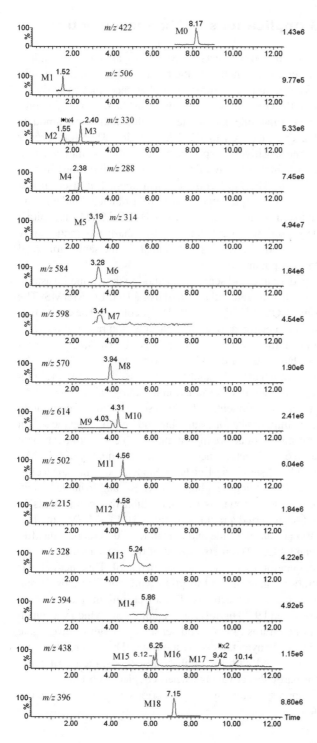

Figure 10.4   UHPLC-MS/MS chromatograms of mosapride and its metabolites in rats after oral administration.

10.3.4 UHPLC-MS Applications in Pharmacokinetics

Pharmacokinetics describes how the body affects a specific drug after administration, and it can be divided into several areas including the extent and rate of absorption, distribution, metabolism, and excretion. The subjects of pharmacokinetics study are usually biological samples, so the concentration of drugs must be measured in complex biological matrix (plasma, urine, bile, and tissue), which contain large amounts of endogenous substances that can interfere with the determination. Therefore, proper bioanalytical methods should be sensitive and specific because of the complex nature of the matrix and the need for high sensitivity to observe concentrations after a low dose and a long time period. The most common instrumentation used in this application is UHPLC-MS with a triple quadrupole mass spectrometer. Tandem mass spectrometry is usually employed for added specificity. UHPLC-MS/MS technique has become the most widely used analytical platform for pharmacokinetic study because it can simultaneously determine various components in the same biological sample and perform high-throughput analysis.

Wang et al. (47) developed a rapid and sensitive method to determine lovastatin in human plasma with sinvastatin as an internal standard using UHPLC-MS/MS. The analysis was carried out on an ACQUITY UPLC^{TM} BEH C$_{18}$ column. The calibration curve of this method was found to be linear over the concentration range of 0.025 to 50 ng/mL. The lower limit of quantification for lovastatin was 0.025 ng/mL. The interday and intraday precision (relative standard deviation) were less than 11%, and the accuracy (relative error) was within 6.0%. The analysis time was shorter than 1.7 min per sample, which met a high-throughput determination of biological samples.

Determination of multiple components of Chinese traditional compound medicines (CTCMs) is of great importance for evaluating and investigating the action mechanism. Pharmacokinetic studies of active ingredients in CTCMs will contribute substantially to the elucidation of their action mechanism and the quality of CTCMs. Yanyan tablets, one of the CTCMs listed in a Ministerial Standard of China, is a clinical medicine commonly used to treat angina, pharynx dryness, and irritable cough. Harpagoside and cinnamic acid are known as the main bioactive components in Yanyan tablets. Xiong et al. (48) reported a rapid, selective UHPLC-MS/MS method for simultaneous determination of harpagoside and cinnamic acid in rat plasma after oral administration of Yanyan tablets. The pretreatment of the plasma sample was a simple one-step protein precipitation by the addition of methanol. This method was validated over the concentration range of 8.00 to 960 ng/mL for harpagoside and 9.90 to 992 ng/mL for cinnamic acid, respectively. The lower limit of quantification was 8.00 ng/mL for harpagoside and 9.90 ng/mL for cinnamic acid, which was much lower than those reported in literatures (49, 50) and sensitive enough to investigate the pharmacokinetic behaviors of Yanyan tablets in rat plasma. The total run time per sample was within 2 min. Compared with the published methods, the sharp peaks produced by UHPLC are of particular advantage when coupled with tandem mass spectrometry, reducing ion suppression and offering superior sensitivity, satisfactory selectivity, and short run time.

UHPLC-MS/MS can also be used for the pharmacokinetics study of TCM constituents and their metabolites. *Rhizoma Drynariae* is a well-known TCM that can replenish the kidney and strengthen the bones. Naringin is known as the main active constituent of *Rhizoma Drynariae*, possessing anti-inflammatory, anti-ulcer, and antioxidation activities. Narigenin, the aglycone of naringin, is the metabolite of naringin and also exhibits anti-ulcer activity. Therefore, both naringin and narigenin levels in plasma are important pharmacokinetic parameters in assessing their efficacy. It is essential to develop a specific, sensitive, and rapid method to determine concentration of naringin and narigenin simultaneously. Li et al. (51) developed a rapid UHPLC-MS/MS method to investigate the pharmacokinetics of naringin and its metabolite narigenin in rat plasma after oral administration of *Rhizoma Drynariae* extract. The analysis was performed on an ACQUITY UPLC^{TM} BEH C$_{18}$ column with acetonitrile-0.4% acetic acid (80:20, *V/V*) as mobile phase. All the results of this method were conformed to the criteria for the analysis of biological samples according to the guidance of the FDA. The pharmacokinetic parameters would be a suitable reference in clinical application for *Rhizoma Drynariae*.

10.4 CONCLUSIONS

This review highlights many benefits of UHPLC using columns packed with sub-2 μm particles, including the high resolution, superior sensitivity, high peak capacity, and high-throughput analysis. These features are further exerted when UHPLC is coupled with MS. UHPLC-MS are particularly useful in the quality control and action mechanism of TCM, in the metabonomic field, for the elucidation of metabolites and metabolism pathways, and in pharmacokinetics study. UHPLC-MS, as a powerful analytical tool, will play a more and more important role in drug discovery and development.

REFERENCES

1. WU, N., CLAUSEN, A.M. *J. Sep. Sci.* 2007, 30(8): 1167–1182.
2. GUILLARME, D., HEINISCH, S., ROCCA, J.L. *J. Chromatogr. A.* 2004, 1052(1–2): 39–51.
3. YANG, B., ZHAO, J., BROWN, J.S., CARR, P.W. *Anal. Chem.* 2000, 72(6):1253–1262.
4. LI, J., HU, Y., CARR, P.W. *Anal. Chem.* 1997, 69(19): 3884–3888.
5. GUILLARME, D., RUTA, J., RUDAZ, S., VEYTHEY, J.L. *Anal. Bioanal. Chem.* 2010, 397(3): 1069–1082.
6. HEINISCH, S., PUY, G., BARRIOULET, M.P., ROCCA, J.L. *J. Chromatogr. A.* 2006, 1118(2): 234–243.
7. THOMPSON, J.D., CARR, P.W. *Anal. Chem.* 2002, 74(5): 1017–1023.
8. TEUTENBERG, T., TUERK, J., HOLZHAUSER, M., GIEGOLD, S. *J. Sep. Sci.* 2007, 30(8): 1101–1114.
9. CABRERA, K. *J. Sep. Sci.* 2004, 27(10–11): 843–852.
10. GUILLARME, D., NGUYEN, D.T.T, RUDAZ, S., VEUTHEY, J.L. *J. Chromatogr. A.* 2007, 1149 (1): 20–29.
11. VAN DE MERBEL, N.C., POELMAN, H. *J. Pharm. Biomed. Anal.* 2003, 33(3): 495–504.
12. GIDDINGS, J.C. *Anal. Chem.* 1965, 37(1): 60–63.
13. KNOX, J.H. *J. Chromatogr. Sci.* 1977, 15: 352.

14. POPPE, H. *J. Chromatogr. A.* 1997, 778(1–2): 3–21.
15. MAJORS, R.E. *LC-GC N. Am.* 2005, 23: 1248–1256.
16. MACNAIR, J.E., LEWIS, K.C., JORGENSON, J.W. *Anal. Chem.* 1997, 69(6): 983–989.
17. GRUMBACH, Eric S., WHEAT, Thomas, E., KELE, Marianna, MAZZEO, Jeffrey R. *LC-GC N. Am. (Suppl.).* 2005: 40–44.
18. Ultra Performance LC™ by design, 2004. Waters Corporation, USA, 720000880EN LL&LW-UL.
19. MELLORS, J.S., JORGENSON, J.W. *Anal. Chem.* 2004, 76(18): 5441–5450.
20. GRUMBACH, E.S., MAZZEO, J.R., DIEHL, D.M. *Proceedings of the 25th International Symposium on Chromatography*, France, 2004, 34.
21. WYNDHAM, K.D., O'GARA, J.E., WALTER, T.H., GLOSE, K.H., LAWRENCE, N.L., ALDEN, B.A., IZZO, G.S., HUDALLA, C.J., IRANETA, P.C. *Anal. Chem.* 2003, 75(24): 6781–6788.
22. NGUYEN, D.T., GUILLARME, D., RUDAZ, S., VEUTHEY, J.L. *J. Sep. Sci.* 2006, 29(12): 1836–1848.
23. NOVÁKOVÁ, L., SOLICHOVÁ, D., SOLICH, P. *J. Sep. Sci.* 2006, 29(16): 2433–2443.
24. NOVÁKOVÁ, L., MATYSOVÁ, L., SOLICH, P. *Talanta.* 2006, 68(3): 908–918.
25. GUILLARME, D., SCHAPPLER, J., RUDAZ, S., VEUTHEY, J.L. *Trend Anal. Chem.* 2010, 29(1): 15–27.
26. WALTER, A.K. *DDT.* 2005, 10: 1357–1367.
27. KAMEL, A., PRAKASH, C. *Curr. Drug Metab.* 2006, 7(8): 837–852.
28. NIESSEN, W.M. *J. Chromatogr. A.* 2003, 1000(1–2): 413–436.
29. LIM, C.K., LORD, G. *Biol. Pharm. Bull.* 2002, 25(5): 547–557.
30. DAWSON, P.H. *Quadrupole Mass Spectrometry and Its Applications.* New York: AIP Press, 1995.
31. MCLUCKEY, S.A., VAN BERKEL, G.J., GOERINGER, D.E., GLISH, G.L. *Anal. Chem.* 1994, 66(14): 737A–743A.
32. LOBVI, E., MATAMOROS, F. *Carbohyd. Polym.* 2007, 68(4):797–807.
33. CHUNYAN, H., RAYMOND, E. *Int. J. Mass Spectrom.* 2001, 212(1–3): 337–357.
34. MAMYRIN, B.A. *Int. J. Mass Spectrom.* 2001, 206: 251–266.
35. HARS, G., MAROS, I. *Int. J. Mass Spectrom.* 2003, 225(2): 101–114.
36. LI, X.Q., XIONG, Z.L., YING, X.X., CUI, L.C., ZHU, W.L., LI, F.M. *Anal. Chim. Acta.* 2006 (580): 170–180.
37. State Administration of Traditional Medicine of China. *Shanghai Scientific and Techonological Publisher*, 1996: 1791.
38. HUANG, W.Z., LIANG, X. Determenation of two flavone glycosides in Trollius ledebourli by HPLC. *Chin Pharm J.* 2000, 35(10):658-659.
39. LIU, Z., WANG, L., LI, W., HUANG, Y., XU, Z.C. *China J. Chin. Mater. Med.* 2004, 29(11): 1049–1051.
40. LI, X.Q., SUN, X.H., CAI, S., YING, X.X., LI, F.M. *Acta Pharmaceutica Sinica.* 2009, 4(8): 895–904.
41. LI, N., CUI, M., LU, X.M., et al. *Biomed. Chromatogr.* 2010, 24(11): 1255–1261.
42. NICHOLSON, J.K. Metabonomics: Nature Drug discovery. 2002, 1(12): 153–161.
43. LI, F.M., LU, X.M., LIU, H.P., LIU, M., XIONG, Z.L. *Biomed. Chromatogr.* 2007, 21(4): 397–405.
44. HUO, T.G., CAI, S., LU, X., SHA, Y., YU, M.Y., LI, F.M. *J. Pharma. Biomed. Anal.* 2009, 49(4): 976–982.
45. SUN, X.H., NIU, L.L., LI, X.Q., LU, X.M., LI, F.M. *J. Pharma. Biomed. Anal.* 2009, 50(1): 27–34.
46. JIANG, K., LI, K.J., QIN, F., LU, X.M., LI, F.M. *Toxicol. In Vitro.* 2011, 25(5): 1033–1038.
47. WANG, D., WANG, D.M., QIN, F., CHEN, L.Y., LI, F.M. *Biomed. Chromatogr.* 2008, 22(5): 511–518.
48. XIONG, Z.L., FU, Y.H., LI, J.J., QIN, F., LI, F.M. *Chromatographia.* 2010, 72: 163–169.
49. WANG, S.J., RUAN, J.X., ZHAO, Y.H., ZHANG, Z.Q. *Biomed. Chromatogr.* 2008, 22(1): 50–57.
50. LI, P., ZHANG, Y., XIAO, L., JIN, X., YANG, K. *Anal. Bioanal. Chem.* 2007, 389(7–8): 2259–2264.
51. LI, X.H., XIONG, Z.L., LU, S., ZHANG, Y., LI, F.M. *Chinese J. Nat. Med.* 2010, 8(1): 40–46.

Index

Ultra-High Performance Liquid Chromatography and Its Applications, First Edition. Edited by Quanyun Alan Xu.
© 2013 John Wiley & Sons, Inc. Published 2013 by John Wiley & Sons, Inc.